Respiratory Disease Series: Diagnostic Tools and Disease Managements

Series Editors

Hiroyuki Nakamura
Ibaraki Medical Center
Tokyo Medical University
Ibaraki, Japan

Kazutetsu Aoshiba
Ibaraki Medical Center
Tokyo Medical University
Ibaraki, Japan

This book series cover a variety of topics in respiratory diseases, with each volume providing an overview of the current state of knowledge, recent discoveries and future prospects for each disease. In each chapter the editors pose critical questions, which are often unresolved clinical issues. These are then discussed by the authors, providing insights and suggestions as to which developments need to be addressed. The series offers new information, which will inspire innovative ideas to further develop respiratory medicine.This collection of monographs is aimed at benefiting patients across the globe suffering from respiratory disease.
Edited by established authorities in the field and written by pioneering experts, this book series will be valuable to those researchers and physicians working in respiratory medicine. The series is aimed at a broad readership, and the books will also be a valuable resource for radiologists, emergency medicine physicians, pathologists, pharmacologists and basic research scientists.

More information about this series at http://www.springer.com/series/15152

Kazuhiro Yamaguchi
Editor

Structure-Function Relationships in Various Respiratory Systems

Connecting to the Next Generation

Editor
Kazuhiro Yamaguchi
Department of Respiratory Medicine
Tokyo Medical University
Tokyo
Japan

Division of Respiratory Medicine
Tohto Clinic, Kenko-Igaku Association
Tokyo
Japan

ISSN 2509-5552 ISSN 2509-5560 (electronic)
Respiratory Disease Series: Diagnostic Tools and Disease Managements
ISBN 978-981-15-5595-4 ISBN 978-981-15-5596-1 (eBook)
https://doi.org/10.1007/978-981-15-5596-1

This Springer imprint is published by the registered company Springer Nature Singapore Pte Ltd.
The registered company address is: 152 Beach Road, #21-01/04 Gateway East, Singapore 189721, Singapore

Preface

It is my great pleasure to dedicate this fabulous book to the memory of two prestigious scientists in the field of pulmonology and their adorable wives. One is Professor Dr. Johannes Piiper and Mrs. Ilse Piiper in Göttingen, Germany. The other is Professor Dr. Ewald Rudolf Weibel and Mrs. Anna Verena Weibel in Bern, Switzerland. Johannes and Ewald inspired me, with enthusiasm, about the importance of always recognizing the structure and functional relationships in the lung. Johannes guided me from the physiological (functional) perspectives, while Ewald from the morphological (anatomical) perspectives. Thus, their instructions have laid my clinical basis today.

Johannes Piiper was born in 1924 in Tartu, Estonia, and deceased in 2012. Although he spent a peaceful time in his adolescence, this abruptly came to an end due to the annexation of Estonia by the Soviet Union in the summer of 1940. At the end of 1946, he went to Göttingen because a large camp for Baltic refugees was set up there. Fortunately, the University of Göttingen accepted him as a student of medical course. In 1958, he visited the United States to work with Prof. Hermann Rahn at the University of Buffalo. In 1960, he finalized the habilitation for physiology at the University of Göttingen. In 1973, he was appointed as the Director of the Department of Physiology of the Max Planck Institute for Experimental Medicine in Göttingen. In 1992, he retired from the institute. The research field of Piiper's group was very wide, including special gas exchange mechanisms in many animals, including humans, dogs, small mammalians, high-altitude mammalians, birds, fishes, and so forth. Of them, the diffusion limitation in alveolar gas phase and alveolar septa in the human lung is the most appreciated work of Piiper. He developed the well-known concept of simultaneous distribution of ventilation-perfusion (V_A/Q) and diffusion capacity-perfusion (D/Q) in the lung.

Ewald R. Weibel was born in 1929 in Buchs AG, Switzerland, and died in the spring of 2019. He started his academic career at the Medical School of Zürich, Göttingen, and Paris. He studied under Prof. Averill A. Liebow at Yale University, Profs. Andre F. Cournand and Dickinson W. Richards at the University of Columbia, and Prof. George E. Palade at the Rockefeller University of New York. In 1966, he was promoted to the Professor and Chairman, Institute of Anatomy, University of

Bern, in Switzerland. In 1994, he stepped down from the position. He acted as the Editor-in-Chief of *Journal of Microscopy* and that of *American Journal of Physiology*: *Lung Cellular and Molecular Physiology*. The most prominent achievement of Weibel would be the publication of the book entitled *Morphometry of the Human Lung* (Springer Verlag and Academic Press, Berlin, New York, 1963). This book still plays a biblical role in the field of pulmonary anatomy even today. In 2010, Weibel and colleagues published an official research policy statement of ATS/ERS concerning standards for quantitative assessment of lung structure. The morphometric estimate of pulmonary diffusing capacity determined by Weibel's group is the monument in this field.

This book is prepared for the young physicians and researchers who are actively engaged in clinical, morphological, or physiological works in the field of respiration. The morphological and physiological approaches are both equally important for upraising the quality of basic science and that of clinical medicine targeting the human respiratory system. This book sheds light on the morphological backgrounds of various functional parameters, including respiratory control system, dynamics and gas transport in upper and lower airways, acinar gas exchange, and gas transfer through alveolar septa. Finally, I would like to offer my sincere gratitude to all of them who contributed to the writing of this book and to Springer Nature Ltd. for cleverer decision on publishing this book.

Tokyo, Japan Kazuhiro Yamaguchi

Contents

Part I
Respiratory Control System

Chapter 1
Anatomy and Physiology of Respiratory Control System: How Are Respiratory Controlling Cells Communicating in the Brain?

Yasumasa Okada, Shigefumi Yokota, and Isato Fukushi

Abstract Autonomic respiratory rhythm is generated in the neuronal networks in the lower brainstem, and the generated neural respiratory output is transmitted to respiratory muscles via cranial and spinal cord motor nerves to control upper airway patency and intrathoracic pressure. Among the respiratory neuronal networks, the pre-Bötzinger Complex located in the bilateral ventrolateral medulla is the kernel for respiratory rhythm generation. In this chapter, we overview the localization of the neuronal networks and anatomical and physiological connectivity among the respiratory neurons. We also introduce the influential theories on the physiological mechanisms of respiratory rhythm generation and discuss how respiratory rhythm is disturbed in pathological conditions.

Keywords Astrocyte · Brainstem · Hypothalamus · Medulla oblongata · Pons
Respiratory control · Respiratory rhythm · Spinal cord

Y. Okada (✉)
Clinical Research Center, Murayama Medical Center, Tokyo, Japan
e-mail: yasumasaokada@1979.jukuin.keio.ac.jp

S. Yokota
Department of Anatomy and Morphological Neuroscience,
Shimane University School of Medicine, Izumo, Japan

I. Fukushi
Clinical Research Center, Murayama Medical Center, Tokyo, Japan

Faculty of Health Sciences, Uekusa Gakuen University, Chiba, Japan

K. Yamaguchi (ed.), *Structure-Function Relationships in Various Respiratory Systems*, Respiratory Disease Series: Diagnostic Tools and Disease Managements, https://doi.org/10.1007/978-981-15-5596-1_1

1 Introduction

Neural respiratory output drives respiratory muscles and maintains ventilation. Thus, persistent generation of adequately regulated neural respiratory output, i.e., respiratory rhythm and pattern, is essential for a healthy life. Although respiratory rhythm and pattern could be changed intentionally, they are usually maintained unconsciously. Autonomic respiratory rhythm and pattern are generated in the neuronal networks in the lower brainstem and spinal cord, and the generated neural respiratory output is transmitted to respiratory muscles via cranial and spinal cord motor nerves to control upper airway patency and intrathoracic pressure [1]. The neuronal network in the brainstem also receives information from various central and peripheral sensing mechanisms detecting the internal and external environmental changes. We overview the anatomy and physiology of the respiratory rhythm generating cellular networks and review how these respiratory rhythm generating cells are communicating in the brain and spinal cord.

2 Localization and Anatomy of Respiratory Rhythm Generating Neuronal Populations

Although it has been known that basic respiratory rhythm is generated in the brainstem, it had been mistakenly believed that the midbrain, especially the corpus quadrigemina, is the most important site [2], until Lumsden (1923) found that the midbrain does not play a vital role and that the medulla oblongata produces essential basic respiratory rhythm, although the pons is necessary for the generation of normal breathing pattern, on the basis of brainstem sectioning experiments in cats [3]. Since those times, the brainstem mechanisms of respiratory rhythm generation have been studied using in vivo animals. However, in 1984 a new type of experimental preparation was developed, i.e., the isolated brainstem–spinal cord preparation that produces respiratory rhythm in vitro [4]. This preparation caused a revolutionary change in the study of central respiratory control. By maximally utilizing the advantage of in vitro preparations, researchers could have conducted various experiments using brainstem sectioning and lesioning, pharmacological, electrophysiological, and imaging techniques that are impossible in in vivo animals [5–7]. On the basis of these in vivo and in vitro studies, we summarize the localization and anatomy of the respiratory rhythm generating neuronal populations. Respiratory neurons, i.e., neurons that fire in synchrony with respiration and are involved in respiratory control, are located in various brain and spinal cord regions, but neurons that play vital roles in respiratory rhythm generation are distributed in the pons and medulla oblongata. Localization of major

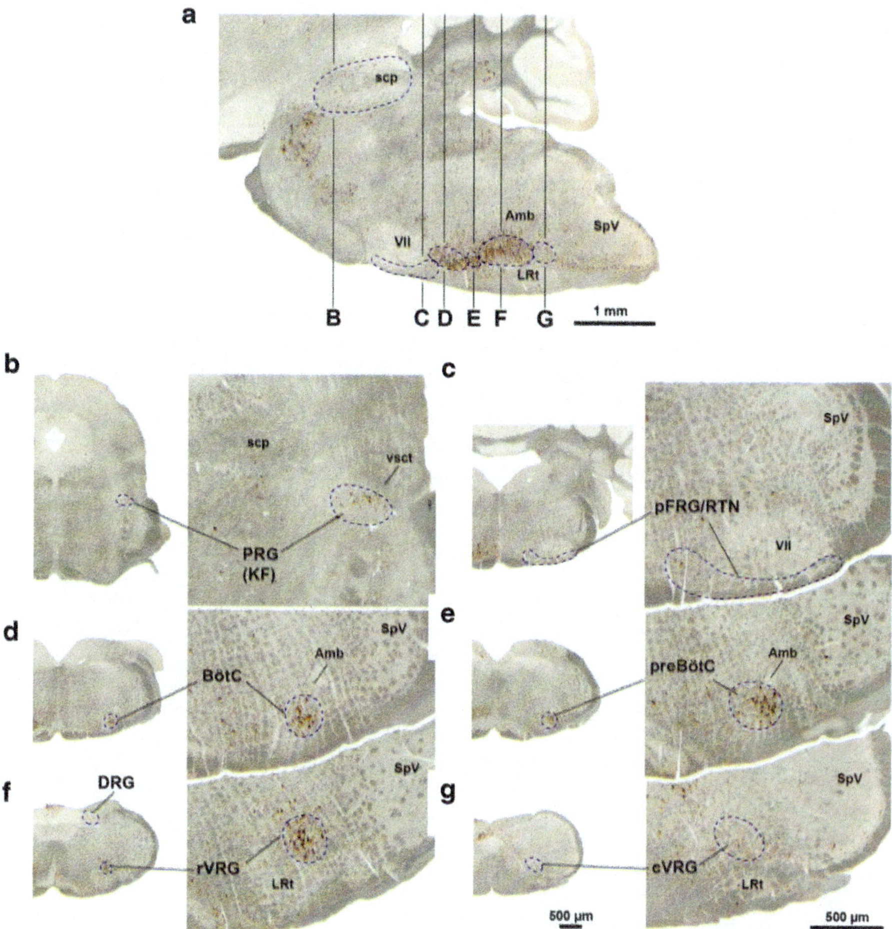

Fig. 1.1 Localization of respiratory neuron populations in the mouse brainstem. (**a**) Sagittal view. B, C, D, E, F, and G denote the levels corresponding to the transversely sectioned pictures **b**, **c**, **d**, **e**, **f**, and **g**, respectively. *Amb* ambiguous nucleus, *BötC* Bötzinger Complex, *cVRG* caudal ventral respiratory group, *DRG* dorsal respiratory group, *KF* Kölliker-Fuse nucleus, *LRt* lateral reticular nucleus, *pFRG* parafacial respiratory group, *PRG* pontine respiratory group, *RTN* retrotrapezoid nucleus, *rVRG* rostral ventral respiratory group, *scp* superior cerebellar peduncle, *SpV* spinal trigeminal nucleus, *vsct* ventral spinocerebellar tract, *VII* facial nucleus

respiratory neuron groups in the pons and medulla oblongata is shown as histological pictures (Fig. 1.1). It has been widely known that there are clusters of respiratory neurons distributed bilaterally in the ventrolateral medulla as longitudinal columns and they are called the ventral respiratory group (VRG) or ventral respiratory column (VRC) [1]. Below are the especially important neuron populations.

2.1 Pre-Bötzinger Complex (preBötC)

In 1991 Smith et al. conducted brainstem transverse sectioning experiments using the isolated brainstem–spinal cord preparation, principally with the same experimental concept with the historical in vivo experiments by Lumsden (1923) [3], gradually and stepwise removed either rostral or caudal portion of the brainstem by monitoring neural respiratory output from cranial or C4 spinal cord motor nerves, and found that a distinct level of the medulla oblongata contains the region where the kernel of the respiratory rhythm generating population exists (Fig. 1.2) [8]. Then, they cut out a slice with several hundred micrometer thickness containing this level from the medulla and confirmed that thus cut out thin slice preparation still generates respiration-like rhythmic activity in vitro. This slice preparation is often called "breathing slice" or "rhythmically active slice" and has been widely applied to the study of respiratory control [8–10].

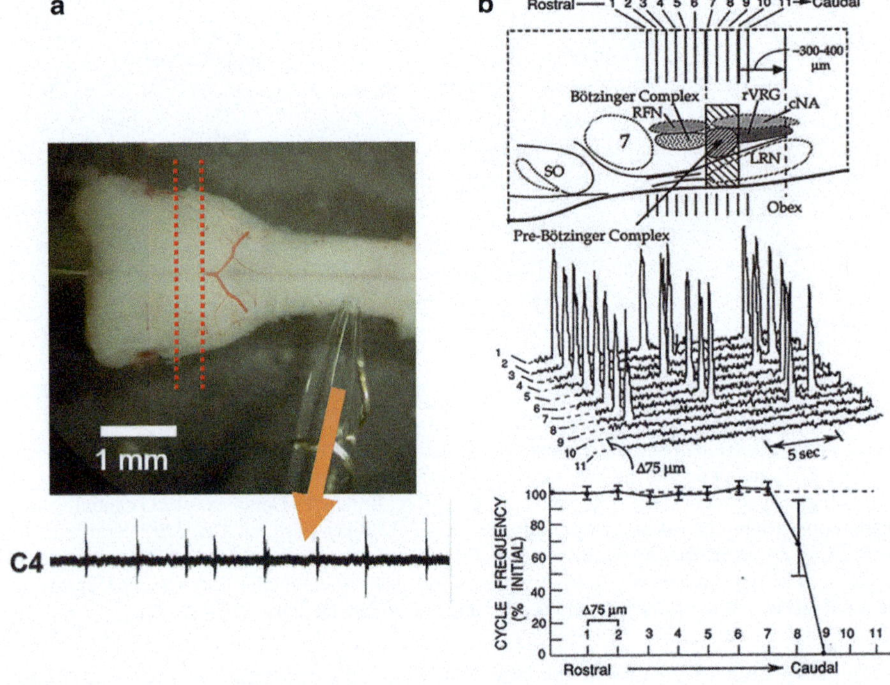

Fig. 1.2 (**a**) Isolated brainstem–spinal cord preparation with its ventral side up. Neural respiratory output can be monitored through a glass suction electrode from the fourth ventral roots of the cervical spinal cord (C4). The medullary layer between the two red dotted lines corresponds to the level of breathing slice. (**b**) Top. Sagittal view of the brainstem. Middle. The brainstem was sectioned from the rostral side 75 µm-stepwise by monitoring neural respiratory output from spinal C4 motor nerves. Bottom. Respiratory rhythm was reduced and finally eliminated after sectioning of the preBötC (levels 8–10) [8]

Furthermore, they identified a limited region in a part of the VRG that contains neurons essential for rhythmogenesis and named it the pre-Bötzinger Complex (preBötC) [8]. They analyzed the function of preBötC neurons and demonstrated that these neurons have voltage-dependent pacemaker-like properties and the intrinsic bursting frequency of these pacemaker neurons depends on the baseline membrane potential, suggesting that rhythm generating network in the preBötC consists of synaptically coupled pacemaker neurons [8, 9]. Then we have analyzed detailed anatomical and functional connectivity to and from the preBötC [10]. Anatomical tract-tracing demonstrated that neurokinin-1 receptor- and somatostatin-immunoreactive preBötC neurons, which are putatively involved in respiratory rhythm generation, are embedded in the plexus of axons originating in the contralateral preBötC. By voltage imaging in rhythmically active slices of neonatal rats, we visually analyzed origination and propagation of inspiratory neural activity as depolarizing wave dynamics on the entire transverse plane as well as within the preBötC, and for the first time demonstrated the inter-preBötC conduction process of inspiratory action potentials. Collectively, we comprehensively elucidated the anatomical pathways to and from the preBötC and dynamics of inspiratory neural information propagation: (1) From the preBötC in one side to the contralateral preBötC, which synchronizes the bilateral rhythmogenic kernels, (2) from the preBötC directly to the bilateral hypoglossal premotor and motor areas as well as to the nuclei tractus solitarius (nTS), and (3) from the hypoglossal premotor areas toward the hypoglossal motor nuclei (Fig. 1.3) [10]. Although the anatomy and physiology of the preBötC have been studied mainly

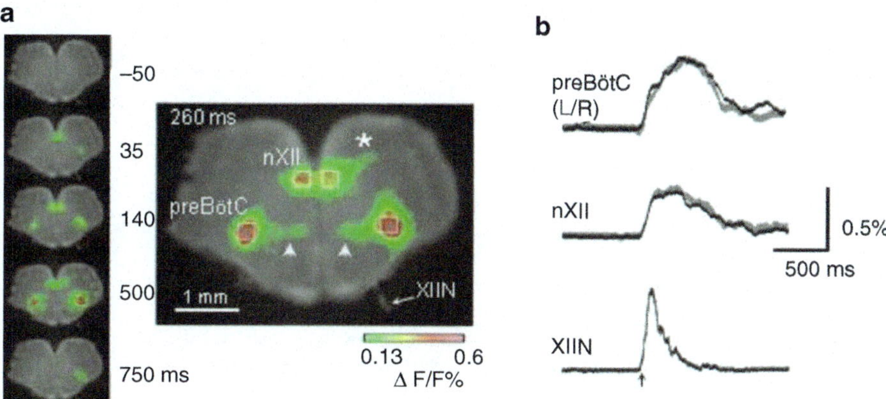

Fig. 1.3 Spontaneous inspiratory activities in the entire plane of a rhythmically active slice visualize by voltage imaging. (**a**) Membrane potential images are shown from the onset (**b**, arrow) of the hypoglossal nerve (XIIN) inspiratory activities. The bilateral preBötC and nXII showed local peaks. Time courses of fluorescence intensity changes in the regions (squares, 260 ms) are shown in **b**, left (L, gray) and right (R, black) along with the integrated XIIN activity. Inspiratory activities of the nTS (*) and the inter-preBötC tract (arrowheads) were visualized (**a**, 260 ms) [10]

in rats and mice, the preBötC has been identified also in humans [11]. The human preBötC is characterized by an aggregation of loosely scattered small neurons, which contain neurokinin 1 receptor as well as somatostatin (Fig. 1.4).

2.2 Parafacial Respiratory Group (pFRG)

In 2003 another respiratory oscillator was found in the most rostral portion of the ventrolateral medulla, superficial to the facial nucleus, by voltage imaging, and it was referred to as the parafacial respiratory group (pFRG) [12]. The population of pre-inspiratory neurons were identified in the pFRG and it was proposed to be involved in the primary respiratory rhythm generation [12]. We evaluated developmental changes of the pFRG and preBötC functions, and demonstrated that in the very neonatal stage the pFRG mainly drives inspiration and significant reorganization of the respiratory neuronal network, characterized by a reduction of pre-inspiratory activity in the pFRG, occurs at P1-P2 in rats [13]. It is also believed that further later, i.e., in a juvenile stage, the pFRG functions as a rhythm generator for active expiration [14]. In any case, the pFRG and preBötC are functionally coupled, but their coupling status changes according to the developmental stage, physiological condition, and neuronal excitability within each oscillator [15, 16].

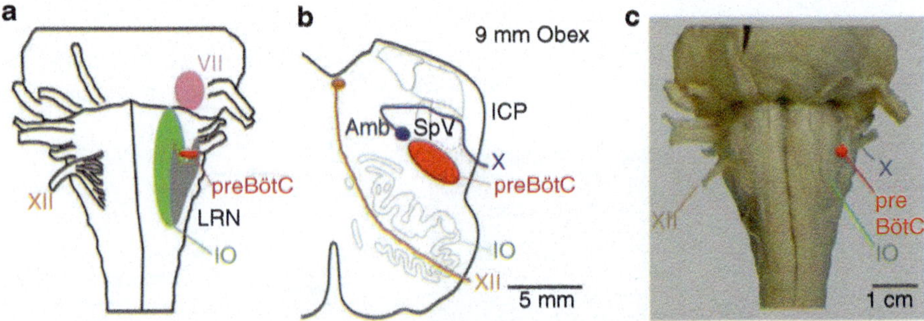

Fig. 1.4 The preBötC identified in the human medulla. (**a**) Schema of the ventral medulla. The preBötC is localized in the ventrolateral medulla, caudal to the facial nucleus (VII) and rostral to the lateral reticular nucleus (LRN). (**b**) On a transversal section at + 9 mm to the obex, the preBötC is located lateral to the inferior olive (IO) ventral to the compact part of the ambiguous nucleus (Amb) and medial to the spinal trigeminal tract (SpV). (**c**) On the ventrolateral surface of the brainstem, the preBötC is positioned in the retro-olivary sulcus lateral to the IO and the rostral rootlets of the hypoglossal nerve (XII) and medial to the ventral rootlets of the vagal nerve (X). ICP, inferior cerebellar peduncle [11]

2.3 *Post-inspiratory Complex (PiCo)*

Recently, an excitatory network that generates the post-inspiratory neuronal activity in mice has been reported [17]. Glutamatergic–cholinergic neurons form the basis of this network, and GABA-mediated inhibition determines the timing and coordination relative to inspiration. This network was named the post-inspiratory complex (PiCo). It has been argued that the PiCo has autonomous rhythm-generating properties, is necessary and sufficient for post-inspiratory activity in vivo, and also shows distinct responses to neuromodulators as compared to other excitatory brainstem networks [17]. In mammals, breathing is composed of three phases, i.e., inspiration, post-inspiration, and active expiration, and it has been proposed by the group of Ramirez that each of the three phases of breathing is generated by a distinct excitatory network: the preBötC, which has been linked to inspiration; the PiCo, for the neuronal control of post-inspiration; and the parafacial respiratory group (pFRG), which has been associated with active expiration, a respiratory phase that is recruited during high metabolic demand, respectively [18].

2.4 *Pontine Respiratory Group (PRG)*

Although the medulla is the center for respiratory rhythm generation, there are several groups of neurons in the pons involved in respiratory control, and they are called pontine respiratory group (PRG) [19–22]. Although sensory feedback from pulmonary stretch receptors (Breuer–Hering reflex) is the major mechanism for the determination of the inspiratory/expiratory phase transition, pontine-based control of inspiratory/expiratory phase transition by both the pontine Kölliker-Fuse nucleus (KF) and parabrachial complex is also important [19–21]. The report on the PiCo is not consistent with this classical concept above, regarding the initiation mechanism of expiration (i.e., inspiratory/expiratory phase transition) [17, 21], and further studies are necessary to clarify the precise mechanism of the inspiratory/expiratory phase transition.

2.5 *High Cervical Respiratory Group (HCRG)*

Although it is without a doubt that autonomic respiratory rhythm is primarily generated in the lower brainstem, it has been reported that when the neuraxis is transected at the level of medullo-spinal cord junction (but not at the level caudal to the C3 spinal cord segment), spontaneous rhythmic breathing occurs transiently in cats, suggesting that there are accessory oscillators in the upper cervical spinal cord [23]. Capability of respiration-like rhythm generation by C1-C2 spinal cord neurons was reported also in the isolated brainstem–spinal cord preparation of neonatal mice

[24]. We identified a novel region that extends from the medullo-spinal cord junction to the C2 spinal cord segment, ~300 μm deep from the ventral surface in neonatal rats, and termed it the high cervical respiratory group (HCRG). We recorded pre-inspiratory and inspiratory neurons in this region. HCRG neurons are distinct from motoneurons, because they are small and spindle-like in shape, and have only two or three long processes, and maybe interneurons involved in respiratory rhythmogenesis in the cervical spinal cord [25, 26].

2.6 Respiration-Related Oscillators in Higher Brain Regions

It is possible that there exist respiration-related oscillators in brain regions further rostral to the pons. We found respiration-coupled rhythmic activity in the lateral hypothalamic area (LHA) by voltage imaging [27]. We consider that the hypothalamic regions including the LHA and dorsomedial hypothalamus (DMH) drive respiratory rhythm, e.g., by secreting awake-promoting substance orexin to wide areas of the brain including the pontine and medullary respiratory regions (Fig. 1.5) [27]. Onimaru and Homma (2007) found that neurons in the limbic regions, i.e., in the piriform–amygdala regions, show spontaneous rhythmic activities synchronized

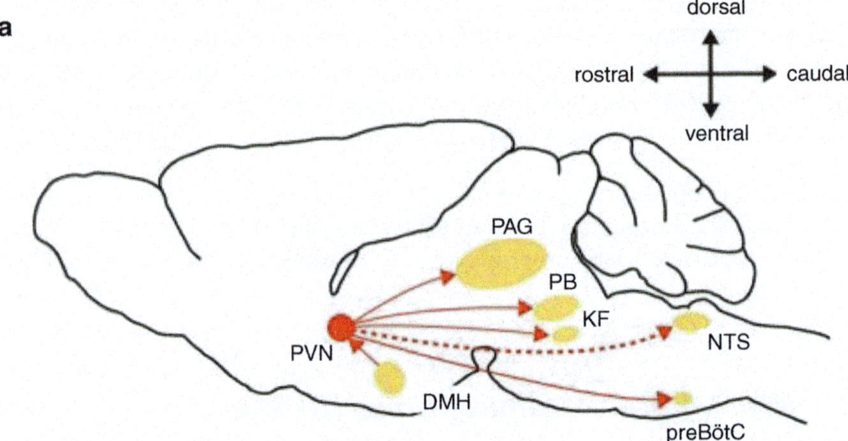

Fig. 1.5 Neuronal populations in the hypothalamus make complex communications within the hypothalamus and with nuclei in the midbrain, pons, medulla, and other brain areas. These pathways and circuits enable the hypothalamus to play roles in the regulation of breathing. (**a**) The projections from/to the PVN. (**b**) The projections from/to the PFA. (**c**) The projections from/to the DMH. *DMH* dorsomedial hypothalamus, *PFA* perifornical area, *PVN* paraventricular nucleus, *PAG* periaqueductal gray, *KF* Kölliker-Fuse nucleus, *PBN* parabrachial nucleus, *preBötC* pre-Bötzinger complex, *RVMM* rostral ventromedial medulla, *NTS* nucleus tractus solitarius [27]

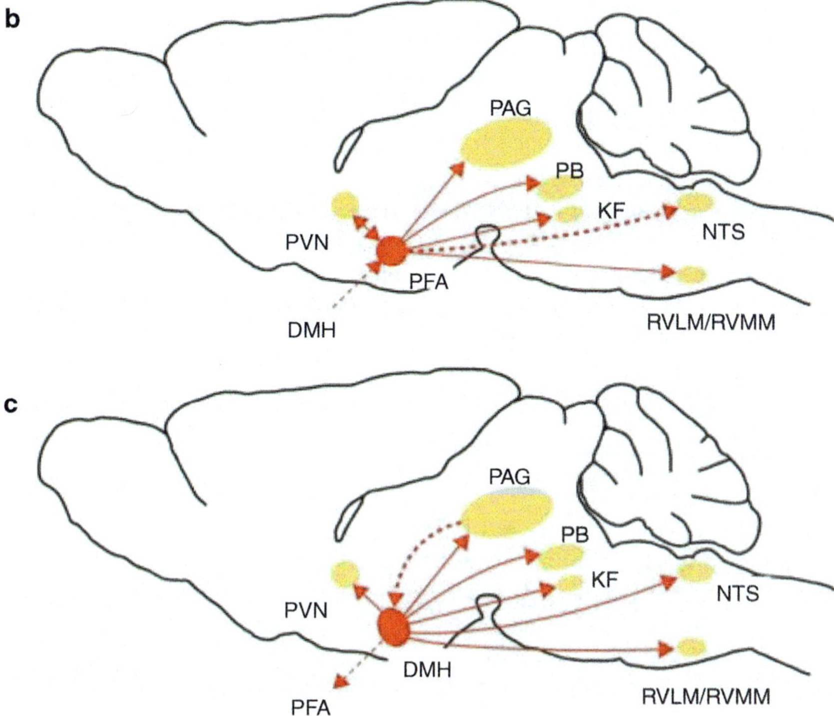

Fig. 1.5 (continued)

with respiratory output [28]. These respiration-synchronized neuronal activities may serve as an auxiliary respiratory rhythm generator, especially in an awake and stress-induced condition, but their precise physiological significance awaits further investigation.

3 Mechanisms of Respiratory Rhythm Generation

The mechanism of respiratory rhythm generation has long been one of the most important research topics in the field of respiratory physiology. There are various types of neurons in the ventrolateral medulla where respiratory rhythm is generated. Each type of neuron shows a distinctive respiration-related firing pattern, and at least some of them should interact and be actively involved in respiratory rhythm generation. Although it has been extensively studied, yet the precise rhythmogenic mechanism has not been fully clarified. There have been a couple of theories and hypotheses proposed; however, none has been fully accepted. Below we introduce currently proposed influential theories.

3.1 Pacemaker Theory

There indeed exist neurons with pacemaker properties in the respiratory network in the rodent brainstem. There are two kinds of pacemaker currents, one is a persistent sodium current (I_{NaP}) and the other is a nonspecific cation current (I_{CAN}). When both I_{NaP} and I_{CAN} are suppressed, respiratory rhythm disappears. However, it must be noted that when the respiratory network activity is boosted by the application of substance-P, once disappeared respiratory rhythm emerges again, suggesting that respiratory rhythm could be generated even without pacemaker current when the activity level of the respiratory network is sufficiently high and that there is also a pacemaker current-independent mechanism in respiratory rhythm generation (Fig. 1.6) [29].

3.2 Network Theory

Historically the network theory has been long supported by researchers especially as the model of in vivo respiratory rhythm generation [30–32]. In the representative model, respiratory rhythm is generated by network interaction among excitatory and inhibitory neurons in which an excitatory pre-inspiratory/inspiratory (pre-I/I) neuron with a pacemaker property is a key inspiratory generator that initiates inspiration, drives inspiratory premotor neurons and motoneurons and receives

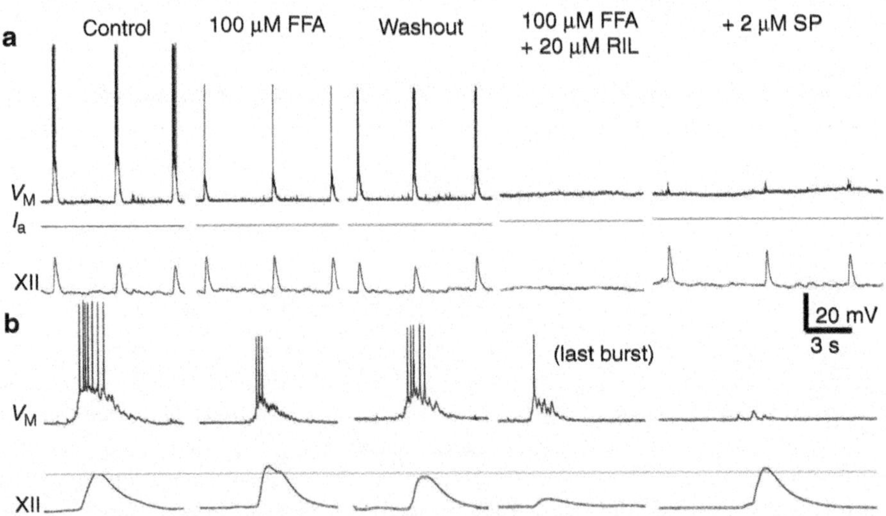

Fig. 1.6 Rhythm generation in the presence of riluzole (RIL) and flufenamic acid (FFA), blockers of pacemaker currents, I_{NaP}, and I_{CAN}, respectively. Coapplication of RIL and FFA abolishes respiratory rhythm. However, the rhythm is rescued with substance-P (SP), which depolarizes cells and boosts excitability but does not directly cause pacemaker properties. (**a**) Continuous segments of the experiment. (**b**) Examples of cellular respiratory drive and XII output plotted with greater time resolution. This finding suggests that a bursting-pacemaker activity is not essential for rhythmogenesis and normally pacemaker currents contribute to rhythmogenesis only by enhancing the general excitability of the network [29]

inhibition from inhibitory neuron network ring composed of mutually inhibiting early-inspiratory (early-I), augmenting-expiratory (aug-E) and post-inspiratory (post-I) neurons (Fig. 1.7) [31, 32]. The network hypothesis of respiratory rhythm

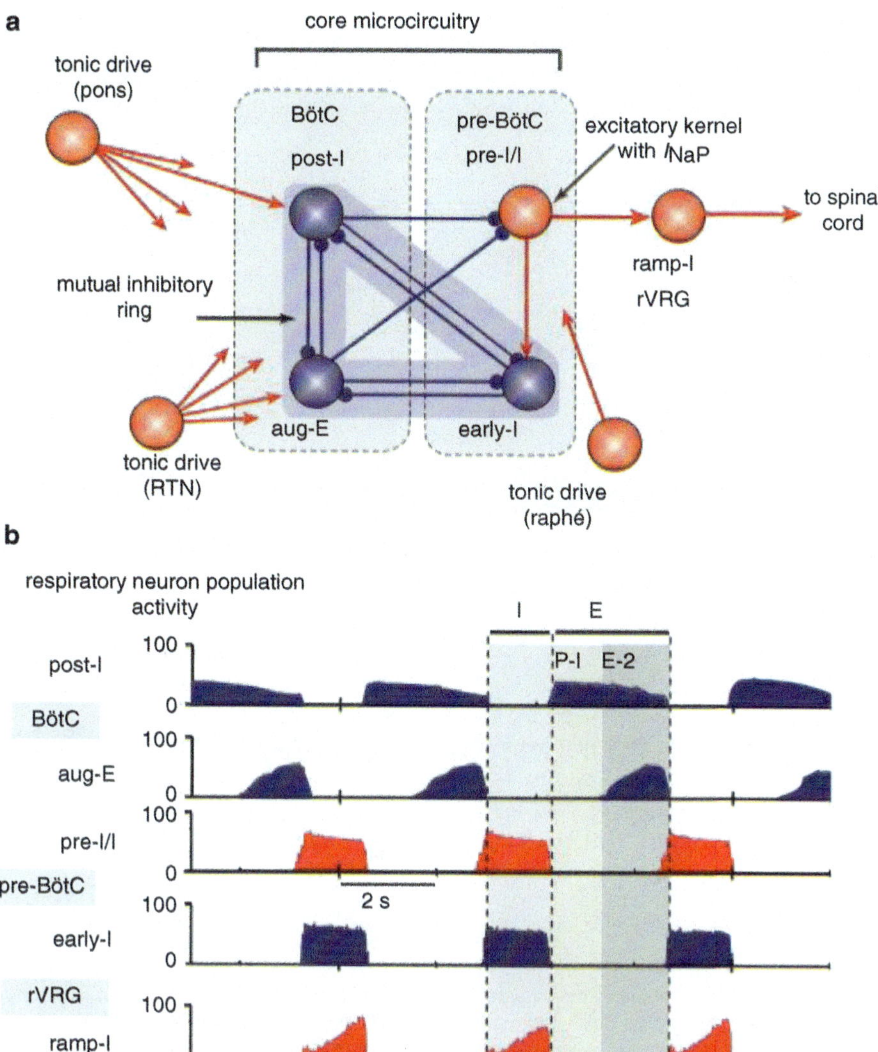

Fig. 1.7 The core microcircuitry within BötC and preBötC respiratory compartments and activity patterns of respiratory neuron populations of the core circuitry that schematically represents the network theory. (**a**) The main interacting neuronal populations in these compartments operating under the control of external excitatory drives from the pons, RTN, and raphe nuclei. Spheres represent neuronal populations (excitatory, red, including tonic drives to the different populations; inhibitory, blue). Red arrows, excitatory connection, small blue circles, inhibitory connection. Inhibitory neurons distributed in the BötC and preBötC compartments are interconnected in a mutual inhibitory ring-like circuit. (**b**) Respiratory neuron population activity patterns by the color-filled activity profiles representing integrated population activity [31]

generation is supported by our observation that there are abundant GABAergic inhibitory respiratory neurons in the mouse preBötC as well as by our experiments using GABA-synthesizing enzyme (GAD67) knockout mice in which the most important inhibitory neurotransmitter GABA was highly reduced and the mice could not generate normal regular respiratory rhythm (Fig. 1.8) [33, 34]. On the other hand, the network hypothesis is not consistent with the experimental findings in which even after the inhibitory synaptic transmission, an essential element in the network mechanism, was eliminated, animals could still maintain respiratory

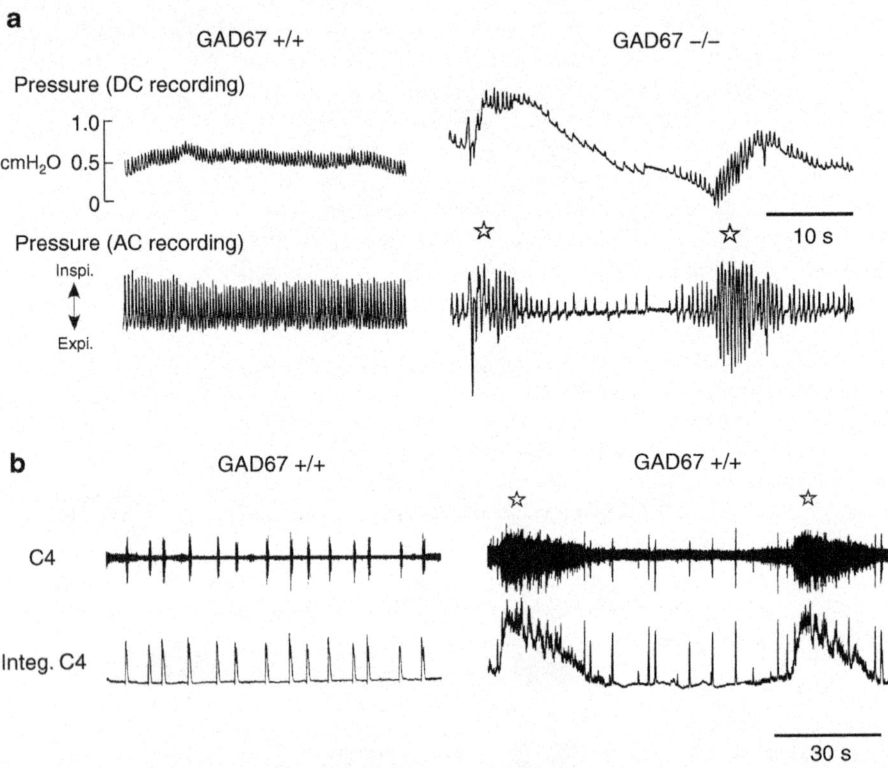

Fig. 1.8 (**a**) In vivo ventilatory movement recorded by whole-body plethysmography. Respiratory rhythm in GAD67+/+ mice (left) was regular and tidal volume was relatively uniform. Irregular respiratory rhythm and apnea were observed in GAD67-/- mice (right). Large variations in tidal volume and gasp-like ventilation with body movement (indicated by star symbols) were observed in GAD67-/- mutant mice. (**b**) Rhythmic activity recorded from C4 ventral roots in in vitro brainstem–spinal cord preparations. Typical recordings obtained from GAD67+/+ (left) and GAD67-/- (right) mice. Rhythm and amplitude of integrated C4 (Integ.C4) bursts of GAD67+/+ mice were regular and constant in contrast to those of GAD67-/- mice, which often showed a large amount of nonrespiratory activity with long durations (indicated by star symbols). These findings indicate the importance of inhibitory neurotransmitter GABA in respiratory rhythm generation [34]

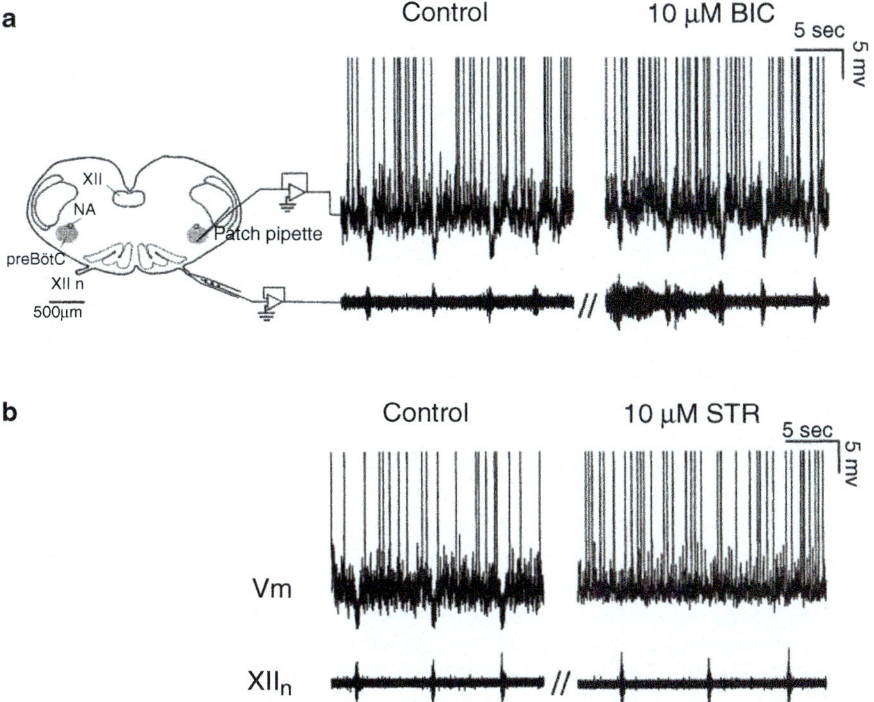

Fig. 1.9 Effects of bicuculline (BIC) and strychnine (STR) on inspiratory-phase inhibitory postsynaptic potentials (IPSPs) of a current clamped preBötC expiratory neuron and on the respiration-related rhythmic motor outflow of the hypoglossal nerve root (XII n). The findings that BIC and STR failed to abolish respiration-related rhythm indicate that inhibitory neurotransmission is not essential for respiratory rhythm generation, contradictorily to the network theory [35]. (**a**) Slice anatomy, electrophysiological recording setup and effects of BIC. XII, hypoglossal nucleus; NA, nucleus ambiguus. (**b**) Effects of STR

rhythm, suggesting that the network mechanism is not essential for respiratory rhythm generation (Fig. 1.9) [35]. It has been also believed by a number of researchers that the respiratory rhythm generating mechanism changes with development from a pacemaker-driven at the neonatal stage to a network-dependent requiring synaptic inhibition at the more matured stage [36].

3.3 Group Pacemaker and Burstlet Theory

The group of Feldman and Del Negro is criticizing both the pacemaker and network theories, and alternatively proposing a group pacemaker and burstlet theory, in which respiratory rhythm is generated as follows. Inspiration starts with a mostly silent network status after preceding inspiration. Then, spontaneous activity in a few preBötC neurons induces postsynaptic activity in other preBötC neurons by mutual excitation. This mutual excitation induces inward currents in preBötC neurons,

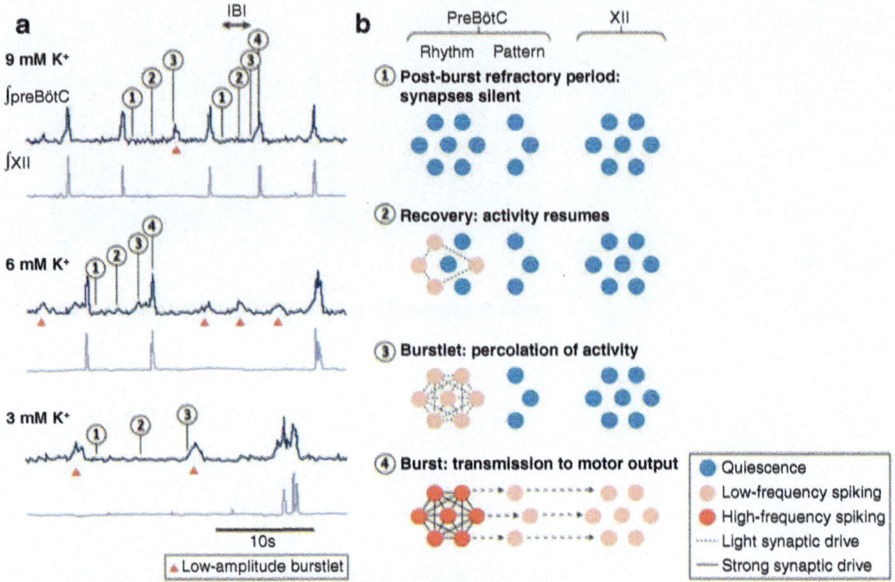

Fig. 1.10 Schema representing the group pacemaker and burstlet theory. (**a**) In vitro recordings of preBötC field potentials and hypoglossal nerve (XII) activity showing rhythmic inspiration-related output at several levels of excitability determined by changes in the extracellular K^+ concentration. (**b**) Schematic of network activity underlying burstlets, bursts, and XII output. Numerals indicate different stages of the trajectories in part a: 1, refractory state following inspiration; 2, spontaneous spiking resumes in some neuronal constituents; 3, active neurons are mutually reinforcing owing to recurrent synaptic interconnections, resulting in low amplitude pre-inspiratory activity (burstlet); and 4, the burstlet triggers a full burst, which propagates to pattern-related preBötC neurons, premotor neurons, and motoneurons, generating XII motor output [38]

causing a network-wide burst. This burst is followed by a transient refractory period possibly due to intrinsic activity-dependent outward currents and synaptic depression that does not require postsynaptic inhibition. After a refractory period, neuronal activities occur spontaneously again and the next cycle starts. In this model pre-inspiratory and inspiratory activity are not always linked; reductions in neural excitability cause occasional failure of inspiratory burst formation and produce only a low amplitude pre-inspiratory activities (burstlet). The low amplitude rhythmogenic pre-inspiratory burstlet activities in the preBötC do not always trigger bursts. Thus, the bursts themselves are not essential for rhythmogenesis, and the neuronal activities that can produce burstlet are the rhythm generator (Fig. 1.10) [37, 38]. The validity of the group pacemaker and burstlet theory must be tested by further experimental investigation.

3.4 Astrocyte-Driven Theory

Astrocytes, a kind of glial cells, had been long considered to play only passive roles in the functioning of the central nervous system, e.g., maintenance of extracellular ionic environment and mechanical and structural support of neurons in the nervous tissue. However, recent progress in glial physiology has clarified that astrocytes play active roles by closely communicating with neurons in information processing. Hülsmann et al. (2000) and Li et al. (2010) reported that blockade of astrocytic metabolism suppresses inspiration-related neural output in rhythmically active slices, suggesting that astrocytic metabolic support is necessary for the maintenance of respiratory rhythm [39, 40]. Also, in an in vivo condition Young et al. (2005) showed that inhibition of astrocytic metabolism reduced respiratory frequency in neonatal rats [41]. Because astrocytes are electrically silent, their activity cannot be electrophysiologically measured but can be analyzed by calcium imaging. We analyzed astrocytic activity in the preBötC of rhythmically active slices by calcium imaging and found that a subset of astrocytes have inspiration-synchronized rhythmic activity and some show pre-inspiratory activity (Fig. 1.11) [42]. Furthermore, we showed that optogenetic selective stimulation of preBötC astrocytes triggers inspiratory busting, suggesting that astrocytes play an active role in respiratory rhythm generation [42]. Our discovery of respiratory modulated astrocytes is not consistent with the report by the group of Hülsmann that spontaneous calcium fluctuations of astrocytes recorded by calcium imaging showed no correlation with respiratory neuronal activity in the preBötC in rhythmically active slices of mice

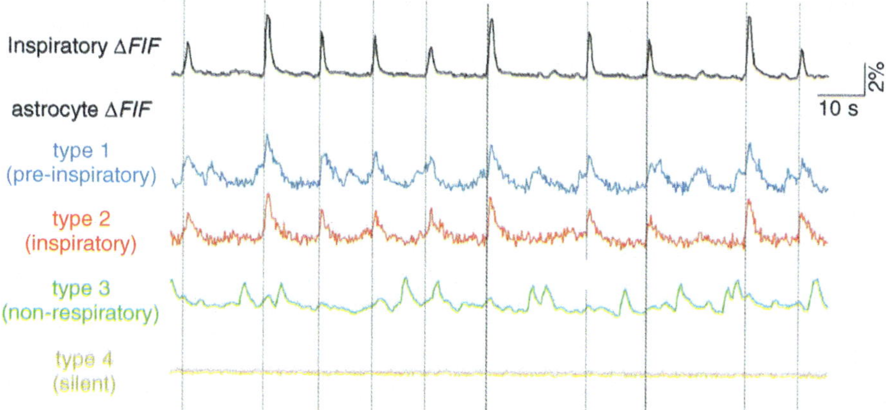

Fig. 1.11 Pre-inspiratory activity of preBötC astrocytes recorded by calcium imaging in a rhythmically active slice. Representative fluorescence calcium activity ($\Delta F/F$) traces of pre-inspiratory, inspiratory, nonrespiratory, and silent astrocytes. Top: bulk $\Delta F/F$ fluctuations of the entire preBötC, representing inspiratory activity [42]

[43]. However, when they reanalyzed the same set of calcium imaging data with a more advanced mathematical technique, they confirmed the presence of pre-inspiratory activities in a subset of astrocytes indicating a functional coupling between astrocytes and neurons in the preBötC [44]. We demonstrated in conscious mice that pharmacological blockade of astrocytic activation attenuates hypoxic ventilatory increase which initially appears when animals are exposed to hypoxia and augments subsequent hypoxic ventilatory depression (with reduction of respiratory frequency), suggesting that astrocytes contribute to the facilitation of respiratory rhythm generation especially under a hypoxic condition [45, 46]. Sheikhbahaei et al. (2018) also analyzed the role of astrocytes in the in vivo rat and demonstrated that blockade of vesicular release in preBötC astrocytes reduces respiratory frequency and decreases rhythm variability [47]. Also, regarding the generation of oscillatory orofacial movement other than breathing, it was reported that masticatory rhythm is driven by astrocytes [48]. Collectively, these reports suggest that astrocytes are actively involved in respiratory rhythm generation, and this notion is supported by the anatomical observation by us and another group that astrocytes in the preBötC are structurally more complex as compared to those in the neighboring medullary regions and are closely coupled with neurons [1, 49].

4 Pathophysiology of Respiratory Rhythm Generator

Clinically it has been well known that acute disturbance of the medulla oblongata due to various pathological events, e.g., injury, hemorrhage, and herniation, induces respiratory arrest and death. Indeed, when the bilateral preBötC is acutely lesioned, respiratory rhythm is eliminated in many cases, although occasionally spontaneous expiratory abdominal muscle activity persists in awake goats [50]. This phenomenon could be explained by considering that when inspiratory rhythm generator in the preBötC does not function respiratory rhythm generation ceases with occasional back up by the expiratory rhythm generator in the pFRG. On the other hand, when the bilateral preBötC are gradually destroyed over weeks by injection of neurotoxin in goats, respiratory rhythm turns to irregular but does not stop due to compensatory activities of other oscillators, indicating that the respiratory rhythm generating mechanism is highly plastic and robust (Fig. 1.12) [51]. Such plasticity and robustness of respiratory rhythm generation are also clinically experienced in patients with gradually progressing large lesions occupying the mostly entire medullary structure, e.g., with cancer metastasis to the medulla [52] or with syringobulbia accompanied by a large cavity in the medulla [53], who show disturbed breathing patterns but persistently generate respiratory rhythm. Pathological lesions disturb respiratory rhythm generation and generally cause hypoventilation. However, brainstem astrocytoma is not the case and could cause severe neurogenic hyperventilation [54], suggesting that tumorous astrocytes secret gliotransmitters that pathologically facilitate respiratory rhythm generation. Such a mechanism is compatible with the astrocyte-driven theory for respiratory rhythm generation.

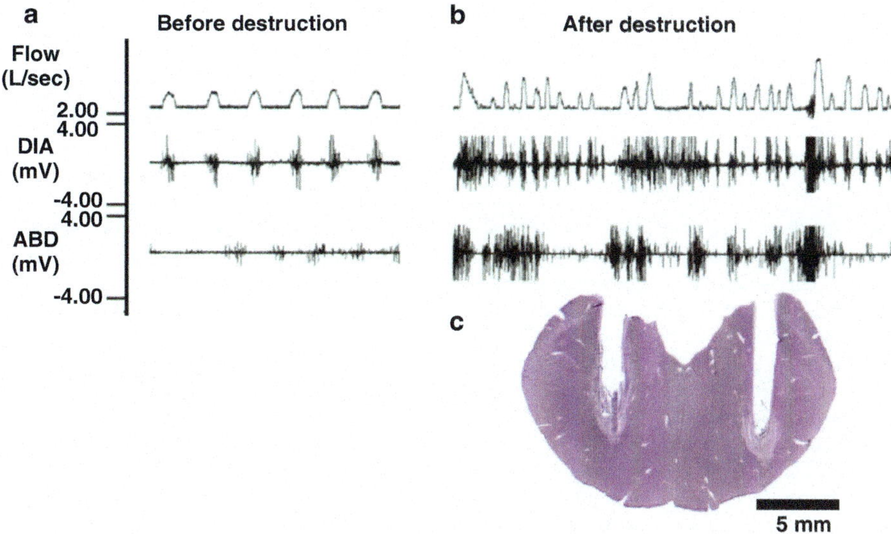

Fig. 1.12 Effect of chronic destruction of preBötC on breathing in an awake goat. Respiratory flow (Flow), diaphragmatic muscle activity (DIA), and abdominal expiratory muscle activity (ABD) before (**a**) and after (**b**) destruction of the preBötC by injection of neurotoxin are illustrated. After the destruction of the preBötC, breathing became irregular, and DIA and ABD showed simultaneous contraction. (**c**) Nearly total destruction of the preBötC by injection of neurotoxin was histologically confirmed [51]

5 Conclusion

The fundamental respiratory rhythm generating networks are located in the pons and medulla oblongata in which the preBötC is the kernel. It has been elucidated that the preBötC rhythm generator is connected with other oscillators and neuron networks in the more rostral brainstem and the spinal cord. However, in a conscious state these oscillators should be communicating also with further higher brain regions including the cerebral cortex. These widely distributed and interconnected regional networks work synchronously, receive information from peripheral and central sensory receptors on internal and external homeostatic changes, and determine the respiratory rhythm and pattern by breath-by-breath. However, the anatomy and physiology of the communication mechanisms between the higher brain and lower brainstem remain totally unelucidated and is the future research topics.

Since the introduction of in vitro preparations to the field of respiratory rhythm generation, studies on the respiratory rhythm generating mechanism have significantly advanced but the rhythmogenic theories remain controversial. Further extensive studies are necessary to fully clarify the anatomical basis and physiological and pathophysiological mechanisms of the respiratory rhythm generation, especially focusing on the role of astrocytes, of which roles are far from elucidated. For this purpose, multilevel advanced experimental approaches at molecular, cellular, network, and system levels should be applied.

References

1. Ikeda K, Kawakami K, Onimaru H, Okada Y, Yokota S, Koshiya N, Oku Y, Iizuka M, Koizumi H. The respiratory control mechanisms in the brainstem and spinal cord: integrative views of the neuroanatomy and neurophysiology. J Physiol Sci. 2017;67:45–62.
2. Marckwald M. The movements of respiration and their innervation in the rabbit. London: Blackie; 1888.
3. Lumsden T. Observations on the respiratory centres in the cat. J Physiol. 1923;57:153–60.
4. Suzue T. Respiratory rhythm generation in the in vitro brain stem-spinal cord preparation of the neonatal rat. J Physiol. 1984;354:173–83.
5. Onimaru H, Arata A, Homma I. Primary respiratory rhythm generator in the medulla of brainstem-spinal cord preparation from newborn rat. Brain Res. 1988;445:314–24.
6. Ballanyi K, Onimaru H, Homma I. Respiratory network function in the isolated brainstem-spinal cord of newborn rats. Prog Neurobiol. 1999;59:583–634.
7. Shinozaki Y, Yokota S, Miwakeichi F, Pokorski M, Aoyama R, Fukuda K, Yoshida H, Toyama Y, Nakamura M, Okada Y. Structural and functional identification of two distinct inspiratory neuronal populations at the level of the phrenic nucleus in the rat cervical spinal cord. Brain Struct Funct. 2019;224:57–72.
8. Smith JC, Ellenberger HH, Ballanyi K, Richter DW, Feldman JL. Pre-Bötzinger complex: a brainstem region that may generate respiratory rhythm in mammals. Science. 1991;254:726–9.
9. Koshiya N, Smith JC. Neuronal pacemaker for breathing visualized in vitro. Nature. 1999;400:360–3.
10. Koshiya N, Oku Y, Yokota S, Oyamada Y, Yasui Y, Okada Y. Anatomical and functional pathways of rhythmogenic inspiratory premotor information flow originating in the pre-Bötzinger complex in the rat medulla. Neuroscience. 2014;268:194–211.
11. Schwarzacher SW, Rüb U, Deller T. Neuroanatomical characteristics of the human pre-Bötzinger complex and its involvement in neurodegenerative brainstem diseases. Brain. 2011;134:24–35.
12. Onimaru H, Homma I. A novel functional neuron group for respiratory rhythm generation in the ventral medulla. J Neurosci. 2003;23:1478–86.
13. Oku Y, Masumiya H, Okada Y. Postnatal developmental changes in activation profiles of the respiratory neuronal network in the rat ventral medulla. J Physiol. 2007;585:175–86.
14. Janczewski WA, Feldman JL. Distinct rhythm generators for inspiration and expiration in the juvenile rat. J Physiol. 2006;570:407–20.
15. Okada Y, Masumiya H, Tamura Y, Oku Y. Respiratory and metabolic acidosis differentially affect the respiratory neuronal network in the ventral medulla of neonatal rats. Eur J Neurosci. 2007;26:2834–43.
16. Lal A, Oku Y, Hülsmann S, Okada Y, Miwakeichi F, Kawai S, Tamura Y, Ishiguro M. Dual oscillator model of the respiratory neuronal network generating quantal slowing of respiratory rhythm. J Comput Neurosci. 2011;30:225–40.
17. Anderson TM, Garcia AJ 3rd, Baertsch NA, Pollak J, Bloom JC, Wei AD, Rai KG, Ramirez JM. A novel excitatory network for the control of breathing. Nature. 2016;536:76–80.
18. Anderson TM, Ramirez JM. Respiratory rhythm generation: triple oscillator hypothesis. F1000Res. 2017;6:139.
19. Chamberlin NL. Functional organization of the parabrachial complex and intertrigeminal region in the control of breathing. Respir Physiol Neurobiol. 2004;143:115–25.

20. Mörschel M, Dutschmann M. Pontine respiratory activity involved in inspiratory/expiratory phase transition. Philos Trans R Soc Lond Ser B Biol Sci. 2009;364:2517–26.
21. Poon CS, Song G. Bidirectional plasticity of pontine pneumotaxic postinspiratory drive: implication for a pontomedullary respiratory central pattern generator. Prog Brain Res. 2014;209:235–54.
22. Yokota S, Kaur S, VanderHorst VG, Saper CB, Chamberlin NL. Respiratory-related outputs of glutamatergic, hypercapnia-responsive parabrachial neurons in mice. J Comp Neurol. 2015;523:907–20.
23. Aoki M, Mori S, Kawahara K, Watanabe H, Ebata N. Generation of spontaneous respiratory rhythm in high spinal cats. Brain Res. 1980;202:51–63.
24. Kobayashi S, Fujito Y, Matsuyama K, Aoki M. Spontaneous respiratory rhythm generation in in vitro upper cervical slice preparations of neonatal mice. J Physiol Sci. 2010;60:303–7.
25. Oku Y, Okabe A, Hayakawa T, Okada Y. Respiratory neuron group in the high cervical spinal cord discovered by optical imaging. Neuroreport. 2008;19:1739–43.
26. Okada Y, Yokota S, Shinozaki Y, Aoyama R, Yasui Y, Ishiguro M, Oku Y. Anatomical architecture and responses to acidosis of a novel respiratory neuron group in the high cervical spinal cord (HCRG) of the neonatal rat. Adv Exp Med Biol. 2009;648:387–94.
27. Fukushi I, Yokota S, Okada Y. The role of the hypothalamus in modulation of respiration. Respir Physiol Neurobiol. 2019;265:172–9.
28. Onimaru H, Homma I. Spontaneous oscillatory burst activity in the piriform-amygdala region and its relation to in vitro respiratory activity in newborn rats. Neuroscience. 2007;144:387–94.
29. Del Negro CA, Morgado-Valle C, Hayes JA, Mackay DD, Pace RW, Crowder EA, Feldman JL. Sodium and calcium current-mediated pacemaker neurons and respiratory rhythm generation. J Neurosci. 2005;25:446–53.
30. von Euler C. On the central pattern generator for the basic breathing rhythmicity. J Appl Physiol Respir Environ Exerc Physiol. 1983;55:1647–59.
31. Smith JC, Abdala AP, Rybak IA, Paton JF. Structural and functional architecture of respiratory networks in the mammalian brainstem. Philos Trans R Soc Lond Ser B Biol Sci. 2009;364:2577–87.
32. Richter DW, Smith JC. Respiratory rhythm generation in vivo. Physiology (Bethesda). 2014;29:58–71.
33. Kuwana S, Tsunekawa N, Yanagawa Y, Okada Y, Kuribayashi J, Obata K. Electrophysiological and morphological characteristics of GABAergic respiratory neurons in the mouse pre-Bötzinger complex. Eur J Neurosci. 2006;23:667–74.
34. Kuwana S, Okada Y, Sugawara Y, Tsunekawa N, Obata K. Disturbance of neural respiratory control in neonatal mice lacking GABA synthesizing enzyme 67-kDa isoform of glutamic acid decarboxylase. Neuroscience. 2003;120:861–70.
35. Shao XM, Feldman JL. Respiratory rhythm generation and synaptic inhibition of expiratory neurons in pre-Bötzinger complex: differential roles of glycinergic and GABAergic neural transmission. J Neurophysiol. 1997;77:1853–60.
36. Hayashi F, Lipski J. The role of inhibitory amino acids in control of respiratory motor output in an arterially perfused rat. Respir Physiol. 1992;89:47–63.
37. Feldman JL, Del Negro CA. Looking for inspiration: new perspectives on respiratory rhythm. Nat Rev Neurosci. 2006;7:232–42.
38. Del Negro CA, Funk GD, Feldman JL. Breathing matters. Nat Rev Neurosci. 2018;19:351–67.
39. Hülsmann S, Oku Y, Zhang W, Richter DW. Metabolic coupling between glia and neurons is necessary for maintaining respiratory activity in transverse medullary slices of neonatal mouse. Eur J Neurosci. 2000;12:856–62.
40. Li GC, Zhang HT, Jiao YG, Wu ZH, Fang F, Cheng J. Glial cells are involved in the exciting effects of doxapram on brainstem slices in vitro. Cell Mol Neurobiol. 2010;30:667–70.
41. Young JK, Dreshaj IA, Wilson CG, Martin RJ, Zaidi SI, Haxhiu MA. An astrocyte toxin influences the pattern of breathing and the ventilatory response to hypercapnia in neonatal rats. Respir Physiol Neurobiol. 2005;147:19–30.

42. Okada Y, Sasaki T, Oku Y, Takahashi N, Seki M, Ujita S, Tanaka KF, Matsuki N, Ikegaya Y. Preinspiratory calcium rise in putative pre-Bötzinger complex astrocytes. J Physiol. 2012;590:4933–44.
43. Schnell C, Fresemann J, Hülsmann S. Determinants of functional coupling between astrocytes and respiratory neurons in the pre-Bötzinger complex. PLoS One. 2011;6:e26309.
44. Oku Y, Fresemann J, Miwakeichi F, Hülsmann S. Respiratory calcium fluctuations in low-frequency oscillating astrocytes in the pre-Bötzinger complex. Respir Physiol Neurobiol. 2016;226:11–7.
45. Fukushi I, Takeda K, Yokota S, Hasebe Y, Sato Y, Pokorski M, Horiuchi J, Okada Y. Effects of arundic acid, an astrocytic modulator, on the cerebral and respiratory functions in severe hypoxia. Respir Physiol Neurobiol. 2016;226:24–9.
46. Fukushi I, Takeda K, Uchiyama M, Kurita Y, Pokorski M, Yokota S, Okazaki S, Horiuchi J, Mori Y, Okada Y. Blockade of astrocytic activation delays the occurrence of severe hypoxia-induced seizure and respiratory arrest in mice. J Comp Neurol. 2020;528(8):1257–64.
47. Sheikhbahaei S, Turovsky EA, Hosford PS, Hadjihambi A, Theparambil SM, Liu B, Marina N, Teschemacher AG, Kasparov S, Smith JC, Gourine AV. Astrocytes modulate brainstem respiratory rhythm-generating circuits and determine exercise capacity. Nat Commun. 2018;9:370.
48. Morquette P, Verdier D, Kadala A, Féthière J, Philippe AG, Robitaille R, Kolta A. An astrocyte-dependent mechanism for neuronal rhythmogenesis. Nat Neurosci. 2015;18:844–54.
49. SheikhBahaei S, Morris B, Collina J, Anjum S, Znati S, Gamarra J, Zhang R, Gourine AV, Smith JC. Morphometric analysis of astrocytes in brainstem respiratory regions. J Comp Neurol. 2018;526:2032–47.
50. Wenninger JM, Pan LG, Klum L, Leekley T, Bastastic J, Hodges MR, Feroah TR, Davis S, Forster HV. Large lesions in the pre-Bötzinger complex area eliminate eupneic respiratory rhythm in awake goats. J Appl Physiol (1985). 2004;97:1629–36.
51. Krause KL, Forster HV, Kiner T, Davis SE, Bonis JM, Qian B, Pan LG. Normal breathing pattern and arterial blood gases in awake and sleeping goats after near total destruction of the presumed pre-Botzinger complex and the surrounding region. J Appl Physiol (1985). 2009;106:605–19.
52. Corne S, Webster K, McGinn G, Walter S, Younes M. Medullary metastasis causing impairment of respiratory pressure output with intact respiratory rhythm. Am J Respir Crit Care Med. 1999;159:315–20.
53. Nogués M, Gené R, Benarroch E, Leiguarda R, Calderón C, Encabo H. Respiratory disturbances during sleep in syringomyelia and syringobulbia. Neurology. 1999;52:1777–83.
54. Rodriguez M, Baele PL, Marsh HM, Okazaki H. Central neurogenic hyperventilation in an awake patient with brainstem astrocytoma. Ann Neurol. 1982;11:625–8.

Chapter 2
Anatomy and Physiology of Chemical Regulatory Mechanisms in Respiration: How Does Chemical Regulation Work with Various Acid–Base Disorders?

Yoshitaka Oyamada

Abstract Acidemia is one of the major excitatory stimuli for respiration. It is sensed by both the peripheral and the central chemoreceptors. In the peripheral chemoreceptor, the carotid body, acidemia inhibits potassium channels (TASK-1/TASK-3 and Ca^{2+}-dependent K^+ channel) on type I cells, then activates them. Activated type I cells release synaptic transmitters, which stimulate postsynaptic nerve endings of afferent fibers. The elicited impulses ascend through the carotid sinus nerve, a branch of the glossopharyngeal nerve, to the respiratory center in the brainstem. Meanwhile, much about the central chemoreception is still to be determined. However, TASK-2 and a proton-sensing G protein-coupled receptor, GPR4, on neurons in the retrotrapezoid nucleus are likely to be implicated in the process. Hypercapnia is also an excitatory stimulus, independent of acidemia. Connexin 26 (Cx26) is expressed on sub-pial astrocytes in the ventrolateral medulla. It is directly sensitive to CO_2. Pharmacological blockade of Cx26 reduced hypercapnic ventilatory response in vivo, suggesting that Cx26 could play a crucial role in the central chemoreception.

Keywords Acidemia · Hypercapnia · Peripheral chemoreceptor · Central chemoreceptor

1 Introduction

Respiration in humans is affected by various factors, such as behavior (e.g., speech, swallowing, and exercise), emotional overload, mechanical stimuli to the broncho-pulmonary system, and imbalances in chemical homeostasis. In this chapter, the

Y. Oyamada (✉)
Respiratory Medicine, National Tokyo Medical Center, Tokyo, Japan
e-mail: oyamada.yoshitaka.fe@mail.hosp.go.jp

K. Yamaguchi (ed.), *Structure-Function Relationships in Various Respiratory Systems*, Respiratory Disease Series: Diagnostic Tools and Disease Managements, https://doi.org/10.1007/978-981-15-5596-1_2

regulatory mechanisms of respiration in chemical imbalances, especially in acid–base imbalance, will be described.

2 Overview of Chemical Regulatory Mechanisms of Respiration

Among various chemical imbalances, hypoxia, hypercapnia, and acidemia (with or without hypercapnia) are major excitatory stimuli for respiration. These stimuli are sensed by two types of chemoreceptors. One is the peripheral chemoreceptor, and the other is the central (i.e., intracranial). Hypoxia elicits ventilatory augmentation mainly by stimulating the peripheral chemoreceptor, while so does hypercapnia mainly via the central chemoreceptor. However, these receptors work in concert. For example, the degree of ventilatory augmentation caused by hypercapnia is substantially influenced by the degree of activation of the peripheral chemoreceptor [1].

Acidemia is sensed by both receptors. When exposed to hypercapnic conditions, an increase in CO_2 eventually results in an increase in H^+ via accelerated production of H_2CO_3 by carbonic anhydrase (CA).

$$CO_2 + H_2O \overset{CA}{\leftrightarrow} H_2CO_3 \leftrightarrow H^+ + HCO_3^-$$

This might lead you to an idea that increased H^+ (i.e., acidemia) is the substantial stimulus under hypercapnic conditions. However, this is wrong. Several prior studies suggest that H^+ is not the unique stimulus [2–5]. For example, in experimental animals, the addition of hypercapnia to acidemia was able to cause stronger ventilatory augmentation than the acidemia alone [4]. And even when accompanied changes in pH were offset by adding HCO_3^-, hypercapnia increased ventilation [5].

3 Peripheral Chemoreceptor

There are two types of peripheral chemoreceptors. One is the carotid body, and the other is the aortic body. The peripheral chemoreceptor "for respiration" is the carotid body. It is considered that the aortic body is not implicated in the control of respiration, but in cardiovascular reflexes, although details in its physiological role are still to be determined.

3.1 The Carotid Body

The carotid body is bilaterally located at the bifurcation of the external and the internal carotid arteries. It is as small as 5 mm in its diameter. The carotid body is composed of type I (glomus) and type II (sustentacular) cells, afferent and efferent nerve fibers, vessels, and connective tissues. Type I cells show similar morphology to adrenal chromaffin cells, while type II cells look like glial cells. Each type II cell encircles 4–6 type I cells. The carotid body receives extremely high blood flow, so that its flow rate per unit weight overwhelms that of other organs (e.g., 15–30 times more than the brain).

Activation of the carotid body causes ventilatory augmentation. Other resultant physiological responses include cardiovascular and sympathomimetic reflexes. Major stimuli for the carotid body are hypoxemia (reduced PaO_2) and acidemia. However, various other factors, such as hypercapnia, hypotension, reduced blood flow, hyperkalemia, hyperthermia, hypoglycemia, and some hormones (catecholamine, angiotensin II, antidiuretic hormone) also stimulate the carotid body [6]. Meanwhile, anemia does not stimulate the carotid body [7]. This is attributed to the high blood flow which the carotid body receives. Because of the high blood flow, dissolved oxygen (which is proportional to PaO_2) is enough to meet the oxygen demand of the carotid body. Reduced PaO_2 results in a decrease in dissolved oxygen, while reduced CaO_2 by anemia or decrease in oxygenated hemoglobin due to CO intoxication does not affect the level of dissolved oxygen.

Figure 2.1 shows the underlying mechanism as to how hypoxemia and acidemia stimulate the carotid body. Hypoxemia and acidemia elicit depolarization in type I cells, mainly by inhibiting Twik-related acid-sensitive K^+ channel (TASK) 1 and 3 heterodimers, which normally set the resting membrane potential of these

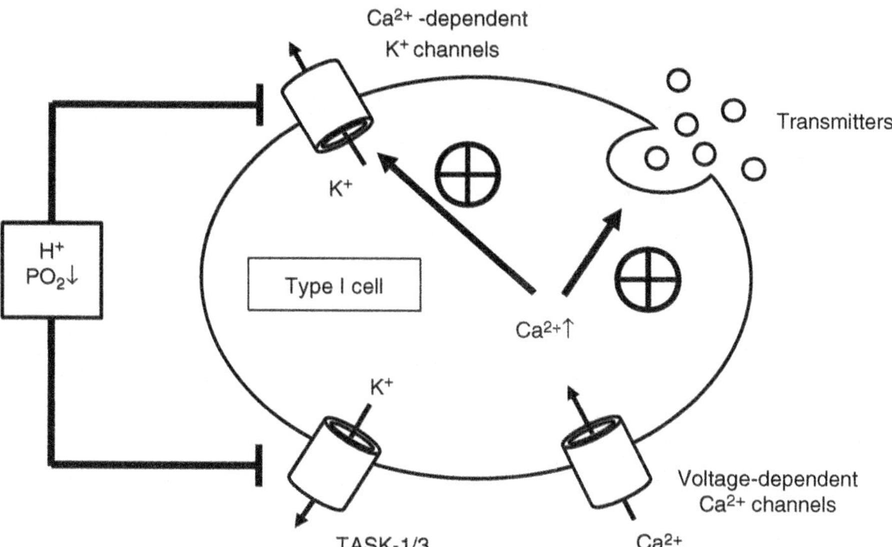

Fig. 2.1 A schema as to how acidemia and hypoxemia activate type I cells. Acidemia and hypoxemia elicit depolarization in type I cells by inhibiting TASK 1 and 3 heterodimers which normally set the resting membrane potential of these cells. The depolarization generates action potentials. This results in the opening of voltage-gated Ca^{2+} channels and subsequent Ca^{2+} entry through the channels. Normally, the resultant increase in intracellular Ca^{2+} opens Ca^{2+}-dependent K^+ channels, so that the cell repolarizes. However, acidemia and hypoxemia also inhibit these channels and, therefore, block repolarization. The increase in intracellular Ca^{2+} triggers the release of chemical transmitters

cells [8]. Acidemia is thought to directly inhibit the heterodimeric channel TASK-1/TASK-3 [8]. Meanwhile, there is still some controversy on how hypoxemia is sensed by the channel. The major putative mechanism is oxygen sensing through changes in mitochondrial energy metabolism. Furthermore, activation of AMP kinase is also considered important (For details, see reference [8]). The depolarization generates action potentials. This results in the opening of voltage-gated Ca^{2+} channels and subsequent Ca^{2+} entry through the channels. Normally, the resultant increase in intracellular Ca^{2+} opens Ca^{2+}-dependent K^+ channels, so that the cell repolarizes. However, hypoxemia and acidemia also inhibit these channels and, therefore, block repolarization [9]. This functions to keep type I cells depolarized. The increase in intracellular Ca^{2+} also triggers the release of chemical transmitters. Released transmitters stimulate postsynaptic nerve endings of afferent fibers. The elicited impulses ascend through the carotid sinus nerve, a branch of the glossopharyngeal nerve, to the respiratory center in the brainstem.

It is still unknown whether CO_2 directly stimulates type I cells or not.

3.2 *The Aortic Body*

The aortic body is a small cluster of vessel-rich tissues located along the aortic arch. It is also composed of type I and type II cells. Its afferent fibers run in the aortic nerve. Like the carotid body, the aortic body is activated by hypoxemia. But its response is weaker [10]. Meanwhile, in contrast to the carotid body, the aortic body responds to reduced CaO_2 [7].

4 Central Chemoreceptor

4.1 *Overview*

Since the 1950s, so many attempts have been made and some of them are still ongoing to identify the central chemoreceptor(s) which primarily sense(s) changes in pH/PCO_2. Experimental techniques evolving with the times (from subarachnoid space perfusion in the dawn to genetic recombination of specific molecules in recent years) have found out several "chemosensitive" areas/nuclei in the central nervous system (CNS). Although evidence is accumulating for some of those areas/nuclei, much about key structures and molecules for the central chemoreception, mechanisms to sense changes in pH/PCO_2, and how the output from each putative chemoreceptor is integrated and finally alters respiration is still to be clarified.

Described below is a summary of pivotal studies on central chemoreception. However, readers should be aware that experimental models (i.e., in vivo, in vitro, or ex vivo), animals used, and their developmental stage (i.e., neonates or adults) often differ among the studies.

4.2 Chemosensitive Areas in the CNS

4.2.1 The Ventrolateral Medulla (VLM)

The VLM is a kernel of respiratory rhythm generation. From the 1960s to 1970s, experiments using topical application of acid [11, 12] or focal lesioning [13] onto the ventral medullary surface were performed in anesthetized, artificially ventilated animals. These studies demonstrated that the acidification of the VLM induced ventilatory augmentation, while the lesioning eliminated the hypercapnic ventilatory response, suggesting that the VLM is chemosensitive. Later, the ventilatory augmentation induced by focal acidification was reconfirmed with a more sophisticated method, microinjection of acetazolamide within the VLM [14]. Studies using neuronal C-FOS, a transcription factor coded by an immediate-early gene *C-FOS*, as a marker of activation by hypercapnia demonstrated immunoreactivity of C-FOS in the VLM including the retrotrapezoid nucleus (RTN) in animals exposed to hypercapnia [15, 16].

The RTN shows connections to the pre-Bötzinger complex (PBC), a putative respiratory rhythm generator in the rostral VLM [17]. And it has been demonstrated that the RTN neurons uniformly express a homeodomain transcription factor PHOX2b whose abnormality causes congenital central hypoventilation syndrome [18]. Therefore, the RTN is considered to be critically involved in the control of respiration.

4.2.2 The Nucleus Tractus Solitarii (NTS)

The NTS is located superficially in the dorsal medulla. This nucleus is a part of the respiratory control network commonly called DRG (dorsal respiratory group). As in the VLM, focal acidification in the NTS caused respiratory augmentation in animals [14, 19], while patients with unilateral focal lesions of rostrolateral medulla including the NTS showed reduced hypercapnic responses [20]. The NTS also showed C-FOS immunoreactivity in animals exposed to hypercapnia [15, 16].

4.2.3 The Locus Coeruleus (LC)

The LC is a nucleus located bilaterally in the dorsal pons adjacent to the base of the fourth ventricle and consists of noradrenergic neurons. The afferent fibers to the LC arise from relatively restricted areas, while the LC sends its efferent to a wide range of the CNS, including the brainstem [21]. Although this nucleus has traditionally been thought to be associated with arousal response, state of vigilance, fear, and anxiety, a series of studies suggests that it might relay inspiratory neural output to higher CNS [22, 23]. As in the VLM and the NTS, focal acidification of the LC augmented breathing [14] and C-FOS immunoreactivity was also increased under hypercapnic conditions [16].

4.2.4 Others

Other chemosensitive areas/nuclei include the medullary raphe, the tuberomammillary nucleus, the lateral hypothalamus, the hypothalamic paraventricular nucleus, and the amygdala [24]. However, these areas/nuclei seem to be more important for other CO_2-dependent physiological responses (i.e., arousal, vigilance, fear, and blood pressure control) rather than control of respiration [24].

4.3 Putative Structures for the Central Chemoreception

4.3.1 Neurons

An electrophysiological ex vivo study using intracellular recordings demonstrated that certain types of respiratory neurons in the VLM exhibited a depolarizing (i.e., excitatory) response to hypercapnic acidosis (Fig. 2.2a) [25]. The response was retained even after the blockade of chemical synaptic transmission either by lowering extracellular Ca^{2+} (i.e., inhibition of transmitter release from presynaptic terminals) or by addition of tetrodotoxin (TTX) to the artificial cerebrospinal fluid (CSF) (i.e., by blocking fast Na^+ channel) (Fig. 2.2b) [25], indicating that the excitatory response could be an intrinsic property of these respiratory neurons. In the study, morphological analysis of recorded neurons was also made. The analysis demonstrated that many chemosensitive neurons extended their dendritic projections to just beneath the

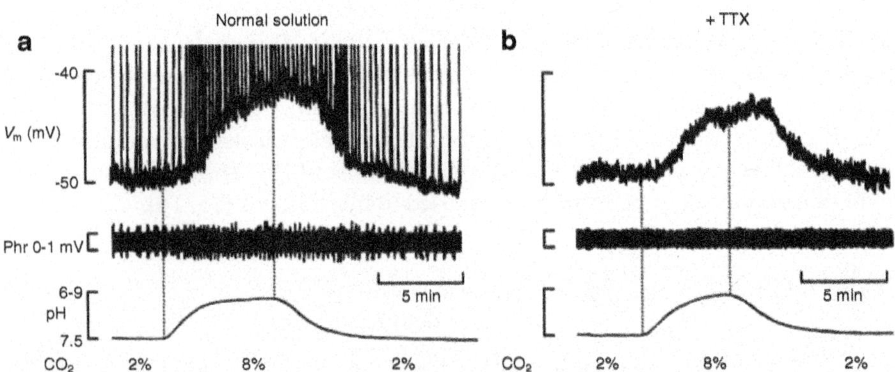

Fig. 2.2 (**a** and **b**) Depolarizing response of an inspiratory neuron in the VLM of the isolated brainstem–spinal cord preparation of the neonatal rat. *Vm*: membrane potential. Phr: phrenic nerve activity. The neuron exhibits periodic bursts of action potentials (shown in a truncated form) synchronized to the phrenic bursts. Increasing CO_2 concentration in the gas bubbling the artificial CSF from 2 to 8% results in an acidic shift in pH at the surface of the medulla. This causes an increase in phrenic burst frequency and a depolarizing shift of baseline membrane potential of the neuron (**a**). The depolarizing response is retained after the blockade of chemical synaptic transmission by adding TTX to the artificial CSF (**b**). (**c**) Morphology of the recorded neuron. A broad arrow indicates dendritic projections, while thin arrows indicate an axon and its collateral [25]

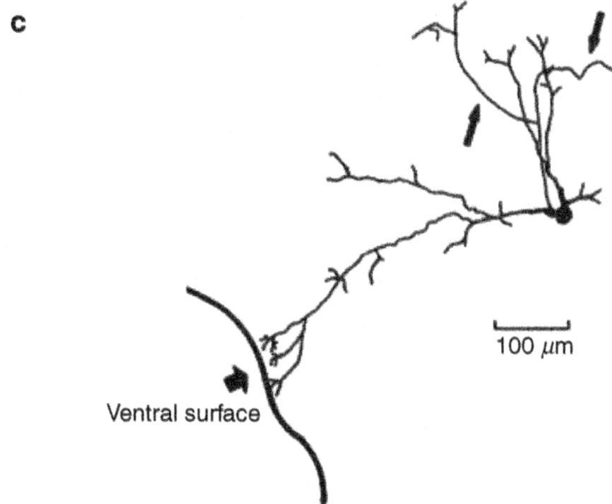

Fig. 2.2 (continued)

medullary surface (Fig. 2.2c) and suggested that the dendrites could be sensors for changes in pH/PCO_2 of surrounding environments [25]. Similar depolarizing response of neurons was observed in the PBC [26], the LC (Figs. 2.3a, b) [22, 23], and in the NTS [27]. It has also been demonstrated that RTN neurons, which are probably described as chemosensitive respiration-modulated or tonic neurons near the facial nucleus in a former study [25], are intrinsically sensitive to acidification and CO_2 [17].

These findings suggest that neurons themselves could be chemoreceptors. Such an idea seems reasonable and attractive. However, nowadays, it cannot be excluded that some attached structures whose communication with neurons is independent of chemical synaptic transmission might affect the excitatory response of these "chemosensitive" neurons. For example, transmitter release from glial cells is not dependent on either extracellular Ca^{2+} or TTX-sensitive Na^+ channels [24].

4.3.2 Glial Cells

The primary cell type at the VLM surface is astrocytes and they compose the marginal glial layer (MGL). Several studies demonstrated that glial cells in the layer show depolarizing responses to hypercapnia [28–30]. Activated astrocytes in the ventral surface of the medulla release ATP [31] and may activate neighboring neurons via a P2-receptor-dependent mechanism [31]. Furthermore, the topical application of a glial toxin, fluorocitrate, into the VLM surface reduced the hypercapnic ventilatory response in anesthetized rats [32]. And blocking the release of ATP reduced the hypercapnic ventilatory response [33], while the application of ATP on the VLM replicated the hypercapnic ventilatory response [33]. These evidences suggest that astrocytes (at least in the VLM) could be the primary chemoreceptor and they might affect ventilatory response to CO_2 by interacting with neuronal dendrites extended within the glial layer via ATP release.

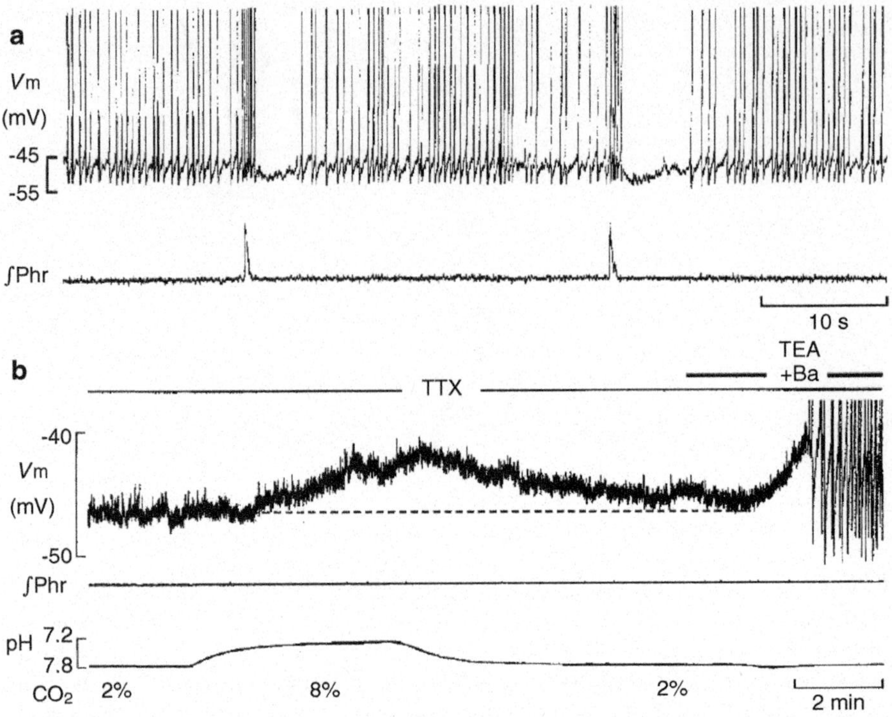

Fig. 2.3 (**a**) Membrane potential trajectory of an LC neuron in the isolated brainstem–spinal cord preparation of the neonatal rat. *Vm*: membrane potential. ∫Phr: integrated signal of phrenic nerve activity. The neuron shows periodic bursts of action potentials and subsequent hyperpolarization synchronized to the phrenic bursts. (**b**) Depolarizing response of an LC neuron to hypercapnic acidosis is retained in the presence of TTX. Subsequent depolarization is due to the addition of K^+ channel blockers, tetraethylammonium (TEA) + barium (Ba) [22]

Okada et al. demonstrated that intrinsically CO_2-sensitive "cells" (not stated as "neurons") defined by expression of C-FOS when exposed to hypercapnia under blockade of chemical synaptic transmission were located inside the MGL [34]. And they surrounded fine vessels [34]. It has also been shown that the medullary surface had an indentation with thickened MGL covered by surface vessels, suggesting that these characteristic anatomical structures may be necessary for the central chemoreception [34].

4.4 Putative Molecules Involved in the Central Chemoreception

4.4.1 Potassium Channels

Baseline membrane potential of neurons is mainly set by potassium conductance. There is a variety of potassium channels and their expression differs among cell types. Expressed channels act in combination and regulate potassium conductance.

From the aspect that depolarizing response underlies the excitability of the cell, some acid-sensitive potassium channels have been investigational targets.

Kir (pH sensitive inwardly rectifying potassium channels) 4.1/5.1 heteromer has been shown to underlie the pH sensitivity of some neurons and glial cells [30, 35]. In sub-pial astrocytes in the RTN, the closure of the channel leads the cells to depolarization, resulting in ATP release [30]. Since ATP is important in hypercapnic ventilatory response (as described in Sect. 4.3.2), Kir expressed on astrocytes in the RTN could play an important role in central chemosensitivity. However, in the brainstem–spinal cord preparations isolated from Kir5.1(-/-) mice, resting phrenic nerve activity and phrenic nerve response to respiratory acidosis or isohydric hypercapnia were similar in those obtained from wild-type mice, suggesting that the role of Kir4.1/5.1 in central chemosensitivity is questionable [36].

TASK-1 channels contribute to the pH sensitivity in raphe neurons [37], maybe in neurons of the LC [38] and the PBC [39]. However, as in the case of Kir5.1, targeted deletion of TASK-1 and TASK-3 genes did not affect the hypercapnic ventilatory response [37]. Meanwhile, the genetic deletion of TASK-2 blunted RTN neuronal pH sensitivity in vitro and diminished the ventilatory response to CO_2/H^+ in vivo [40]. TASK-2 channel has a more restricted brainstem distribution and was localized to the PHOX2b-expressing neurons in the retrotrapezoid nucleus (RTN) [40]. Moreover, a recent study in human which investigated associations of genetic variants with the response to hypercapnia has found a suggestive association for hypercapnic ventilatory response with TASK-2 rs2815118 (intronic tagging SNP) and rs150380866 (coding variant Arg → Trp) [41].

4.4.2 GPR4

GPR4 is one of proton-sensing G protein-coupled receptors that are activated by extracellular acidic pH. A recent study revealed that the genetic deletion of GPR4 disrupted acidosis-dependent activation of RTN neurons, increased apnea frequency, and blunted ventilatory response to CO_2 [42]. In the study, the reintroduction of GPR4 into RTN neurons restored CO_2-dependent RTN neuronal activation and rescued the ventilatory phenotype [42]. Furthermore, additional elimination of TASK-2 expressed in RTN neurons essentially abolished the ventilatory response to CO_2 [42].

4.4.3 Connexin 26 (Cx26)

Connexin is a protein that forms gap junctions. Among many connexin families, Cx26 is found to be expressed on the leptomeninges and sub-pial astrocytes, including those at the VML surface [43]. Cx26 is directly sensitive to changes in PCO_2, resulting in ATP release by its opening [43]. Expression of Cx26 made HeLa cells CO_2-sensitive and induced ATP release from those cells [44]. Pharmacological blockade of Cx26 reduced ATP release elicited by hypercapnia in vitro and in vivo [43]. It also reduced the hypercapnic ventilatory response in vivo [43]. These findings suggest that Cx26 could play a crucial role in central chemosensitivity. Studies using genetic manipulation are awaited.

5 Conclusion

Fig. 2.4 summarizes how respiration is chemically regulated in the presence of acidemia (with or without hypercapnia).

An increase in H^+ is perceived by both the peripheral and the central chemoreceptors. In the peripheral chemoreceptor (i.e., the carotid body), TASK-1/TASK-3 heterodimer on type I cells is the key structure for H^+ sensing (see also Fig. 2.1).

On the other hand, there are still many arguments on H^+/CO_2 sensing processes in the central chemoreception. Huckstepp and Dale proposed in their review on central chemosensitivity that "a primary chemosensory nucleus must contain

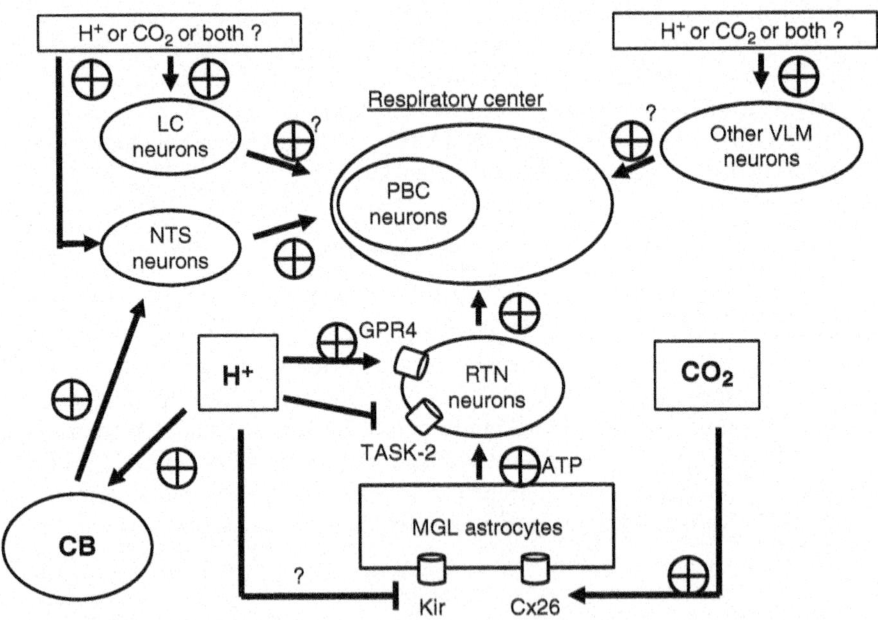

Fig. 2.4 A schema summarizing how respiration is chemically regulated in the presence of acidemia (with or without hypercapnia). RTN neurons are activated by H^+ through inhibition of TASK-2 and activation of GPR4. They are also activated by ATP released from neighboring astrocytes. In astrocytes, CO_2-induced ATP release occurs via the opening of Cx26. H^+ may also activate astrocytes, resulting in their release of ATP. Neurons in the NTS and the LC as well as non-RTN neurons in the VLM may contribute to the central chemosensitivity. However, H^+/CO_2 sensing mechanism in these cells is still to be determined. Acidemia also activates the peripheral chemoreceptor (the carotid body, CB)

intrinsically chemosensitive cells" [24]. And "an intrinsically chemosensitive cells must contain a molecular transducer that enables it to respond directly to the relevant chemosensory stimulus. The necessity of this molecular transducer for chemosensory responses must be demonstrated by either genetic deletion or selective antagonism" [24]. Furthermore, "This cell should then communicate, directly or indirectly, with the physiological centers of breathing" [24]. Taking this proposal into account, it seems reasonable to consider RTN neurons as one of central chemoreceptors. RTN neurons are activated by H^+ through inhibition of TASK-2 and activation of GPR4. They are also activated by ATP released from neighboring astrocytes. In astrocytes, CO_2-induced ATP release occurs via the opening of Cx26. H^+ may also activate astrocytes, resulting in their release of ATP. Neurons in the NTS and the LC as well as non-RTN neurons in the VLM may contribute to the central chemosensitivity. However, H^+/CO_2 sensing mechanism in these cells is still to be determined. Specific anatomical structures such as neuronal dendrites just beneath the medullary surface and indentations in the medullary surface with thickened MGL covered by surface vessels may be necessary for the central chemoreception.

References

1. Smith CA, Blain GM, Henderson KS, Dempsey JA. Peripheral chemoreceptors determine the respiratory sensitivity of central chemoreceptors to CO_2: role of carotid body CO_2. J Physiol. 2015;593(18):4225–43. https://doi.org/10.1113/JP270114.
2. Pappenheimer JR, Fencl V, Heisey SR, Held D. Role of cerebral fluids in control of respiration as studied in unanesthetized goats. Am J Phys. 1965;208:436–50.
3. Loeschke HH. Central chemosensitivity and the reaction theory. J Physiol. 1982;332:1–24.
4. Eldridge FL, Kiley JP, Millhorn DE. Respiratory responses to medullary hydrogen ion changes in cats: different effects of respiratory and metabolic acidosis. J Physiol. 1985;358:285–97.
5. Shams H. Differential effects of CO_2 and H^+ as central stimuli of respiration in the cat. J Appl Physiol. 1985;58:357–64.
6. Kumar P, Bin-Jaliah I. Adequate stimuli of the carotid body: more than an oxygen sensor? Respir Physiol Neurobiol. 2007;157:12–21. https://doi.org/10.1016/j.resp.2007.01.007.
7. Fitzgerald R, Lahiri S. The respiratory system. In: Fishman AP, Cherniack NS, Widdicombe JS, Geiger SR, editors. Handbook of physiology. Bethesda: Lippincott Williams and Wilkins; 1986. p. 316–24.
8. Buckler KJ. TASK channels in arterial chemoreceptors and their role in oxygen and acid sensing. Pflugers Arch - Eur J Physiol. 2015;467:1013–25. https://doi.org/10.1007/s00424-015-1689-1.
9. Peers C, Wyatt CN. The role of maxiK channels in carotid body chemotransduction. Respir Physiol Neurobiol. 2007;157:75–82. https://doi.org/10.1016/j.resp.2006.10.010.
10. Lahiri S, Mokashi A, Mulligan E, Nishino T. Comparison of aortic and carotid chemoreceptor responses to hypercapnia and hypoxia. J Appl Physiol Respir Environ Exerc Physiol. 1981;51:55–61. https://doi.org/10.1152/jappl.1981.51.1.55.
11. Mitchell RA, Loeschcke HH, Massion WH, Severinghaus JW. Respiratory responses mediated through superficial chemosensitive areas on the medulla. J Appl Physiol. 1963;18:523–33. https://doi.org/10.1152/jappl.1963.18.3.523.

12. Schlaefke ME, See WR, Loeschcke HH. Ventilatory response to alterations of H+ ion concentration in small areas of the ventral medullary surface. Respir Physiol. 1970;10:198–212. https://doi.org/10.1016/0034-5687(70)90083-6.
13. Schlaefke ME, Kille JF, Loeschcke HH. Elimination of central chemosensitivity by coagulation of a bilateral area on the ventral medullary surface in awake cats. Pflugers Arch. 1979;378:231–41. https://doi.org/10.1007/bf00592741.
14. Coates EL, Li A, Nattie EE. Widespread sites of brain stem ventilatory chemoreception. J Appl Physiol. 1993;75:5–14. https://doi.org/10.1152/jappl.1993.75.1.5.
15. Sato M, Severinghaus JW, Basbaum AI. Medullary CO_2 chemoreceptor neuron identification by c-fos immunocytochemistry. J Appl Physiol. 1992;73:96–100. https://doi.org/10.1152/jappl.1992.73.1.96.
16. Berquin P, Bodineau L, Gros F, Lamicol N. Brainstem and hypothalamic areas involved in respiratory chemoreflexes: a Fos study in adult rats. Brain Res. 2000;857:30–40. https://doi.org/10.1016/s0006-8993(99)02304-5.
17. Mulkey DK, Stornetta RL, Weston MC, Simmons JR, Parker A, Bayliss DA, Guyenet PG. Respiratory control by ventral surface chemoreceptor neurons in rats. Nat Neurosci. 2004;7:1360–9. https://doi.org/10.1038/nn1357.
18. Stornetta RL, Moreira TS, Takakura AC, Kang BJ, Chang DA, West GH, Brunet JF, Mulkey DK, Bayliss DA, Guyenet PG. Expression of Phox2b by brainstem neurons involved in chemosensory integration in the adult rat. J Neurosci. 2006;26:10305–14. https://doi.org/10.1523/JNEUROSCI.2917-06.2006.
19. Nattie EE, Li A. CO_2 dialysis in nucleus tractus solitaries region of rat increases ventilation in sleep and wakefulness. J Appl Physiol. 2002;92:2119–30. https://doi.org/10.1152/japplphysiol.01128.2001.
20. Morrell MJ, Heywood P, Moosavi SH, Guz A, Stevens J. Unilateral focal lesions in the rostrolateral medulla influence chemosensitivitiy and breathing measured during wakefulness, sleep, and exercise. J Neurol Neurosurg Psychiatry. 1999;67:637–45. https://doi.org/10.1136/jnnp.67.5.637.
21. Aston-Jones G, Shipley MT, Grzanna R. The locus coeruleus, A5 and A7 noradrenergic cell groups. In: Paxinos G, editor. The rat nervous system. San Diego, CA: Academic; 1995. p. 183–213.
22. Oyamada Y, Ballantyne D, Mückenhoff K, Scheid P. Respiration-modulated membrane potential and chemosensitivity of locus coeruleus neurones in the in vitro brainstem-spinal cord of the neonatal rat. J Physiol. 1998;513:381–98. https://doi.org/10.1111/j.1469-7793.1998.381bb.x.
23. Oyamada Y, Andrzejewski M, Mückenhoff K, Scheid P, Ballantyne D. Locus coeruleus neurones in vitro: pH-sensitive oscillations of membrane potential in an electrically coupled network. Respir Physiol. 1999;118:131–47. https://doi.org/10.1016/s0034-5687(99)00088-2.
24. Huckstepp RTR, Dale N. Redefining the components of central CO_2 chemosensitivity – towards a better understanding of mechanism. J Physiol. 2011;589(23):5561–79. https://doi.org/10.1113/jphysiol.2011.214759.
25. Kawai A, Ballantyne D, Mückenhoff K, Scheid P. Chemosensitive medullary neurones in the brainstem-spinal cord preparation of the neonatal rat. J Physiol. 1996;492:277–92. https://doi.org/10.1113/jphysiol.1996.sp021308.
26. Kawai A, Onimaru H, Homma I. Mechanisms of CO_2/H^+ chemoreception by respiratory rhythm generator neurons in the medulla from newborn rats in vitro. J Physiol. 2006;572(2):525–37. https://doi.org/10.1113/jphysiol.2005.102533.
27. Huang RQ, Erlichman JS, Dean JB. Cell-cell coupling between CO_2-excited neurons in the dorsal medulla oblongata. Neuroscience. 1997;80:41–57. https://doi.org/10.1016/s0306-4522(97)00017-1.
28. Fukuda Y, Honda Y, Schläfke ME, Loeschcke HH. Effects of H^+ on the membrane potential of silent cells in the ventral and dorsal surface layer of the rat medulla in vitro. Pflugers Arch. 1978;376:229–35. https://doi.org/10.1007/bf00584955.

29. Ritucci NA, Erlichman JS, Leiter JC, Putnam RW. Response of membrane potential and intracellular pH to hypercapnia in neurons and astrocytes from rat retrotrapezoid nucleus. Am J Physiol Regul Interr Comp Physiol. 2005;289:R851–61. https://doi.org/10.1152/ajpregu.00132.2005.
30. Wenker JC, Kréneisz O, Nishiyama A, Mulkey DK. Astrocytes in the retrotrapezoid nucleus sense H^+ by inhibition of a Kir4.1-5.1-like current and may contribute to chemoreception by a purinergic mechanism. J Neurophysiol. 2010;104:3042–52. https://doi.org/10.1152/jn.00544.2010.
31. Gourine AV, Kasymov V, Marina N, Tang F, Figueiredo MF, Lan S, Teschemacher AG, Spyer KM, Deisseroth K, Kasparov S. Astrocytes control breathing through pH-dependent release of ATP. Science. 2010;329:571–5. https://doi.org/10.1126/science.1190721.
32. Erlichman JS, Li A, Nattie EE. Ventilatory effects of glial dysfunction in a rat brain stem chemoreceptor region. J Appl Physiol. 1998;85:1599–604. https://doi.org/10.1152/jappl.1998.85.5.1599.
33. Gourine AV, Llaudet E, Dale N, Spyer KM. ATP is a mediator of chemosensory transduction in the central nervous system. Nature. 2005;436:108–11. https://doi.org/10.1038/nature03690.
34. Okada Y, Chen Z, Jiang W, Kuwana S, Eldridge FL. Anatomical arrangement of hypercapnia activated cells in the superficial ventral medulla of rats. J Appl Physiol. 2002;93:427–39. https://doi.org/10.1152/japplphysiol.00620.2000.
35. Wu J, Xu H, Shen W, Jiang C. Expression and coexpression of CO_2-sensitive Kir channels in brainstem neurons of rats. J Membr Biol. 2004;197:179–91. https://doi.org/10.1007/s00232-004-0652-4.
36. Trapp S, Tucker SJ, Gourine AV. Respiratory responses to hypercapnia and hypoxia in mice with genetic ablation of Kir5.1 (Kcnj16). Exp Physiol. 2011;96:451–9. https://doi.org/10.1113/expphysiol.2010.055848.
37. Mulkey DK, Talley EM, Stornetta RL, Siegel AR, West GH, Chen X, Sen N, Mistry AM, Guyenet PG, Bayliss DA. TASK channels determine pH sensitivity in select respiratory neurons but do not contribute to central respiratory chemosensitivity. J Neurosci. 2007;27:14049–58. https://doi.org/10.1523/JNEUROSCI.4254-07.2007.
38. Filosa JA, Putnam RW. Multiple targets of chemosensitive signaling in locus coeruleus neurons: role of K^+ and Ca^{2+} channels. Am J Physiol Cell Physiol. 2003;53:C145–55. https://doi.org/10.1152/ajpcell.00346.2002.
39. Koizumi H, Smerin SE, Yamanishi T, Moorjani BR, Zang R, Smith JC. TASK channels contribute to the K^+-dominated leak current regulating respiratory rhythm generation in vitro. J Neurosci. 2010;30:4273–84. https://doi.org/10.1523/JNEUROSCI.4017-09.2010.
40. Bayliss DA, Barhanin J, Gestreau C, Guyenet PG. The role of pH-sensitive TASK channels in central respiratory chemoreception. Pflugers Arch. 2015;467:917–29. https://doi.org/10.1007/s00424-014-1633-9.
41. Goldberg S, Ollila HM, Lin L, Sharifi H, Rico T, Andlauer O, Aran A, Bloomrosen E, Faraco J, Fang H, Mignot E. Analysis of hypoxic and hypercapnic ventilatory response in healthy volunteers. PLoS One. 2017;12:e0168930. https://doi.org/10.1371/journal.pone.0168930.
42. Kumar NN, Velic A, Soliz J, Shi Y, Li K, Wang S, Weaver JL, Sen J, Abbott SB, Lazarenko RM, Ludwig MG, Perez-Reyes E, Mohebbi N, Bettoni C, Gassmann M, Suply T, Seuwen K, Guyenet PG, Wagner CA, Bayliss DA. Regulation of breathing by CO_2 requires the proton-activated receptor GPR4 in retrotrapezoid nucleus neurons. Science. 2015;348:1255–60. https://doi.org/10.1126/science.aaa0922.
43. Huckstepp RT, id Bihi R, Eason R, Spyer KM, Dicke N, Willecke K, Marina N, Gourine AV, Dale N. Connexin hemichannel-mediated CO_2-dependent release of ATP in the medulla oblongata contributes to central respiratory chemosensitivity. J Physiol. 2010;588:3901–20. https://doi.org/10.1113/jphysiol.2010.192088.
44. Huckstepp RT, Eason R, Sachdev A, Dale N. CO_2-dependent opening of connexin 26 and related βconnexins. J Physiol. 2010;588:3921–31. https://doi.org/10.1113/jphysiol.2010.192096.

Chapter 3
Coordination of Swallowing and Breathing: How Is the Respiratory Control System Connected to the Swallowing System?

Yoshitaka Oku

Abstract Coordination of swallowing and breathing is an important airway-protecting mechanism. Although it is regulated by the interaction between central pattern generators of swallowing and breathing within the brainstem, various factors modulate the coordination of swallowing and breathing. Swallowing normally occurs during expiration, and respiration after swallowing resumes with expiration. However, swallowing can occur immediately following inspiration (I-SW pattern), and respiration can resume with inspiration (SW-I pattern). Such atypical breathing–swallowing coordination occurs when the timing of swallows is inappropriate and tends to increase due to diseases and aging. In patients with chronic obstructive pulmonary disease (COPD), swallowing physiology is altered, and breathing–swallowing coordination is impaired. COPD patients with a higher frequency of I-SW and/or SW-I patterns have a higher frequency of exacerbations. Thus, we suggest that the impairment of breathing–swallowing coordination can cause exacerbations and may influence the prognosis of diseases. Two modalities, swallowing rehabilitation and low-level continuous positive airway pressure (CPAP), may ameliorate breathing–swallowing coordination. More studies are needed to elucidate whether such interventions improve outcomes.

Keywords Breathing–swallowing coordination · Chronic obstructive pulmonary disease · Continuous positive airway pressure · Exacerbation · Aspiration

Y. Oku (✉)
Department of Physiology, Hyogo College of Medicine, Nishinomiya, Hyogo, Japan
e-mail: yoku@hyo-med.ac.jp

K. Yamaguchi (ed.), *Structure-Function Relationships in Various Respiratory Systems*, Respiratory Disease Series: Diagnostic Tools and Disease Managements, https://doi.org/10.1007/978-981-15-5596-1_3

1 Introduction

Pneumonia is a major cause of death in the elderly worldwide. In a case-control study, the incidence of aspiration pneumonia was 18% in nursing home patients and 15% in community-acquired pneumonia patients [1]; however, the incidence of aspiration pneumonia is thought to be much greater in the elderly population. Aspiration pneumonia occurs recurrently because it is due to the functional impairment of swallowing. Chronic obstructive pulmonary disease (COPD) is another major cause of death worldwide, and exacerbation seriously affects its morbidity and mortality [2]. A number of studies suggest that the functional impairment of swallowing is associated with exacerbations of COPD [3–5]. Therefore, understanding the physiology and pathophysiology underlying swallowing malfunction is essential for clinicians who treat these patients. A variety of pathological and physiological factors cause swallowing disorder, but the two major causes of swallowing disorder in the elderly are muscle weakness and the inappropriate timing of swallowing relative to the respiratory cycle. Thus, in this chapter, we will focus on how the timing of swallowing and the coordination between swallowing and breathing are regulated and discuss the clinical consequences of the alteration of coordination due to aging, diseases, and interventions.

2 Swallowing Physiology

2.1 Swallowing as an Airway Protecting Reflex

The pharynx is the common pathway of breath and swallowed food. During swallowing, the airway is securely protected against aspiration by the epiglottis, vocal cords, and vestibular folds that cover the laryngeal orifice. In addition, the airway is protected physiologically by "deglutition apnea," the phenomenon in which the respiration is stopped during the event of swallowing. Deglutition apnea is regulated by the interaction between central pattern generators (CPGs) for swallowing and respiration in the brainstem [6].

During swallowing, the hyoid bone first slowly moves in the rostral and posterior direction and then rapidly moves toward the anterior direction. By this movement, the larynx elevates, the epiglottis falls to cover the larynx, and the upper esophageal sphincter opens to let pass a food bolus. After the completion of swallowing, the hyoid bone descends to its original position. The elevation of the larynx often starts before the provisional arrest of respiratory flow. This phenomenon is one of the remarkable characteristics of patients with stroke [7] and can be interpreted as a behavior to compensate for the delayed triggering of the swallowing reflex. The restoration of respiration usually begins with expiration (SW-E pattern), and expiration begins earlier than the completion of the laryngeal descent to its original position [8]. This can also be interpreted as an airway-protecting behavior to avoid aspirating a food residue within the pharynx.

2.2 *Importance of Subglottic Pressure in Swallowing Efficiency*

Gross et al. [9, 10] postulated that subglottic pressure plays an important role in swallowing efficiency. The reduction of subglottic pressure prolongs the pharyngeal contraction in healthy subjects [9]. In patients with tracheostomy, the reduction of subglottic pressure lengthens the pharyngeal transit time and increases the chance of pharyngeal residue and aspiration. Closing the tracheostomy tube during swallowing improves swallowing dynamics and reduces aspiration. The mechanisms by which subglottic pressure affects swallowing efficiency remain to be elucidated.

2.3 *Swallowing Frequency*

Spontaneous swallow frequency rates (swallows per minute: SPM) in healthy subjects are 1.01 in young subjects and 0.58 in elderly subjects [11]. SPM is diminished in patients with acute stroke, and the reduction is correlated with the severity of dysphagia [12]. Healthy subjects swallow 5–6 times per hour during sleep [13]. Swallows occur when the subject arises, which accompanies body movement and changes in breathing pattern, or in an arousal state, as identified by an electroencephalogram, that does not accompany body movement or changes in breathing pattern [14]. The frequency of swallows is highest at Stage I sleep and decreases as the depth of sleep progresses [14]. Therefore, the clearance of the upper airway dramatically deteriorates during sleep, particularly in elderly subjects. Saliva secretion also decreases, which may result in bacterial growth and accumulation of gastric juice and refluxed gastric contents.

3 Neural Substrates of Swallowing Reflex

3.1 *Swallowing CPG and Its Interaction with Respiratory CPG*

The swallowing CPG consists of the dorsal swallowing group (DSG) located within the nucleus of the solitary tract (NTS) and the ventral swallowing group (VSG) located in the ventral reticular formation adjacent to the nucleus ambiguus (AMB) [15]. These swallowing neuron groups anatomically overlap with the dorsal and ventral respiratory groups, and CPGs for swallowing and respiration functionally share a common neuronal network [16, 17].

Figure 3.1 illustrates the interaction between CPGs of swallowing and breathing in a simplified model. Here, we consider a respiratory CPG model that consists of preinspiratory/inspiratory (pre-I/I) neurons in the pre-Bötzinger complex (pre-BötC); expiratory-decrementing (E-DEC), expiratory-augmenting (E-AUG), and inspiratory-decrementing (I-DEC) neurons in the Bötzinger complex (BötC); and

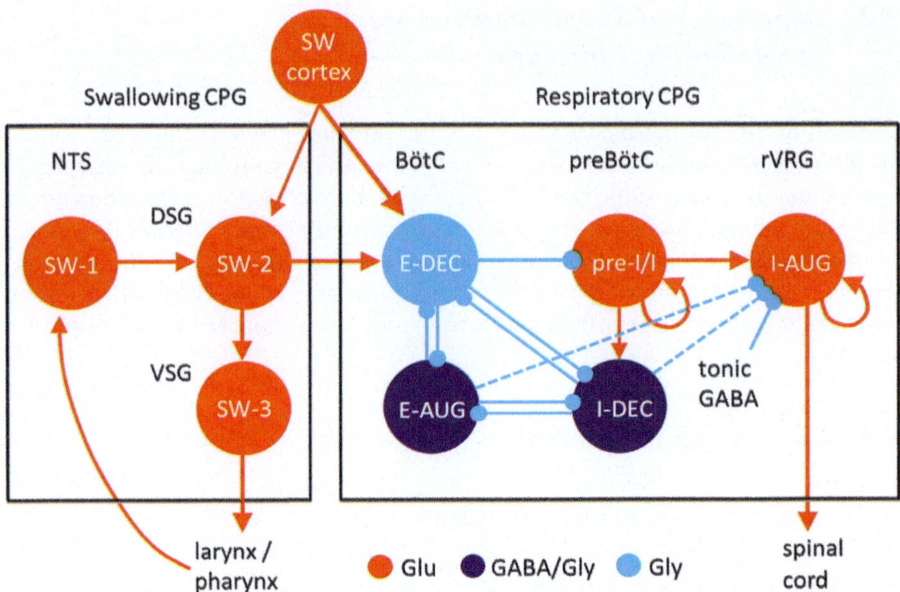

Fig. 3.1 Interaction between respiratory and swallowing CPGs. Excitatory neuron pools and projections are colored in red (Glu; glutamatergic neurons), whereas inhibitory neuron pools and projections are colored either in blue (Gly: glycinergic neurons) or in purple (GABA/Gly: GABA/glycine-coreleasing neurons). See text for detailed explanation

inspiratory augmenting (I-AUG) neurons in the rostral ventral respiratory group (rVRG). The preBötC is the kernel of the inspiratory rhythm generation, whereas E-DEC, E-AUG, and I-DEC neurons in the BötC shape the respiratory pattern by constituting an "inhibitory ring." I-AUG neurons are inspiratory premotor neurons. For simplicity, the swallowing CPG is composed of two interneuron groups (SW-1 and SW-2) and one premotor neuron group (SW-3).

When a liquid or a food bolus reaches the pharynx (the valleculae or the pyriform recess), the signal is transmitted to the primary afferent relay neurons in the NTS (SW-1) via the superior laryngeal nerve (SLN) and the glossopharyngeal nerve, then activates swallowing interneurons (SW-2) in the DSG that gate the afferent information and trigger the swallowing reflex by a burst of firings. The swallowing reflex is a programmed sequence of muscle activities, where premotor neurons (SW-3) in the VSG generate a stereotypic motor pattern almost irrespective of the strength of the afferent stimulus. When a swallow is triggered, E-DEC neurons are activated to suppress respiratory activity (deglutition apnea). During a volitional swallow, swallowing-related cortical areas (SW cortex) activate (presumably) E-DEC neurons to suppress respiration (both inspiration and expiration). Thus, in the case of a volitional swallow, respiration stops before the swallowing reflex occurs. Simultaneously, the SW cortex lowers the threshold of SW-2 neurons to facilitate the swallowing reflex.

Since the coupling of swallowing and respiratory rhythms occurs in decerebrate animals, the coordination of swallowing and breathing is primarily regulated by the interaction between these CPGs of swallowing and respiration. When the synaptic transmission of the Kölliker–Fuse nucleus in the pons is disrupted, the airway-protecting laryngeal adduction during swallowing and the coordination between swallowing and breathing is impaired [18], suggesting that the Kölliker–Fuse nucleus is required for these functions.

Electrical stimulation of the SLN elicits swallows in cats and rats [16, 19, 20]. During continuous electrical SLN stimulation, swallows preferentially occur at respiratory phase transitions (the phase transition between stage II expiration and inspiration, the transition between inspiration and stage I expiration, and the transition between stage I and II of expiration), where E-DEC neurons are disinhibited. This is physiologically reasonable because E-DEC neurons are inhibitory neurons that have broad projections to other respiratory neurons and thus can contribute to deglutition apnea when disinhibited.

3.2 *Swallowing-Related Cortical and Subcortical Areas*

Functional brain mapping has revealed that swallowing activates cortical areas (sensorimotor cortex, premotor cortex, cingulate cortex, and insulate cortex) and subcortical areas (internal capsule, hypothalamus, amygdala, substantia nigra, and midbrain reticular formation) [15, 21]. These cortical and subcortical areas not only facilitate the swallowing reflex but also modulate swallow-related muscles by receiving sensory feedback [22].

The incidence of dysphagia after stroke varies between 41% and 78% [23–25]. Therefore, patients with stroke have a high incidence of dysphagia and a risk of aspiration pneumonia. Swallowing-related cortical areas display interhemispheric asymmetry, and right hemisphere strokes tend to produce pharyngeal dysphagia, whereas left hemisphere strokes tend to produce oral dysphagia [26]. Recovery of swallowing function after stroke is associated with an increase in the activation of swallow-related cortical areas in the unaffected hemisphere [27], suggesting that functional recovery depends on the reorganization of neuronal circuits, i.e., projections from the unaffected hemisphere to the swallowing CPG. A short-term pharyngeal sensory stimulation causes a persistent augmentation of swallowing motor representation in the cortex [27], which might be applicable to the neurorehabilitation of dysphagia after stroke.

4 Physiological Basis of Coordination Between Breathing and Swallowing

In conscious humans, swallows preferentially occur during the early-middle phase of expiration and respiration after the swallow is resumed with expiration (E-SW-E pattern) [28, 29]. In relation to the lung volume, swallows preferentially occur at a specific range of lung volume (200–370 ml above the functional residual capacity) that is optimal for fundamental swallowing functions—laryngeal elevation, upper airway closure, the opening of the upper and lower esophageal sphincters [30]. However, other types of coordination are observed, i.e., swallowing occurring immediately following inspiration (I-SW pattern) and respiration resuming with inspiration (SW-I pattern), resulting in additional three (I-SW-E, E-SW-I, I-SW-I) patterns [31]. We recruited 269 healthy elderly subjects with a 10-item eating assessment tool (EAT-10) score less than 3, as well as 30 dysphagic patients, to evaluate breathing–swallowing coordination [32]. Figure 3.2 shows the relative frequency of I-SW/SW-I patterns in healthy elderly subjects and dysphagic patients. In general, healthy subjects had a lower occurrence rate of I-SW/SW-I patterns; however, even in healthy subjects, 20 (7.5%) subjects had a high (>40%) I-SW rate, and 25 (9.4%) subjects had a high SW-I rate in water swallows.

Swallowing is a physiological perturbation of the respiratory rhythm [33]. Swallowing causes phase resetting of the respiratory cycle and changes the interval to the next breath depending on the timing in the respiratory cycle when the swallow is initiated. Swallows initiated at late times in the respiratory cycle have the shortest interval to the next breath and tend to happen with the SW-I pattern, whereas swallows initiated at early times naturally tend to happen in the I-SW pattern [32] (Fig. 3.3).

Swallows with the I-SW-E pattern accompany the prolonged latency to the onset of swallowing from the onset of deglutition apnea and the prolongation of the duration of deglutition apnea, suggesting that this pattern is an adaptive response to compensate for the delayed onset of the swallowing reflex [32]. In contrast, the E-SW-I pattern tends to occur when the timing of swallows in the respiratory cycle

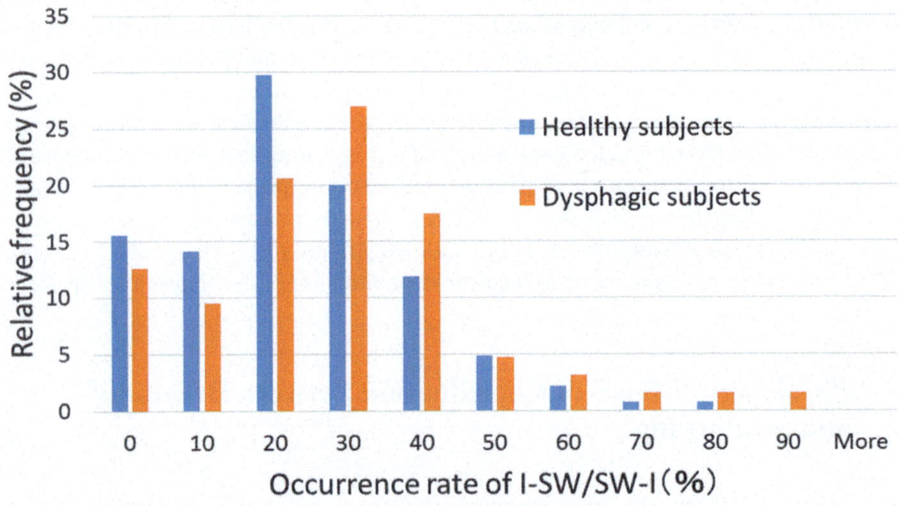

Fig. 3.2 The relative frequency of I-SW/SW-I patterns in healthy elderly subjects and dysphagic patients

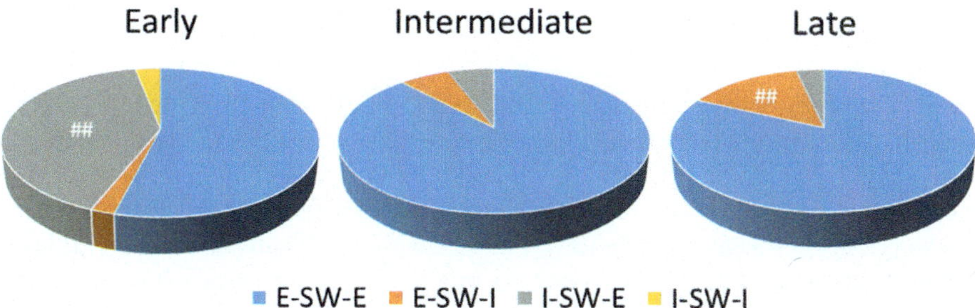

Fig. 3.3 Frequency distributions of different breathing–swallowing coordination patterns in early (old phase <0.4), intermediate (old phase between 0.4 and 1.0), and late (old phase >1.0) timings of swallowing. The old phase represents the timing of swallowing in the respiratory cycle, which is expressed as the time from the preceding inspiration, where the mean length of the respiratory cycle is normalized to 1 [33]. ##: P value vs. Intermediate <0.0001 (chi-square test followed by Haberman's residual test)

is delayed. Patients with dysphagia often have difficulty initiating swallows, and the timing of swallows is delayed, which may result in E-SW-I-pattern swallows.

5 Impaired Swallowing in COPD: Its Pathophysiology and Consequences

5.1 Swallowing Function in COPD

COPD is a lung disease characterized by chronic obstruction of lung airflow that interferes with normal breathing and is not fully reversible. Since breathing and swallowing cannot be done simultaneously, a prolonged time to empty the inspired air due to an expiratory flow limitation may interfere with the act of swallowing. Therefore, it is reasonable to hypothesize that swallowing is impaired in patients with COPD, though such a phenomenon is poorly recognized by clinicians. Coelho [34] was the first to report that patients with COPD have impaired swallowing function: "The overall picture was one of reduced strength in all aspects of the swallow, coupled with a reduced ability to use pulmonary air to clear the larynx and ensure airway protection." Subsequent studies revealed a higher (17–20%) prevalence of dysphagia in patients with COPD [35] compared with control subjects.

The impairment of swallowing function can be evaluated by the latency from water injection into the pharynx to the onset of the swallowing reflex (e.g., the simple two-step swallowing-provocation test; STS-SPT) [36] or the repetitive saliva swallowing test (RSST) [37]. In moderate and severe stages of COPD, a higher incidence of abnormal swallowing function is observed by both tests; however, even in the mild stages of COPD, the incidence of abnormal RSST count (<3 voluntary swallows within 30 s) is higher than in controls [38].

5.2 *Altered Swallowing Physiology in COPD*

Swallowing physiology in patients with COPD is different from that of healthy subjects. The maximal laryngeal elevation during swallowing is reduced in patients with COPD [35]. Some patients use voluntary airway-protecting maneuvers during swallowing, such as prolonged airway closure and earlier laryngeal closure relative to the cricopharyngeal opening [35]. Patients with COPD are more likely to have either penetration of pharyngeal contents into the larynx or actual aspiration when swallowing large fluid volumes and favor an I-SW-E pattern with larger boluses [39]. Further, patients with COPD have a longer pharyngeal swallowing phase than normal subjects, which is associated with a decrease in the difference between the duration of maximal laryngeal elevation and the duration of pharyngeal transit [40]. As described in Sect. 4, the latency to the onset of swallowing from the onset of deglutition apnea and the duration of deglutition apnea are lengthened in swallows with the I-SW-E pattern compared to the E-SW-E pattern [32]. Therefore, the I-SW-E pattern observed in patients with COPD while swallowing large boluses is interpreted as an airway-protecting behavior to compensate for the delayed triggering of the swallowing reflex and a longer pharyngeal transit time.

5.3 *Influences of Smoking on Swallowing Function*

Smoking, the leading cause of COPD, also causes swallowing abnormalities. Smoking increases both the threshold volume for triggering the pharyngo-upper esophageal sphincter contractile response and the threshold volume for reflexive pharyngeal swallowing [41]. In addition, smoking lowers the upper and lower esophageal sphincter pressures [42]. Therefore, smoking triggers or aggravates gastroesophageal reflux and adversely affects the clearance of refluxed gastric contents. This may be related to a decreased laryngopharyngeal sensitivity in patients with COPD [43].

5.4 *Gastroesophageal Reflux Disease (GERD) and COPD Exacerbations*

GERD is a common comorbidity of COPD, with a prevalence of 10–29% in the Western population and 5–14% among Japanese adults [44]. Terada, Muro et al. [45] first reported that GERD symptoms are an important factor associated with COPD exacerbation. Later, large-scale cohort studies verified that GERD is an independent predictor of frequent exacerbations [46–48]. In the Copenhagen City Heart

Study, in which 1259 patients with COPD were enrolled, the association between GERD and COPD exacerbation was found only in those individuals who did not use acid-inhibitory treatment regularly [47]. Interestingly, another cohort study with 4483 participants [48] showed that the use of a proton-pump inhibitor (PPI) was associated with frequent exacerbations but did not meaningfully influence the GERD–exacerbation association. A recent cohort study with 638 participants revealed that therapy with PPIs for GERD did not reduce the risk for severe exacerbations in COPD [49]. These results may suggest that regurgitation itself, rather than acid, causes COPD exacerbations. Patients with COPD who had experienced exacerbations in the previous year had prolonged latencies from water injection into the pharynx to the onset of the swallowing reflex compared to patients without exacerbations [3]. A more recent prospective study showed that a prolonged latency to initiate the swallowing reflex predisposed patients with COPD to exacerbations [4]. They suggested that abnormal swallowing reflexes in COPD might be affected by the comorbidity of GERD and cause bacterial colonization. Alternatively, impairment of swallowing function may contribute to the manifestation of GERD symptoms by lowering of upper airway clearance and microaspiration. Tsuzuki et al. [5] showed that RSST is also a useful test for predicting exacerbations in patients with COPD.

6 Coordination of Swallowing and Breathing in Aging and Diseases

6.1 Discoordination Between Swallowing and Breathing May Predict Aspiration, Exacerbation, and Prognosis of Diseases

I-SW and SW-I patterns increase with aging [29] and in patients with stroke [50], head and neck cancer after treatment [51], Parkinson's disease [52], or COPD [53]. I-SW-pattern swallows in combination with a delayed onset of the swallowing reflex may cause aspiration before the swallow, particularly in the case of liquid swallows. On the other hand, SW-I-pattern swallows may cause aspiration after the swallow when food residue is in the pharynx after the first swallowing reflex. Healthy subjects rarely aspirate even on such occasions; however, SW-I-pattern swallows and the shortening of the duration of deglutition apnea are correlated with laryngeal penetration and aspiration in patients with Parkinson's disease [54]. Further, frequent I-SW and/or SW-I patterns (>25% occurrence rate) are associated with COPD exacerbation [55]. Therefore, identifying such subjects with a high I-SW/SW-I rate and treating them to reduce the I-SW/SW-I rate may prevent exacerbations of these diseases.

6.2 Coordination of Swallowing and Breathing in Patients with Obstructive Sleep Apnea Syndrome (OSAS)

CPAP prolongs the latency and reduces the frequency of swallows in healthy subjects [56]. Animal experiments suggest that these effects are mediated by bronchopulmonary receptors rather than upper airway receptors [57]. In patients with OSAS, swallows occur in periods of microarousal, and swallows with the SW-I pattern are increased compared with healthy subjects [14]. CPAP restores the frequency of SW-I-pattern swallows to the normal level in patients with OSAS [14], and the frequency and duration of GERD symptoms are also reduced [58].

7 Factors Affecting Coordination Between Swallowing and Breathing

7.1 Coordination of Swallowing and Breathing During Anesthesia

As described in Sect. 4, in conscious subjects, swallows preferentially occur during the early–middle phase of expiration, when the lung volume falls within a specific range. However, in unconscious subjects, e.g., during anesthesia, such a preference has not been observed [59]. It is well known that anesthesia attenuates airway defensive reflexes and heightens the risk of aspiration. During anesthesia, when swallows occur during inspiration or at the expiratory-to-inspiratory phase transition, coughing and apnea are often evoked due to laryngeal stimulation, resulting in a risk of aspiration [60]. Further, anesthesia affects the coordination of swallowing and breathing. Sevoflurane and propofol decrease the frequency of spontaneous swallows and increase the frequency of I-SW and SW-I pattern swallows [61].

7.2 Coordination of Swallowing and Breathing During Loaded Breathing

The timing of swallows in the respiratory cycle is modified during loaded breathing even in conscious subjects. For example, hypercapnia increases the frequency of I-SW and SW-I swallows in both the conscious and unconscious states [60, 61]. However, hypercapnia decreases the frequency of swallows in the conscious state, whereas it augments the frequency in the unconscious state.

When resistive or elastic loading is applied to the respiratory system, the frequency of swallows during inspiration and at the expiratory-to-inspiratory phase transition increases simultaneously with changes in breathing patterns [62].

7.3 Coordination of Swallowing and Breathing During Continuous Positive Airway Pressure and Assisted Ventilation

A brief (~100 ms) inspiratory flow is often recorded immediately after swallowing. This is not a true inspiration but a negative pressure associated with the relaxation of the pharyngeal constrictor muscle, known as the swallow noninspiratory flow (SNIF) [63]. When a bilevel positive pressure ventilator is used in S (spontaneous) or ST (spontaneous/timed) mode, SNIF may trigger inspiratory support, resulting in an SW-I pattern swallow [64, 65]. Terzi et al. [64] proposed that a device equipped with an off switch is effective to prevent swallowing-induced autotriggering of inspiration. In contrast, continuous positive airway pressure (CPAP) ventilation reduces the frequency of SW-I-pattern swallows in patients with OSAS [14].

Recently, we found that low-pressure (4 cm H_2O) CPAP decreases the SW-I frequency, increases the SNIF occurrence, and normalizes the timing of swallowing, suggesting that low-pressure CPAP alleviates the risk of aspiration in patients with COPD [66]. Low-pressure CPAP may affect the lung volume during quiet breathing by relieving the expiratory flow limitation and thereby shift the timing of swallowing toward early expiration. Therefore, although CPAP reduces the frequency of swallows [56], it has a beneficial effect from the viewpoint of airway protection.

Nasal high-flow oxygen therapy (NHF) has recently been widely used for hypoxemic respiratory failure. Since NHF generates low-level positive airway pressure during expiration with the mouth closed (2.7 cm H_2O at 35 L/min [67] and 5.5 cm H_2O at 40 L/min [68]), it is expected to ameliorate the breathing–swallowing coordination, similar to low-level CPAP. However, in healthy volunteers, swallows during the inspiratory phase and at the expiratory-inspiratory transition tended to increase, although the occurrence rates were not significantly different [69]. On the other hand, NHF shortened the latency for inducing the swallowing reflex [69].

8 Swallowing Rehabilitation to Improve the Coordination Between Swallowing and Breathing

Several methods potentially ameliorate the coordination between swallowing and breathing. The first method is the supraglottic swallow, a maneuver to ensure laryngeal closure during swallowing, in which each subject is instructed to inhale and hold his breath before and during swallowing and exhale after swallowing. By the use of this maneuver, the subject is expected to swallow at a high lung volume and naturally resume his respiration with expiration after swallowing, which would reduce SW-I-pattern swallows. Another method uses biofeedback, where information on the respiratory phase and lung volume is visually presented so that a subject can swallow at an optimal time within the respiratory cycle [70]. Training with this method has improved the swallowing function, such as in the penetration-aspiration score, in patients with head neck cancer.

9 Conclusion

Coordination of swallowing and breathing is regulated by the interaction between central pattern generators of swallowing and breathing within the brainstem. The coordination is impaired in various conditions, e.g., aging, anesthesia, neurological diseases, and respiratory diseases, leading to aspiration and exacerbation of diseases, and it may influence prognosis. Swallowing rehabilitation and low-level CPAP are candidates for therapeutic intervention to normalize the coordination of swallowing and breathing. Studies are needed to elucidate whether such interventions improve the outcome.

References

1. Reza Shariatzadeh M, Huang JQ, Marrie TJ. Differences in the features of aspiration pneumonia according to site of acquisition: community or continuing care facility. J Am Geriatr Soc. 2006;54(2):296–302. https://doi.org/10.1111/j.1532-5415.2005.00608.x.
2. Soler-Cataluna JJ, Martinez-Garcia MA, Roman Sanchez P, Salcedo E, Navarro M, Ochando R. Severe acute exacerbations and mortality in patients with chronic obstructive pulmonary disease. Thorax. 2005;60(11):925–31. https://doi.org/10.1136/thx.2005.040527.
3. Kobayashi S, Kubo H, Yanai M. Impairment of the swallowing reflex in exacerbations of COPD. Thorax. 2007;62(11):1017. https://doi.org/10.1136/thx.2007.084715.
4. Terada K, Muro S, Ohara T, Kudo M, Ogawa E, Hoshino Y, et al. Abnormal swallowing reflex and COPD exacerbations. Chest. 2010;137(2):326–32.
5. Tsuzuki A, Kagaya H, Takahashi H, Watanabe T, Shioya T, Sakakibara H, et al. Dysphagia causes exacerbations in individuals with chronic obstructive pulmonary disease. J Am Geriatr Soc. 2012;60(8):1580–2. https://doi.org/10.1111/j.1532-5415.2012.04067.x.
6. Bautista TG, Sun QJ, Pilowsky PM. The generation of pharyngeal phase of swallow and its coordination with breathing: interaction between the swallow and respiratory central pattern generators. Prog Brain Res. 2014;212:253–75. https://doi.org/10.1016/B978-0-444-63488-7.00013-6.
7. Wang CM, Shieh WY, Chen JY, Wu YR. Integrated non-invasive measurements reveal swallowing and respiration coordination recovery after unilateral stroke. Neurogastroenterol Motil. 2015;27(10):1398–408. https://doi.org/10.1111/nmo.12634.
8. Martin BJ, Logemann JA, Shaker R, Dodds WJ. Coordination between respiration and swallowing: respiratory phase relationships and temporal integration. J Appl Physiol (1985). 1994;76(2):714–23.
9. Gross RD, Atwood CW Jr, Grayhack JP, Shaiman S. Lung volume effects on pharyngeal swallowing physiology. J Appl Physiol (1985). 2003;95(6):2211–7. https://doi.org/10.1152/japplphysiol.00316.2003.
10. Gross RD, Steinhauer KM, Zajac DJ, Weissler MC. Direct measurement of subglottic air pressure while swallowing. Laryngoscope. 2006;116(5):753–61. https://doi.org/10.1097/01.mlg.0000205168.39446.12.

11. Crary MA, Sura L, Carnaby G. Validation and demonstration of an isolated acoustic recording technique to estimate spontaneous swallow frequency. Dysphagia. 2013;28(1):86–94. https://doi.org/10.1007/s00455-012-9416-y.
12. Crary MA, Carnaby GD, Sia I. Spontaneous swallow frequency compared with clinical screening in the identification of dysphagia in acute stroke. J Stroke Cerebrovasc Dis. 2014;23(8):2047–53. https://doi.org/10.1016/j.jstrokecerebrovasdis.2014.03.008.
13. Lichter I, Muir RC. The pattern of swallowing during sleep. Electroencephalogr Clin Neurophysiol. 1975;38(4):427–32.
14. Sato K, Umeno H, Chitose S, Nakashima T. Sleep-related deglutition in patients with OSAHS under CPAP therapy. Acta Otolaryngol. 2011;131(2):181–9. https://doi.org/10.3109/00016489.2010.520166.
15. Jean A. Brain stem control of swallowing: neuronal network and cellular mechanisms. Physiol Rev. 2001;81(2):929–69.
16. Oku Y, Tanaka I, Ezure K. Activity of bulbar respiratory neurons during fictive coughing and swallowing in the Decerebrate Cat. J Physiol Lond. 1994;480:309–24.
17. Sugiyama Y, Shiba K, Mukudai S, Umezaki T, Hisa Y. Activity of respiratory neurons in the rostral medulla during vocalization, swallowing, and coughing in Guinea pigs. Neurosci Res. 2014;80:17–31. https://doi.org/10.1016/j.neures.2013.12.004.
18. Bautista TG, Dutschmann M. Ponto-medullary nuclei involved in the generation of sequential pharyngeal swallowing and concomitant protective laryngeal adduction in situ. J Physiol. 2014;592(Pt 12):2605–23. https://doi.org/10.1113/jphysiol.2014.272468.
19. Dick TE, Oku Y, Romaniuk JR, Cherniack NS. Interaction between central pattern generators for breathing and swallowing in the cat. J Physiol Lond. 1993;465:715–30.
20. Saito Y, Ezure K, Tanaka I, Osawa M. Activity of neurons in ventrolateral respiratory groups during swallowing in decerebrate rats. Brain and Development. 2003;25(5):338–45. https://doi.org/10.1016/S0387-7604(03)00008-1.
21. Kern MK, Jaradeh S, Arndorfer RC, Shaker R. Cerebral cortical representation of reflexive and volitional swallowing in humans. Am J Physiol Gastrointest Liver Physiol. 2001;280(3):G354–60.
22. Miller AJ. The neurobiology of swallowing and dysphagia. Dev Disabil Res Rev. 2008;14(2):77–86. https://doi.org/10.1002/ddrr.12.
23. Falsetti P, Acciai C, Palilla R, Bosi M, Carpinteri F, Zingarelli A, et al. Oropharyngeal dysphagia after stroke: incidence, diagnosis, and clinical predictors in patients admitted to a neurorehabilitation unit. J Stroke Cerebrovasc Dis. 2009;18(5):329–35. https://doi.org/10.1016/j.jstrokecerebrovasdis.2009.01.009.
24. Masrur S, Smith EE, Saver JL, Reeves MJ, Bhatt DL, Zhao X, et al. Dysphagia screening and hospital-acquired pneumonia in patients with acute ischemic stroke: findings from get with the guidelines—stroke. J Stroke Cerebrovasc Dis. 2013;22(8):e301–9. https://doi.org/10.1016/j.jstrokecerebrovasdis.2012.11.013.
25. Sorensen RT, Rasmussen RS, Overgaard K, Lerche A, Johansen AM, Lindhardt T. Dysphagia screening and intensified oral hygiene reduce pneumonia after stroke. J Neurosci Nurs. 2013;45(3):139–46. https://doi.org/10.1097/JNN.0b013e31828a412c.
26. Robbins J, Levin RL. Swallowing after unilateral stroke of the cerebral cortex: preliminary experience. Dysphagia. 1988;3(1):11–7.
27. Hamdy S, Rothwell JC, Aziz Q, Singh KD, Thompson DG. Long-term reorganization of human motor cortex driven by short-term sensory stimulation. Nat Neurosci. 1998;1(1):64–8. https://doi.org/10.1038/264.
28. Nishino T, Yonezawa T, Honda Y. Effects of swallowing on the pattern of continuous respiration in human adults. Am Rev Respir Dis. 1985;132(6):1219–22. https://doi.org/10.1164/arrd.1985.132.6.1219.
29. Shaker R, Li Q, Ren J, Townsend WF, Dodds WJ, Martin BJ, et al. Coordination of deglutition and phases of respiration: effect of aging, tachypnea, bolus volume, and chronic obstructive pulmonary disease. Am J Phys. 1992;263(5 Pt 1):G750–5.

30. McFarland DH, Martin-Harris B, Fortin AJ, Humphries K, Hill E, Armeson K. Respiratory-swallowing coordination in normal subjects: lung volume at swallowing initiation. Respir Physiol Neurobiol. 2016;234:89–96. https://doi.org/10.1016/j.resp.2016.09.004.
31. Martin-Harris B, Brodsky MB, Michel Y, Ford CL, Walters B, Heffner J. Breathing and swallowing dynamics across the adult lifespan. Arch Otolaryngol Head Neck Surg. 2005;131(9):762–70. https://doi.org/10.1001/archotol.131.9.762.
32. Yagi N, Oku Y, Nagami S, Yamagata Y, Kayashita J, Ishikawa A, et al. Inappropriate timing of swallow in the respiratory cycle causes breathing-swallowing discoordination. Front Physiol. 2017;8:676. https://doi.org/10.3389/fphys.2017.00676.
33. Paydarfar D, Gilbert RJ, Poppel CS, Nassab PF. Respiratory phase resetting and airflow changes induced by swallowing in humans. J Physiol. 1995;483(Pt 1):273–88.
34. Coelho CA. Preliminary findings on the nature of dysphagia in patients with chronic obstructive pulmonary disease. Dysphagia. 1987;2(1):28–31.
35. Mokhlesi B, Logemann JA, Rademaker AW, Stangl CA, Corbridge TC. Oropharyngeal deglutition in stable COPD. Chest. 2002;121(2):361–9.
36. Teramoto S, Matsuse T, Fukuchi Y, Ouchi Y. Simple two-step swallowing provocation test for elderly patients with aspiration pneumonia. Lancet. 1999;353(9160):1243. https://doi.org/10.1016/S0140-6736(98)05844-9.
37. Oguchi K, Saitoh E, Mizuno M, Kusudo S, Tanaka T, Onogi K. The repetitive saliva swallowing test (RSST) as a screening test of functional dysphagia (1) Normal values of RSST. Jpn J Rehabil Med. 2000;37:375–82.
38. Ohta K, Murata K, Takahashi T, Minatani S, Sako S, Kanada Y. Evaluation of swallowing function by two screening tests in primary COPD. Eur Respir J. 2009;34(1):280–1. https://doi.org/10.1183/09031936.00016909.
39. Cvejic L, Harding R, Churchward T, Turton A, Finlay P, Massey D, et al. Laryngeal penetration and aspiration in individuals with stable COPD. Respirology. 2011;16(2):269–75. https://doi.org/10.1111/j.1440-1843.2010.01875.x.
40. Cassiani RA, Santos CM, Baddini-Martinez J, Dantas RO. Oral and pharyngeal bolus transit in patients with chronic obstructive pulmonary disease. Int J Chron Obstruct Pulmon Dis. 2015;10:489–96. https://doi.org/10.2147/COPD.S74945.
41. Dua K, Bardan E, Ren J, Sui Z, Shaker R. Effect of chronic and acute cigarette smoking on the pharyngo-upper oesophageal sphincter contractile reflex and reflexive pharyngeal swallow. Gut. 1998;43(4):537–41.
42. Stanciu C, Bennett JR. Smoking and gastro-oesophageal reflux. Br Med J. 1972;3(5830):793–5.
43. Clayton NA, Carnaby-Mann GD, Peters MJ, Ing AJ. The effect of chronic obstructive pulmonary disease on laryngopharyngeal sensitivity. Ear Nose Throat J. 2012;91(9):370, 2, 4 passim.
44. Kang JY. Systematic review: geographical and ethnic differences in gastro-oesophageal reflux disease. Aliment Pharmacol Ther. 2004;20(7):705–17. https://doi.org/10.1111/j.1365-2036.2004.02165.x.
45. Terada K, Muro S, Sato S, Ohara T, Haruna A, Marumo S, et al. Impact of gastro-oesophageal reflux disease symptoms on COPD exacerbation. Thorax. 2008;63(11):951–5. https://doi.org/10.1136/thx.2007.092858.
46. Hurst JR, Vestbo J, Anzueto A, Locantore N, Mullerova H, Tal-Singer R, et al. Susceptibility to exacerbation in chronic obstructive pulmonary disease. N Engl J Med. 2010;363(12):1128–38. https://doi.org/10.1056/NEJMoa0909883.
47. Ingebrigtsen TS, Marott JL, Vestbo J, Nordestgaard BG, Hallas J, Lange P. Gastro-esophageal reflux disease and exacerbations in chronic obstructive pulmonary disease. Respirology. 2015;20(1):101–7. https://doi.org/10.1111/resp.12420.
48. Martinez CH, Okajima Y, Murray S, Washko GR, Martinez FJ, Silverman EK, et al. Impact of self-reported gastroesophageal reflux disease in subjects from COPDGene cohort. Respir Res. 2014;15:62. https://doi.org/10.1186/1465-9921-15-62.
49. Baumeler L, Papakonstantinou E, Milenkovic B, Lacoma A, Louis R, Aerts JG, et al. Therapy with proton-pump inhibitors for gastroesophageal reflux disease does not reduce the risk for

severe exacerbations in COPD. Respirology. 2016;21(5):883–90. https://doi.org/10.1111/resp.12758.
50. Leslie P, Drinnan MJ, Ford GA, Wilson JA. Swallow respiration patterns in dysphagic patients following acute stroke. Dysphagia. 2002;17(3):202–7. https://doi.org/10.1007/s00455-002-0053-8.
51. Gillespie MB, Brodsky MB, Day TA, Sharma AK, Lee FS, Martin-Harris B. Laryngeal penetration and aspiration during swallowing after the treatment of advanced oropharyngeal cancer. Arch Otolaryngol Head Neck Surg. 2005;131(7):615–9. https://doi.org/10.1001/archotol.131.7.615.
52. Gross RD, Atwood CW Jr, Ross SB, Eichhorn KA, Olszewski JW, Doyle PJ. The coordination of breathing and swallowing in Parkinson's disease. Dysphagia. 2008;23(2):136–45.
53. Gross RD, Atwood CW Jr, Ross SB, Olszewski JW, Eichhorn KA. The coordination of breathing and swallowing in chronic obstructive pulmonary disease. Am J Respir Crit Care Med. 2009;179(7):559–65.
54. Troche MS, Huebner I, Rosenbek JC, Okun MS, Sapienza CM. Respiratory-swallowing coordination and swallowing safety in patients with Parkinson's disease. Dysphagia. 2011;26(3):218–24. https://doi.org/10.1007/s00455-010-9289-x.
55. Nagami S, Oku Y, Yagi N, Sato S, Uozumi R, Morita S, et al. Breathing-swallowing discoordination is associated with frequent exacerbations of COPD. BMJ Open Respir Res. 2017;4(1):e000202. https://doi.org/10.1136/bmjresp-2017-000202.
56. Nishino T, Sugimori K, Kohchi A, Hiraga K. Nasal constant positive airway pressure inhibits the swallowing reflex. Am Rev Respir Dis. 1989;140(5):1290–3. https://doi.org/10.1164/ajrccm/140.5.1290.
57. Samson N, Roy B, Ouimet A, Moreau-Bussiere F, Dorion D, Mayer S, et al. Origins of the inhibiting effects of nasal CPAP on nonnutritive swallowing in newborn lambs. J Appl Physiol (1985). 2008;105(4):1083–90. https://doi.org/10.1152/japplphysiol.90494.2008.
58. Kerr P, Shoenut JP, Millar T, Buckle P, Kryger MH. Nasal CPAP reduces gastroesophageal reflux in obstructive sleep apnea syndrome. Chest. 1992;101(6):1539–44.
59. Nishino T, Hiraga K. Coordination of swallowing and respiration in unconscious subjects. J Appl Physiol (1985). 1991;70(3):988–93.
60. Nishino T, Hasegawa R, Ide T, Isono S. Hypercapnia enhances the development of coughing during continuous infusion of water into the pharynx. Am J Respir Crit Care Med. 1998;157(3 Pt 1):815–21. https://doi.org/10.1164/ajrccm.157.3.9707158.
61. D'Angelo OM, Diaz-Gil D, Nunn D, Simons JC, Gianatasio C, Mueller N, et al. Anesthesia and increased hypercarbic drive impair the coordination between breathing and swallowing. Anesthesiology. 2014;121(6):1175–83. https://doi.org/10.1097/ALN.0000000000000462.
62. Kijima M, Isono S, Nishino T. Coordination of swallowing and phases of respiration during added respiratory loads in awake subjects. Am J Respir Crit Care Med. 1999;159(6):1898–902. https://doi.org/10.1164/ajrccm.159.6.9811092.
63. Brodsky MB, McFarland DH, Michel Y, Orr SB, Martin-Harris B. Significance of nonrespiratory airflow during swallowing. Dysphagia. 2012;27(2):178–84. https://doi.org/10.1007/s00455-011-9350-4.
64. Terzi N, Normand H, Dumanowski E, Ramakers M, Seguin A, Daubin C, et al. Noninvasive ventilation and breathing-swallowing interplay in chronic obstructive pulmonary disease. Crit Care Med. 2014;42(3):565–73. https://doi.org/10.1097/CCM.0b013e3182a66b4a.
65. Hori R, Isaka M, Oonishi K, Yabe T, Oku Y. Coordination between respiration and swallowing during non-invasive positive pressure ventilation. Respirology. 2016;21(6):1062–7. https://doi.org/10.1111/resp.12790.
66. Hori R, Ishida R, Isaka M, et al. Effects of noninvasive ventilation on the coordination between breathing and swallowing in patients with chronic obstructive pulmonary disease. Int J Chron Obstruct Pulmon Dis. 2019;14:1485–94.
67. Parke R, McGuinness S, Eccleston M. Nasal high-flow therapy delivers low level positive airway pressure. Br J Anaesth. 2009;103(6):886–90. https://doi.org/10.1093/bja/aep280.

68. Groves N, Tobin A. High flow nasal oxygen generates positive airway pressure in adult volunteers. Aust Crit Care. 2007;20(4):126–31. https://doi.org/10.1016/j.aucc.2007.08.001.
69. Sanuki T, Mishima G, Kiriishi K, Watanabe T, Okayasu I, Kawai M, et al. Effect of nasal high-flow oxygen therapy on the swallowing reflex: an in vivo volunteer study. Clin Oral Investig. 2017;21(3):915–20. https://doi.org/10.1007/s00784-016-1822-3.
70. Martin-Harris B, McFarland D, Hill EG, Strange CB, Focht KL, Wan Z, et al. Respiratory-swallow training in patients with head and neck cancer. Arch Phys Med Rehabil. 2015;96(5):885–93. https://doi.org/10.1016/j.apmr.2014.11.022.

Part II
Upper Airway

Chapter 4
Anatomy and Function of Upper Airway During Sleep: What Are Essential Mechanisms Eliciting Apneas During Sleep?

Satoru Tsuiki

Abstract During sleep, both the tonic and inspiratory-modulated components of upper airway muscle activity are suppressed in parallel with the characteristic postural atonia. In individuals with a constricted upper airway related to anatomical factors such as a large tongue relative to maxilla and mandible (i.e., oropharyngeal crowding), this sleep-related reduction in upper airway muscle activity could precipitate an upper airway occlusion more easily and lead to obstructive sleep apnea. However, in the narrowed upper airway of patients with obstructive sleep apnea, a neuromuscular compensatory mechanism augments the activity of the upper airway dilator muscles in defense of upper airway patency, particularly during inspiration. When the crowded oropharynx is improved by the application of positive airway pressure and/or mandibular advancement, upper airway dilator muscle activity during inspiration is reduced, returning the upper airway toward a normal configuration and pattern of muscle function in patients with obstructive sleep apnea. Accordingly, the upper airway patency in humans is precisely controlled by the close anatomical–functional relationship.

Keywords Upper airway · Obstructive sleep apnea · Anatomical balance · Genioglossus muscle

S. Tsuiki (✉)
Institute of Neuropsychiatry, Shibuya-ku, Tokyo, Japan

Division of Aging and Geriatric Dentistry, Department of Oral Function and Morphology, Tohoku University Graduate School of Dentistry, Aoba-ku, Sendai, Japan

Department of Oral Health Sciences, Faculty of Dentistry, The University of British Columbia, Vancouver, BC, Canada
e-mail: tsuiki@somnology.com

K. Yamaguchi (ed.), *Structure-Function Relationships in Various Respiratory Systems*, Respiratory Disease Series: Diagnostic Tools and Disease Managements, https://doi.org/10.1007/978-981-15-5596-1_4

1 Introduction

1.1 Human Upper Airway and Obstructive Sleep Apnea

Obstructive sleep apnea (OSA), as characterized by repetitive upper airway occlusion during sleep, is a common, underdiagnosed public health problem that predisposes the patient to cardiac complications. Understanding the relationship between anatomy and function of the upper airway increases our knowledge of pathogenesis of OSA and, moreover, might contribute to developing new treatment modalities for OSA.

Anatomically, human upper airway includes the extrathoracic trachea, larynx, pharynx, and nose. Within these segments, the pharynx is a segment bounded rostrally (cranially) by the nasopharynx and caudally by the glottic chink that further consists of four subsegments: nasopharynx, velopharynx, oropharynx, and hypopharynx (Fig. 4.1). Of particular interest, the pharynx is biologically compliant lacking bony or rigid support and the only collapsible segment of the respiratory tract. This is quite reasonable for serving its multifunction including alimentation, phonation, and respiration. It is likely that the evolution of both anatomy and neural control of the pharynx in humans has made its multiple ends. Swallowing requires highly collapsible tube where a coordinated neuromuscular action propels contents in the oral cavity into the esophagus and not into the nose or trachea. Speaking entails rapid movements of collapsible pharyngeal tube controlled by the central nervous system. In contrast, respiration is a physiological behavior involuntarily performed while awake and asleep. Thus, sleep-induced physiological phenomenon and body position change from upright to supine can be triggers increasing the pharyngeal collapsibility, leading to respiratory complications such as snoring and OSA [1–3].

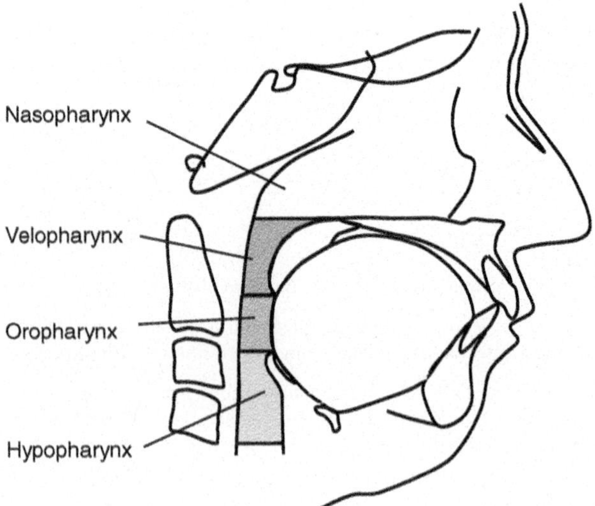

Fig. 4.1 Anatomical features of the human upper airway. The pharynx consists of four anatomical subsegments: nasopharynx, velopharynx, oropharynx, and hypopharynx

1.2 Upper Airway Patency During Wakefulness

During wakefulness, in order to counteract the intraluminal negative pressure during inspiration and to maintain its patency, the pharynx is stabilized by contraction of upper airway dilating muscles including the genioglossus (GG): an extrinsic tongue protruding muscle. Especially, the anterior wall of the pharynx is partly determined by the tongue, and tongue position is controlled by rhythmic activity in pace with inspiration as well as tonic activity throughout inspiration/expiration [1, 2, 4, 5]. Since some GG fibers diverge to the pharynx from the mental spine in the mandible, activation of these muscle fibers may result in (1) advancement of the tongue ventrally, thereby (2) enlargement of the pharyngeal space, and also (3) acting to stiffen the airway wall, assisting to maintain the airway patency [1, 6] (Fig. 4.2).

Another important aspect of the GG function in humans is the antigravitational role [6]. It has been reported that OSA may worsen and/or maybe observed mainly in the supine position while microgravity during spaceflight reduced sleep-disordered breathing [7, 8]. These reports suggest that gravitational pull on the tongue is expected to reduce pharyngeal size. By comparing the amount of reduction in the upper airway size at various levels, Tsuiki et al. [3] found that the velopharynx, not oropharynx, is not only the narrowest site in both upright and supine body positions but also the most changeable site in awake OSA patients. This finding is partly in agreement with a previous work by Miyamoto et al. [9] who reported the oropharyngeal size remained constant when the body position changed from upright to supine in OSA patients. It is speculated that

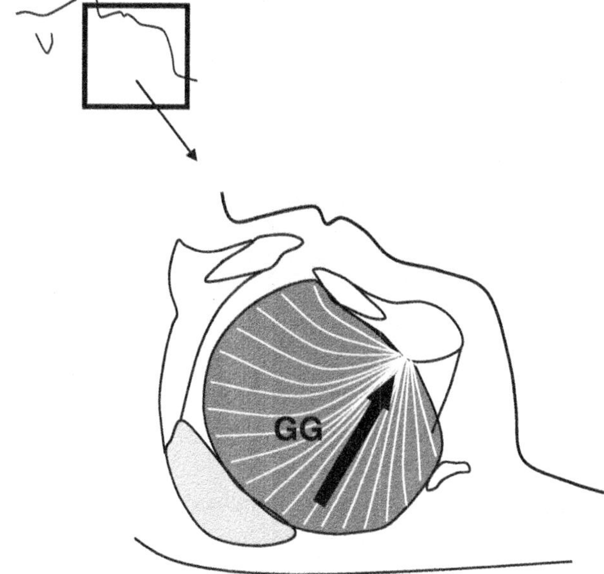

Fig. 4.2 A schematic illustration to describe the upper airway and the genioglossus muscle. GG: genioglossus

activation of the GG muscle can be interpreted to maintain the position of the tongue and oropharyngeal size. Although the palatal musculature exhibits compensation mechanisms similar to the GG muscle, the palatal musculature has been demonstrated to have less reflexive activation than the GG muscle in the supine position [10].

1.3 Factors Controlling Genioglossus Muscle Activity

Before describing how upper airway patency changes from awake to asleep, the factors controlling the GG muscle activity and links between GG activity and respiration should be summarized here. There are three primary neural inputs controlling GG muscle activity [5]. First, the negative pressure in the upper airway reflexively activates mechanoreceptors in the larynx (i.e., superior laryngeal nerve afferent activity) and ultimately leads to an increased hypoglossal output to the GG muscle by way of nucleus of the solitary tract in the brainstem and hypoglossal motor nucleus (i.e., negative-pressure reflex). Second, hypoglossal motor nuclei that control GG muscle receive input from central-pattern generating neurons in the ventral medulla (i.e., central-pattern generator), which produces inspiratory-phasic activity of the GG muscle. This activity is independent of the presence/absence of negative-pressure reflex as the inspiratory activation of GG occurs prior to negative pressure development in the upper airway: 50–100 ms prior to diaphragmatic activation [11]. Accordingly, during inspiration, both respiratory neurons from the central pattern generator and the negative-pressure reflex modulate the GG muscle activity. Third, neurons that modulate arousal (active awake, less active, or inactive asleep), such as serotonergic neurons, provide a tonic excitatory influence on upper airway motoneurons including hypoglossal motoneurons. This has been called the “wakefulness stimulus” and generally increases muscle activity. Apart from these three primary factors, it should be noted tonic excitatory effects on GG muscle, regardless of respiratory phase, cannot be neglected if the GG muscle is lengthened and tensed as seen during mandibular advancement, which would lead to augmentation of the input from the proprioceptors in the GG muscle [12].

2 Upper Airway Collapse During Sleep; Who Develops OSA?

Even in severe OSA individuals with a structurally narrowed and collapsible upper airway, obstructive apnea never occurs during wakefulness because neuromuscular compensation mechanism is augmented in defense of upper airway patency. Indeed, compared to normal subjects, patients with OSA have significantly greater GG muscle activity while awake [12–14]. However, the compensation per se appears unsatisfactory because the increased GG muscle activity of OSA patients merely provides a smaller cross-sectional area of the upper airway and higher supraglottic resistance compared to non-apneic persons [15–18].

During sleep, the GG muscle activity is suppressed in parallel with the characteristic postural atonia both in healthy people and in OSA patients. This sleep-related reduction in GG muscle activity reminds us of either the loss of negative-pressure reflex or respiratory input from central pattern generator. However, Taranto-Montemuro et al. [19] recently reported that both phasic and tonic activity of GG increased in association with changes in the negative pressure level of the airway (i.e., more negative) during sleep. Therefore, the decrease in the GG muscle activity is due to a remarkable reduction in the wakefulness stimuli. This physiological phenomenon further brings to mind two possibilities for the tongue collapse and subsequently result in upper airway narrowing and/or occlusion; the GG muscle activity is either more reduced in patients with OSA compared to normal subjects during sleep or OSA patients exhibiting a normal sleep-related reduction in GG muscle activity in conjunction with the underlying structural upper airway narrowing [20].

Repeatedly, the upper airway cross-sectional area is determined by the interaction between the individual anatomic structure per se and the upper airway muscle activity. The cross-sectional area of the upper airway is determined by transmural pressure (Ptm), the difference between intraluminal pressure (Plumen), and pharyngeal tissue pressures (Ptissue): Ptm = Plumen—Ptissue [20, 21]. Thus, airflow during inspiration, which reduces the Plumen, might decrease the Ptm, leading to a decreased upper airway cross-sectional area. However, during both wakefulness and sleep, this is not always the case under the presence of the activity of GG muscle. An increase in GG muscle activity (Pmuscle) meaning contraction of GG muscle acts to stiffen as well as to enlarge the cross-sectional area, which would result in reducing the Ptissue [12]. An increase in Ptm, caused by a more positive Plumen and/or more negative Ptissue (including a greater Pmuscle), increases the cross-sectional area of the upper airway. Conversely, a decrease in Ptm, caused by a more negative Plumen and/or more positive Ptissue (including a smaller Pmuscle), decreases the cross-sectional area of the upper airway. Accordingly, Ptm is farther expressed as the following equation: Ptm = Plumen-(Ptissue-Pmuscle) [21]. This equation suggests that the cross-sectional area of the upper airway "purely" reflects the intrinsic structural property of the upper airway when the upper airway muscle activity is completely suppressed (i.e., Pmuscle = 0) [20, 21].

3 Anatomical Difference Between OSA and Normals

Isono et al. [20] succeeded in critically separating anatomic factors from neural factors by means of eliminating the upper airway muscle activity using general anesthesia. They further found that OSA patients have an abnormally narrow passive pharynx, which constitutes the principal pathogenic alteration against normals [20–22]. What kind of an anatomical abnormality do OSA patients then possess in the upper airway and surrounding area?

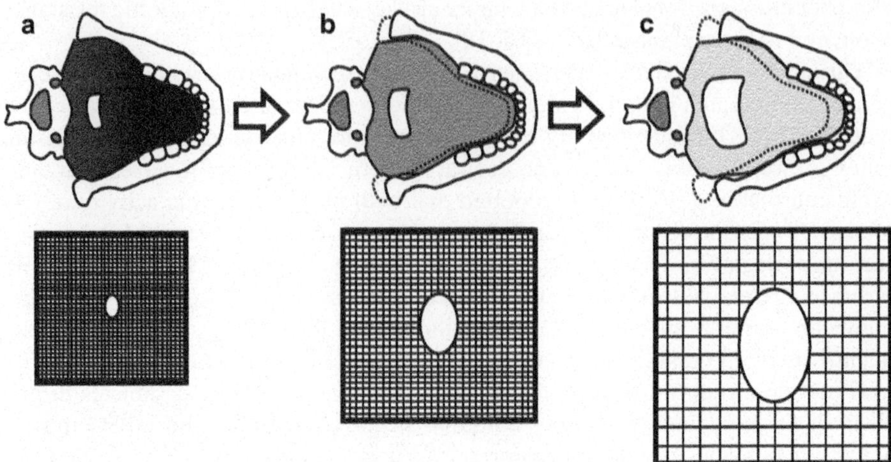

Fig. 4.3 Schematic illustrations of the relationships between the cervical spine, mandible, dental arch, tongue and upper airway, and dose-dependent effects of mandibular advancement on oropharyngeal crowding (left to right). The upper airway is surrounded by oropharyngeal soft tissue, which is enclosed by dentofacial and vertebral hard tissue (top). The upper airway is assumed to be a "collapsible tube" that is encircled by soft material within a rigid box (bottom). A larger tongue associated with obesity readily increases the relative amount of soft tissue inside the rigid box, which would result in the collapse of the tube and the worsening of OSA. An increased shading density indicates increased soft tissue pressure inside the craniofacial bony enclosure. Mandibular advancement restores the increased upper airway collapsibility to normal by enlarging the size of the bony enclosure (**a** to **b** and **c**)

Structurally, the pharyngeal airway is surrounded by soft tissue such as the tongue, which is enclosed by bony structures such as the mandible and cervical vertebrae [22]. Under the suppression of pharyngeal dilator muscle contraction during sleep and anesthesia, the anatomical balance between the soft-tissue volume inside the bony enclosure and the bony enclosure size seems to determine the pharyngeal airway size. Accordingly, the anatomical imbalance between upper airway soft-tissue volume and bony enclosure size results in pharyngeal airway obstruction during sleep and anesthesia when reduction or loss of upper airway muscle activity. Watanabe et al. demonstrated that obesity and craniofacial abnormalities synergistically increase pharyngeal collapsibility during general anesthesia and paralysis [22]. Tsuiki et al. [23] further reported while the tongue was significantly larger in subjects with larger maxillomandibular dimensions, OSA patients had a significantly larger tongue for a given maxillomandibular size. This anatomical abnormality "oropharyngeal crowding" specific to OSA patients is shown in Fig. 4.3.

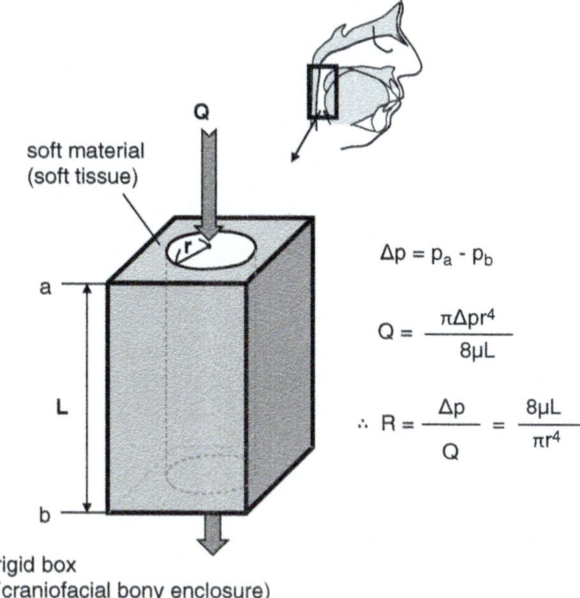

Fig. 4.4 Schematic explanations for the simplified upper airway model as a cylindrical and collapsible tube. L: length of the tube, Δp: the difference between the upstream (p_a) and downstream (p_b) pressures in the tube, Q; volume of airflow per unit time, R; tube resistance, r: radius of the tube. Q and Δp and/or Q and R are proportional to each other (i.e., $Q = \Delta p \cdot R$), R can be expressed as: $R = \Delta p/Q = 8\mu L/\pi r4$, where μ is constant

4 Upper Airway as a Cylindrical and Collapsible Tube

Rheologically, the upper airway could be assumed as a simple cylindrical tube. Airflow dynamics are influenced by a change in the tube radius (r) and longitudinal tube length (L), since the flow rate (volume of airflow per unit time; Q) through the tube is proportional to the fourth power of r and the difference between the upstream and downstream pressures (Δp) in the tube [24]. More precisely, since Q and Δp and/or Q and the tube resistance (R) are proportional to each other (i.e., $Q = \Delta p \cdot R$), R can be expressed as: $R = \Delta p/Q = 8\mu L/\pi r^4$, where μ is constant. It is indicated that the resistance of the tube increases on the condition that the length of the tube becomes longer whereas the tube radius becomes smaller. This increase in the radius, even though the amount of the change is quite small, provides a great impact in reducing the tube resistance. Enlargement of the upper airway dimension, for example, by mandibular advancement can sometimes dramatically results in improving the airflow dynamics, leading to a remarkable improvement of oxygenation and apnea–hypopnea index [25, 26] (Fig. 4.4).

5 Functional Contribution of Advancement of the Mandible to Upper Airway Patency

As noted earlier, the activity of GG muscle is augmented to maintain upper airway patency during wakefulness in patients with a compromised upper airway as in the case with OSA [13]. However, sleep greatly suppresses GG muscle activity, which may lead to partial or complete upper airway collapse. If mandibular advancement, which is aimed to improve the oropharyngeal crowding, increased the activity of the GG muscle, this effect should help to maintain the patency of the airway, since the activation of GG activity could stiffen the compliant airway wall. In contrast, if we consider that mandibular advancement increases the cross-sectional area and decreases the propensity for upper airway occlusion, a larger, less-collapsible upper airway could improve airflow dynamics, which would require less GG activity [12].

With the use of a remotely controlled mandibular positioner, Almeida and colleagues [27] investigated the effects of passive mandibular advancement on GG muscle activity and severity of OSA during sleep. They found that both phasic and tonic activity of GG decreased with protrusion of the mandible in a dose-dependent manner [27]: Enlargement of the airway by mandibular advancement enabled the GG muscle not to activate but to normalize its hyperactivity. Interestingly, the reduction in GG muscle activity was similar to the effect of nasal continuous positive airway pressure (CPAP) therapy, in that successful application of CPAP decreased GG muscle activity in OSA patients [28], although the approach used to improve airway patency was totally different between the two treatment modalities. It would be reasonable to speculate that input from upper airway mechanoreceptors during inspiration decreased phasic GG muscle activity, since protrusion of the mandible reduced the propensity for upper airway collapse. An increase in the bony enclosure size (e.g., mandibular advancement by an oral appliance) had no effect on the volume of the upper airway soft issues in this case. However, it can be interpreted that protrusion of the mandible itself acts to change the position and size of the bony enclosure, thereby reducing Ptissue in OSA patients as reported by Isono et al. who succeeded in decreasing Ptissue in the passive pharynx in patients with OSA [29]. This speculation is also supported by an animal study in which mandibular advancement reduced the anterior and lateral soft tissue pressure in anesthetized rabbits [30]. Therefore, therapeutic outcomes of mandibular advancement in oral appliance therapy are likely to be more anatomically based rather than via activation of upper airway muscle activity. (Fig. 4.5)

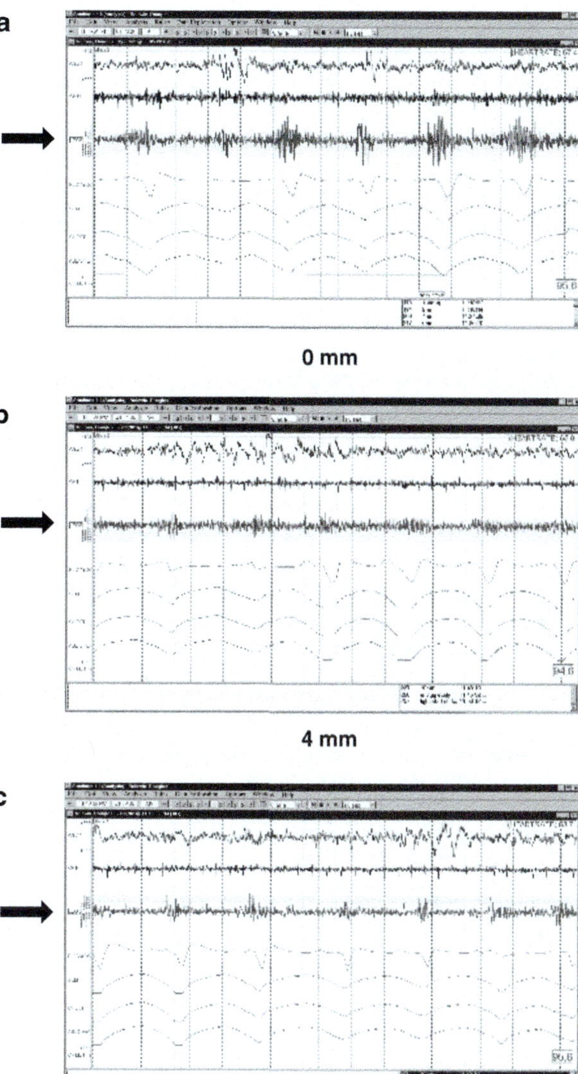

Fig. 4.5 Dose-dependent decreases in respiratory-related genioglossus muscle activity in association with remotely controlled mandibular advancement during sleep in an OSA patient (**a** to **b** and **c**). The third trace (arrow) represents raw genioglossus muscle activity

6 Conclusion

In response to the clinical question "What are essential mechanisms eliciting apneas during sleep?," obstructive apnea occurs if sleep-related physiological reduction of upper airway muscle is added on top of oropharyngeal crowding in humans. When oropharyngeal crowding is normalized, augmented neuromuscular compensatory activation of upper airway muscle is also normalized and then OSA is improved. There certainly exists a close anatomical-functional relationship to control the upper airway patency in humans.

Acknowledgments Research discussed in this chapter was in part supported by grants from the Japanese Society for the Promotion of Science (Grant-in-Aid for Scientific Research; 17K11793, 19K10236, 20K10085).

References

1. Remmers JE, de Groot WJ, Sauerland EK, Anch AM. Pathogenesis of upper airway obstruction during sleep. J Appl Physiol. 1978;44:931–8. https://doi.org/10.1152/jappl.1978.44.6.931.
2. Kuna ST, Remmers JE. Anatomy and physiology of upper airway obstruction. In: Kryger MH, Roth T, Dement WC, editors. Principles and practice of sleep medicine. 3rd ed. Philadelphia: WB Saunders; 2000. p. 840–58.
3. Tsuiki S, Almeida FR, Bhalla PS, Lowe AA, Fleetham JA. Supine-dependent changes in upper airway size in awake obstructive sleep apnea patients. Sleep Breath. 2003;7:43–50. https://doi.org/10.1007/s11325-003-0043-6.
4. Tsuiki S, Ono T, Ishiwata Y, Kuroda T. Functional divergence of human genioglossus motor units with respiratory-related activity. Eur Respir J. 2000;15:906–10. https://doi.org/10.1034/j.1399-3003.2000.15e16.x.
5. White DP. Pathogenesis of obstructive and central sleep apnea. Am J Respir Crit Care Med. 2005;172:1363–70. https://doi.org/10.1164/rccm.200412-1631SO.
6. Kobayashi I, Perry A, Rhymer J, Wuyam B, Hughes P, Murphy K, Innes JA, McIvor J, Cheesman AD, Guz A. Inspiratory coactivation of the genioglossus enlarges retroglossal space in laryngectomized humans. J Appl Physiol. 1996;80:1595–604. https://doi.org/10.1152/jappl.1996.80.5.1595.
7. Joosten SA, O'Driscoll DM, Berger PJ, Hamilton GS. Supine position related obstructive sleep apnea in adults: pathogenesis and treatment. Sleep Med Rev. 2014;18:7–17. https://doi.org/10.1016/j.smrv.2013.01.005.
8. Elliott AR, Shea SA, Dijk DJ, Wyatt JK, Riel E, Neri DF, et al. Microgravity reduces sleep-disordered breathing in humans. Am J Respir Crit Care Med. 2001;164:478–85. https://doi.org/10.1164/ajrccm.164.3.2010081.
9. Miyamoto K, Özbek MM, Lowe AA, Fleetham JA. Effect of body positions on tongue posture in awake patients with obstructive sleep apnoea. Thorax. 1997;52:255–9. https://doi.org/10.1136/thx.52.3.255.
10. Horner RL. Motor control of the pharyngeal musculature and implications for the pathogenesis of obstructive sleep apnea. Sleep. 1996;19:827–53. https://doi.org/10.1093/sleep/19.10.827.
11. van Lunteren E. Muscles of the pharynx: structural and contractile properties. Ear Nose Throat J. 1993;72:27–9. 33
12. Tsuiki S, Ryan CF, Lowe AA, Inoue Y. Functional contribution of mandibular advancement to awake upper airway patency in obstructive sleep apnea. Sleep Breath. 2007;11:245–51. https://doi.org/10.1007/s11325-007-0119-9.

13. Mezzanotte WS, Tangel DJ, White DP. Waking genioglossal electromyogram in sleep apnea patients versus normal controls (a neuromuscular compensatory mechanism). J Clin Invest. 1992;89:1571–9. https://doi.org/10.1172/JCI115751.
14. Suratt PM, McTier RF, Wilhoit SC. Upper airway muscle activation is augmented in patients with obstructive sleep apnea compared with that in normal subjects. Am Rev Respir Dis. 1988;137:889–94. https://doi.org/10.1164/ajrccm/137.4.889.
15. Bohlman ME, Haponik EF, Smith PL, Allen RP, Bleecker ER, Goldman SM. CT demonstration of pharyngeal narrowing in adult obstructive sleep apnea. AJR Am J Roentgenol. 1983;140:543–8. https://doi.org/10.2214/ajr.140.3.543.
16. Horner RL, Shea SA, McIvor J, Guz A. Pharyngeal size and shape during wakefulness and sleep in patients with obstructive sleep apnoea. Q J Med. 1989;72:719–35.
17. Lowe AA, Fleetham JA, Adachi S, Ryan CF. Cephalometric and computed tomographic predictors of obstructive sleep apnea severity. Am J Orthod Dentofac Orthop. 1989;107:589–95.
18. Anch AM, Remmers JE, Bruce H 3rd. Supraglottic airway resistance in normal subjects and patients with occlusive sleep apnea. J Appl Physiol. 1982;53:1158–63. https://doi.org/10.1152/jappl.1982.53.5.1158.
19. Taranto-Montemurro L, Messineo L, Sands SA, Azarbarzin A, Marques M, Edwards BA, Eckert DJ, White DP, Wellman A. The combination of atomoxetine and oxybutynin greatly reduces obstructive sleep apnea severity. A randomized, placebo-controlled, double-blind crossover trial. Am J Respir Crit Care Med. 2019;199:1267–76. https://doi.org/10.1164/rccm.201808-1493OC.
20. Isono S, Remmers JE, Tanaka A, Sho Y, Sato J, Nishino T. Anatomy of pharynx in patients with obstructive sleep apnea and in normal subjects. J Appl Physiol. 1997;82:1319–26. https://doi.org/10.1152/jappl.1997.82.4.1319.
21. Isono S. Upper airway muscle function during sleep. In: Gloughlin GM, Marcus CL, Caroll JL, editors. Sleep and breathing in children: a developmental approach. Marcell Dekker Inc.: New York; 2000. p. 261–91.
22. Watanabe T, Isono S, Tanaka A, Tanzawa H, Nishino T. Contribution of body habitus and craniofacial characteristics to segmental closing pressures of the passive pharynx in patients with sleep-disordered breathing. Am J Respir Crit Care Med. 2002;165:260–5.
23. Tsuiki S, Isono S, Ishikawa T, Yamashiro Y, Tatsumi K, Nishino T. Anatomical balance of the upper airway and obstructive sleep apnea. Anesthesiology. 2008;108(6):1009–15. https://doi.org/10.1164/ajrccm.165.2.2009032.
24. Mathiasen RA, Cruz RM. Asymptomatic near-total airway obstruction by a cylindrical tracheal foreign body. Laryngoscope. 2005;115:274–7. https://doi.org/10.1097/01.mlg.0000154732.16034.83.
25. Kato J, Isono S, Tanaka A, Watanabe T, Araki D, Tanzawa H, et al. Dose-dependent effects of mandibular advancement on pharyngeal mechanics and nocturnal oxygenation in patients with sleep-disordered breathing. Chest. 2000;117:1065–72. https://doi.org/10.1378/chest.117.4.1065.
26. Walker-Engström ML, Ringqvist I, Vestling O, Wilhelmsson B, Tegelberg A. A prospective randomized study comparing two different degrees of mandibular advancement with a dental appliance in treatment of severe obstructive sleep apnea. Sleep Breath. 2003;7:119–30. https://doi.org/10.1007/s11325-003-0119-3.
27. Almeida FR, Tsuiki S, Hattori Y, Takei Y, Inoue Y, Lowe AA. Dose-dependent effects of mandibular protrusion on genioglossus activity in sleep apnoea. Eur Respir J. 2011;37:209–12. https://doi.org/10.1183/09031936.00194809.
28. Strohl KP, Redline S. Nasal CPAP therapy, upper airway muscle activation, and obstructive sleep apnea. Am Rev Respir Dis. 1986;134:555–8.
29. Isono S, Tanaka A, Sho Y, Konno A, Nishino T. Advancement of the mandible improves velopharyngeal airway patency. J Appl Physiol. 1995;79:2132–8. https://doi.org/10.1152/jappl.1995.79.6.2132.
30. Kairaitis K, Stavrinou R, Parikh R, Wheatley JR, Amis TC. Mandibular advancement decreases pressures in the tissues surrounding the upper airway in rabbits. J Appl Physiol. 2006;100:349–56. https://doi.org/10.1152/japplphysiol.00560.2005.

Chapter 5
Instability of Upper Airway During Anesthesia and Sedation: How Is Upper Airway Unstable During Anesthesia and Sedation?

Shiroh Isono

Abstract Interaction between upper airway structures and muscles serves to achieve various physiological functions of the upper airway. Within the upper airway, the pharynx is structurally a collapsible tube, and its luminal size and stiffness are neurally regulated by pharyngeal muscle contraction which significantly decreases in unconscious states such as sleep and anesthesia, and is lost during muscle paralysis. Obstructive sleep apnea patients with structurally collapsible pharynx are at risk for unstable breathing leading to critical respiratory complications during and after anesthesia and sedation. In contrast, the laryngeal airway is usually maintained during anesthesia and sedation unless the laryngeal reflexes occur, and maximized during complete paralysis while precise neural regulation of the vocal cord movements is depressed in unconscious states. Understandings of these fundamental differences are important for exploring the mechanisms of the pharyngeal and laryngeal obstruction and for determining their appropriate treatments to prevent life-threatening severe hypoxemia.

Keywords Upper airway · Pharynx · Obstructive sleep apnea · Larynx

1 Introduction

Physiological functions of the upper airway are performed by an interaction between upper airway structural properties and neural control of upper airway muscle activity. Within the upper airway, both structural properties and physiological functions

S. Isono (✉)
Department of Anesthesiology, Graduate School of Medicine, Chiba University, Chiba, Japan
e-mail: shirohisono@yahoo.co.jp

K. Yamaguchi (ed.), *Structure-Function Relationships in Various Respiratory Systems*, Respiratory Disease Series: Diagnostic Tools and Disease Managements, https://doi.org/10.1007/978-981-15-5596-1_5

significantly differ between the pharyngeal and laryngeal airways aside from a role for a breathing conduit. The pharynx is a collapsible tube in which airway size and stiffness are determined by regulating the contraction of surrounding pharyngeal muscles. The pharyngeal airway patency is significantly impaired in unconscious states such as sleep and anesthesia and lost during muscle paralysis. In contrast, the laryngeal airway is usually maintained during and anesthesia unless the laryngeal reflexes occur, and maximized during complete paralysis while precise neural regulation of the vocal cord movements is depressed in unconscious states. Understandings of these fundamental differences are important for exploring the mechanisms of the pharyngeal and laryngeal obstruction and for determining their appropriate treatments to prevent life-threatening severe hypoxemia.

2 Neural Mechanisms for Stable Pharyngeal Airway Maintenance

2.1 Interaction Between Anatomical and Neural Mechanisms for Stable Upper Airway Maintenance

The neural balance between two antagonistic forces influence pharyngeal airway size [1] (Fig. 5.1). Pharyngeal collapsing forces are produced by contraction of the inspiratory pump muscles (e.g., the diaphragm and external intercostal muscles). The negative airway pressure within the pharyngeal airway during inspiration promotes the collapse of the extra-thoracic airway. The pharyngeal dilating forces are produced by contraction of the pharyngeal airway dilator muscles such as the genioglossus muscle. Activities of both the inspiratory pump and pharyngeal airway muscles are regulated by "neural mechanisms." The level of consciousness [2], chemical stimuli [3], and airway reflexes [4, 5] interactively modulate the final activation of the inspiratory and pharyngeal muscles.

The neural mechanisms operate best during wakefulness assuring a patent pharyngeal airway. However, the reduction of the level of consciousness by anesthesia or sleep significantly decreases activities of the pharyngeal dilator muscles, more than it decreases the activities of the inspiratory pump muscles (e.g., the diaphragm, external intercostals) [2, 6, 7], resulting in a loss of neural balance and therefore a narrowing of the pharyngeal airway. Depressed neural mechanisms during anesthesia and sleep are of crucial importance, particularly in patients with impaired anatomical mechanisms, such as neonates, small infants, and patients with obstructive sleep apnea, who significantly depend importantly on neural mechanisms for maintaining the pharyngeal airway [8, 9]. Accordingly, the pharyngeal airway size is determined by the interaction between the neural and anatomical mechanisms. Impairment of anatomical mechanisms is a primary cause of pharyngeal obstruction, and the neural mechanisms usually operate to compensate for the impairment of anatomical mechanisms.

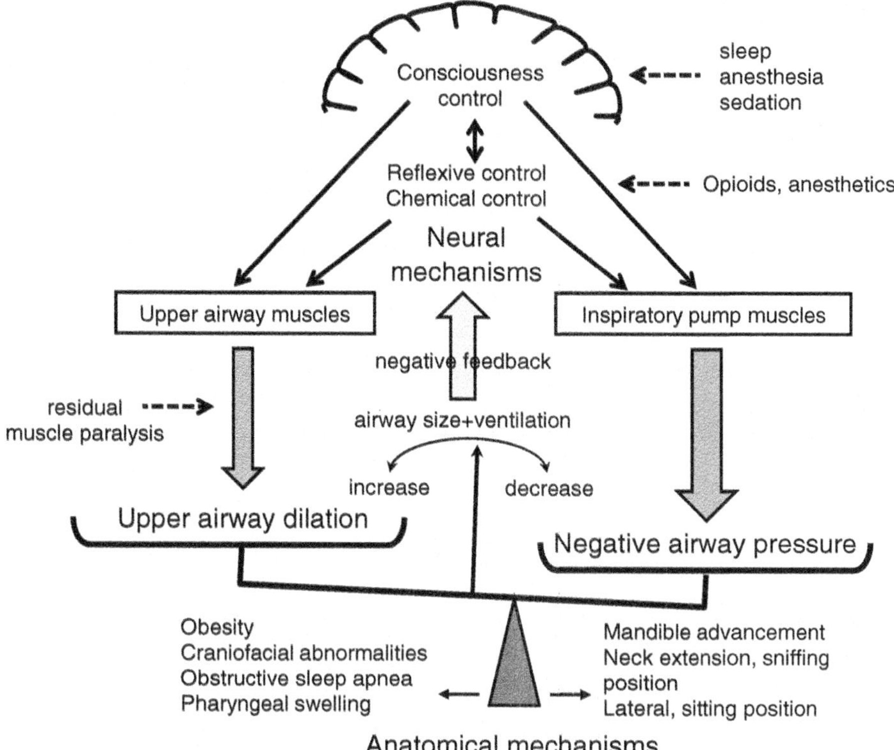

Fig. 5.1 Schematic explanation of the interaction between anatomical and neural mechanisms for stable breathing and upper airway maintenance

2.2 Sequential Interactions Between Anatomical and Neural Mechanisms

There are sequential interactions between anatomical and neural mechanisms during the process of obstruction and reopening of the pharyngeal airway (Fig. 5.2) [10]. The neural mechanisms for the pharyngeal airway maintenance are ensured by the continual supervision of the higher central nervous system during wakefulness. Serotonergic or noradrenergic neurons are considered to be involved in the wakefulness stimulus to the upper airway motor neuron [11] which is significantly depressed during sleep, anesthesia, sedation, and impaired consciousness [2]. Compensatory pharyngeal muscle activation during wakefulness is particularly important for obstructive sleep apnea adults and children with structural abnormalities at the pharyngeal airway [12, 13]. Loss of consciousness is, therefore, an initial key event for inducing various degrees of pharyngeal obstruction depending on both magnitude of depression of pharyngeal muscle contraction and the structural stability of the pharyngeal airway [2, 6, 7, 9, 14].

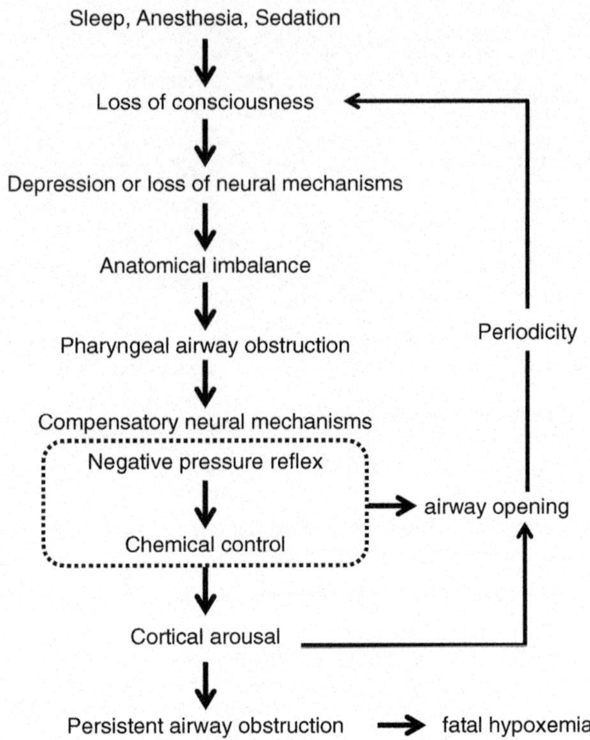

Fig. 5.2 Sequential interactions between anatomical and neural mechanisms during the process of obstruction and reopening of the pharyngeal airway

Once the pharyngeal airway obstructs, the negative pressure reflex induced by negative airway pressure during inspiration operates to reverse the obstruction as a first line of compensative response by re-increasing pharyngeal dilator muscle activities [4, 5]. Contribution of this reflex may vary depending on the muscles involved, sleep stage, anesthesia and sedation level, and presence of obstructive sleep apnea. The level of peak genioglossal activity is directly associated with the magnitude of negative inspiratory airway pressure during both wakefulness and NREM sleep while the slope of the association is greater during wakefulness than NREM sleep [15]. Negative airway pressure is evidenced to regulate the genioglossal activity even within an inspiration, and the responsiveness of the negative pressure reflex is not altered within the physiological range of the chemical stimuli [16, 17]. The trigeminal and superior laryngeal nerves are reported to mediate an important component of the reflex responses as afferent pathways [5]. The involvement of central pattern generator is suggested in the reflex since the responsiveness is significantly attenuated when negative airway pressure is produced by an iron lung ventilator during NREM sleep, which eliminated the inspiratory pump muscle activation [18].

Probably because of greater surface tension on the pharyngeal airway mucosa, the greater pharyngeal muscle activation is necessary to reopen the closed airway [19]. When the augmentation of the pharyngeal dilator muscle activities is not sufficient to reopen the airway, hypercapnia and hypoxemia develop during the pharyngeal obstruction. The pharyngeal muscles' responses to the chemical stimuli differ from those of the inspiratory pump muscles, in which the genioglossus is activated only below a SaO_2 threshold, while the diaphragm activity increases monotonously in response to progressive hypoxia and hypercapnia [3]. Therefore, the imbalance between the pharyngeal dilating forces and the inspiratory collapsing forces favors the maintenance of pharyngeal obstruction [1]. However, it should be also recognized that the increased negative airway pressure progressively increases pharyngeal dilator muscle activities through the augmented negative pressure reflex.

Furthermore, the chemical stimuli also play a significant role in the induction of the arousal response, which used to be considered as the last line of defense for reversal of the occluded airway and the most important survival response [20, 21]. A burst of pharyngeal muscle activation usually occurs upon cortical arousal and overcomes the pharyngeal obstruction [1]. Notably, cortical arousal is not commonly observed upon the restoration of patent pharyngeal airway in neonates and small children [22]. In addition to increased respiratory drive to the inspiratory pump muscles during obstructive apnea, arousal stimuli induce ventilatory overshoot immediately after the restoration of the patent pharyngeal airway. Resultant reduction of chemical drive to motoneurons of both upper airway and inspiratory pump muscles predisposes the reoccurrence of periodic obstructive apnea events.

2.3 Roles of Pharyngeal Muscles in the Neural Mechanisms

Pharyngeal muscles are divided into pharyngeal airway dilators and constrictors based on structural arrangements of the muscle fibers. Major constrictor muscles of the pharynx consist of the superior, middle, and inferior pharyngeal constrictors; however, dilators are numerous. While the net vector balance between dilators and constrictors is generally believed to determine the actual movement of pharyngeal airway structures, both dilators and constrictors also synergistically contribute to the stiffening of the pharyngeal airway wall [23].

At the level of the velopharyngeal airway (between the edge of the soft palate and the end of the nasal septum), the role of each muscle depends on the preference of the breathing route. Contraction of the palatoglossal and palatopharyngeal muscles opens the retropalatal airway, allowing for nasal breathing. These muscles are, therefore, considered to be the dilators during nasal breathing. The tensor veli palatini and the levator veli palatini, which elevate the soft palate, are considered to be the dilators during oral breathing. Normal persons are predominantly nasal breathers during sleep as well as wakefulness (96 vs. 93%) [24]. In response to obstruction within the nasal passage, reflexive switching from nasal to oral breathing route is initiated by stimulation of the mechanoreceptors within the airway mucosa. Nasal topical anesthesia, sleep, and sedation significantly delay the switching reflex [25–27].

At the level of the oropharyngeal airway (between the tip of the epiglottis and the edge of the soft palate, also known as the retroglossal airway), the genioglossus muscle dilates the airway for respiration, while its fan-like projection of muscle fibers produces complicated movements of the tongue. The genioglossus is the most extensively studied muscle, since the reduction of its activity is known to be associated with an increase in upper airway resistance [1].

At the hypopharyngeal airway (between the vocal cords and the tip of the epiglottis), net vector forces of the suprahyoid muscles, which attach superiorly to the hyoid bone (i.e., the geniohyoid and mylohyoid), and the infrahyoid muscles, which attach inferiorly to the hyoid bone (i.e., the sternohyoid, thyrohyoid, and omohyoid), determine the position of the hyoid bone and its function in the dilatation of this segment. The position of the hyoid bone and the direction of muscle fibers significantly influence the net vector forces of these muscles [28]. Anesthesia and muscle paralysis posteriorly displace the hyoid bone and an increase of lung volume caudally displaces the hyoid bone. The hyoid bone position is a marker for changes of upper airway muscle activation and longitudinal pharyngeal wall tension due to lung volume change as well as the balance of the soft tissue volume of the craniofacial size [29].

Many dilator muscles exhibit phasic inspiratory activity, especially during mechanical loading [27, 30, 31] or chemical stimuli [3] since inspiration is the most critical phase for airway maintenance and increase of dilating forces during this period serves to oppose negative intraluminal pressure which acts to narrow the pharyngeal airway [1]. Activation of the alae nasi occurs prior to the onset of inspiratory airflow, especially during sleep (90 ms prior to the onset of flow during wakefulness vs. 196 ms during sleep) [32]. This sequence is considered to act as a stiffener for the collapsible pharynx against the collapsing forces of inspiratory efforts. Electromyogram activity of the diaphragm gradually increases during inspiration and peaks in the late inspiratory phase. In contrast, pharyngeal muscle activity peaks in the early inspiratory phase, and higher activity is maintained throughout the inspiration.

Interestingly, inspiratory activation of the superior pharyngeal constrictor is frequently observed in association with airway reopening on arousal in patients with obstructive sleep apnea [33]. One possible interpretation of their findings is that contraction of the constrictor muscle results in the stiffening of the pharyngeal airway wall, which assists in the maintenance of a patent airway reestablished by a burst of the genioglossus upon arousal. Increases in tonic muscle activity, which is often observed in dilator muscles during mechanical loading and chemical stimuli, may also contribute to the stiffening of the pharyngeal airway wall. Pharyngeal muscles not only act as dilators or constrictors in controlling the pharyngeal airway size but also act as stiffeners in regulating compliance of the pharyngeal airway wall. Recent physiological and clinical studies on electrical stimulation of the upper airway muscles strongly support this concept [23, 34, 35].

3 Effects of Anesthesia and Sedation on the Pharyngeal Neural Mechanisms

3.1 Loss of the Neural Mechanisms During Anesthesia Induction and Airway Maintenance Techniques

Both inspiratory efforts and pharyngeal airway patency are lost during anesthesia induction with intravenous anesthetics, opioids, and muscle relaxants, and anesthesiologists need to ventilate the patients through the anesthesia facemask with airway maneuvers in order to maintain airway patency and oxygenation. Facemask ventilation is difficult in 5% and impossible in 0.15% of general anesthesia population despite proper airway maneuvers [36, 37]. Neck radiation changes, male sex, sleep apnea, Mallampati III or IV, and presence of beard were identified as independent risk factors for impossible mask ventilation.

Sato et al. confirmed difficulty in mask ventilation during anesthesia induction in patients with severe obstructive sleep apnea, and demonstrated the effectiveness of two hands mask ventilation with triple airway maneuvers including mandible advancement, head extension, and mouth opening [38]. Under general anesthesia and complete paralysis, Isono et al. found that the passive closing pressures of patients with obstructive sleep apnea are above the atmospheric pressure (2 ± 3 cm H_2O), whereas those of subjects without obstructive sleep apnea are below the atmospheric pressure (−4 ± 4 cm H_2O) [14]. Accordingly, with using various airway maintenance techniques, anesthesiologists need to reduce the closing pressures by approximately 10 cm H_2O more in sleep apnea patients than in non-sleep apnea subjects [39]. Or awake tracheal intubation is advised in patients with anticipated difficult or impossible mask ventilation [40].

Proper positioning of the patient and adequate denitrogenation before induction of general anesthesia are essential (Fig. 5.3). Inhalation of pure oxygen for more than 3 min with the fitted mask eliminates nitrogen from the lung and maximizes the apnea tolerance period [41]. The sitting position or reversed Trendelenburg position prolongs the apnea tolerance time by increasing the functional residual capacity of the lung [42]. The sitting position also decreases the closing pressures by approximately 6 cm H_2O and is advantageous for mask ventilation [43]. The sniffing position or ramping position in obese patients decreases the closing pressure by approximately 4 cm H_2O and is advantageous for both mask ventilation and tracheal intubation; as such, a standard head and neck position should be used for anesthesia induction [44]. Recent airway management guidelines recommend the use of muscle relaxants for improving both facemask ventilation and tracheal intubation [40]. Succinylcholine, a short-acting depolarizing muscle relaxant, is demonstrated to open the closed pharyngeal airway during pharyngeal muscle fasciculation [45]. Triple airway maneuvers with two hands are more effective than those with one hand [38]. Use of anesthesia ventilator with pressure control mode is advised to

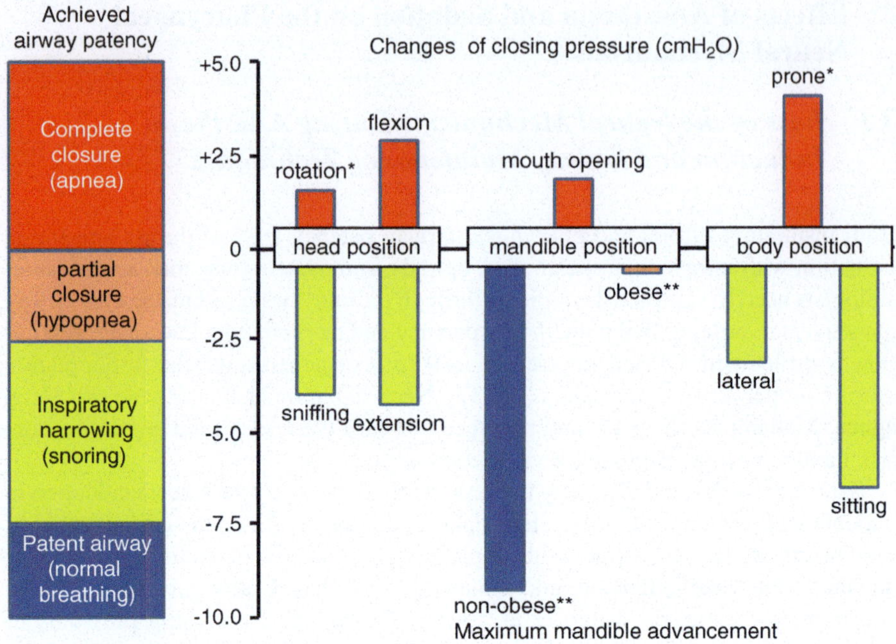

Fig. 5.3 Changes of pharyngeal closing pressures in response to various mechanical airway interventions to subjects in neutral head and neck position with mouth closed in supine posture

apply PEEP to maintain the pharyngeal patency and increase lung volume and to prevent gas-insufflation into stomach by limiting peak inspiratory pressure [46].

3.2 *Differential Effects of Anesthetics and Sedatives on the Inspiratory Pump Muscles and Pharyngeal Muscles*

Anesthetics and sedatives have depressant effects on breathing (Table 5.1). Dose-dependent tidal volume reduction is a typical breathing pattern. Respiratory rate initially increases but eventually central apnea occurs depending on the doses and types of anesthetics and sedatives administered [47]. These drugs also have a significant influence on pharyngeal patency. Nishino et al. demonstrated that halothane and intravenous anesthetics and sedatives, such as thiopental and diazepam, decrease the hypoglossal nerve activity more than the phrenic nerve activity in cat preparation [6]. Notably, the magnitude of differential suppression varies among these drugs, and ketamine presents the least differential effects, indicating that ketamine has the advantage of maintaining a patent airway whereas increased salivation during ketamine anesthesia may increase the airway opening pressure and airflow resistance and offset the beneficial effect on airway patency [6]. While the mechanisms of the selective depression of pharyngeal muscle activation are unclear, anesthetic

Table 5.1 Influences of anesthesia and sleep on control of breathing

	Respiratory rate	Tidal volume	Negative pressure reflex	Hypoxic ventilatory response	Hypercapnic ventilatory response	Behavioral control
Topical upper airway anesthesia	No change	No change	Depressed	Preserved	Preserved	Preserved
Inhalational anesthetics (deep)	Increase	Decrease	Depressed	Depressed	Depressed	Lost
Residual inhalational anesthetics	No change	No change	N/A	Depressed	Preserved	Diminished
Propofol anesthesia	Increase	Decrease	Depressed	Depressed	Depressed	Lost
Propofol sedation	Slightly decrease	Decrease	Diminished	Diminished	Diminished	Depressed
Dexmedetomidine	Slightly decrease	Slightly decrease	N/A	Depressed	Depressed	Diminished
Opioids	Decrease	Slightly increase	N/A	Depressed	Depressed	Diminished
Residual muscle relaxant	No change	No change	N/A	Depressed	Preserved	Preserved
NREM sleep	Slightly decrease	Slightly decrease	Diminished	Diminished	Diminished	Depressed
REM sleep	Irregular	Irregular	Depressed	Depressed	Depressed	Depressed

drugs could directly suppress the respiratory center of the upper airway, as the suppression was dose-dependent under constant chemical (PaO_2, $PaCO_2$, and pH) stimuli [6]. Ochiai et al. reported selective and dose-dependent depression of genioglossus electromyogram activity in adult cats [7] and kittens [48]. They indicated a developmental difference of selective genioglossal depression between kittens and adult cats. These animal studies demonstrate preferential suppression of the pharyngeal dilators to the inspiratory pump muscles (diaphragm > intercostal >> genioglossus), indicating the development of pharyngeal obstruction during anesthesia according to the Neural Balance Model (Fig. 5.1).

3.3 Differential Effects of Non-depolarizing Muscle Relaxants on the Inspiratory Pump Muscles and Pharyngeal Muscles

Non-depolarizing muscle relaxants have differential effects on various muscles [49, 50]. The pharyngeal muscles are more sensitive to a small dose of vecuronium than is the diaphragm; and a low concentration of enflurane facilitates the differential effects of vecuronium [51]. Therefore, according to the balance of forces model, patients partially paralyzed with residual muscle relaxant are at increased risk of pharyngeal airway obstruction, and the risk is further augmented under influences of both residual anesthetics and muscle relaxant during recovery from anesthesia [52]. Eikermann et al. demonstrated depression of inspiratory pharyngeal airway dilation in awake healthy volunteers partially paralyzed with rocuronium at the level of train of four ratio 0.8 while the expiratory pharyngeal airway size and ventilation were maintained [53].

In fact, residual muscle paralysis was evident in patients with postoperative critical respiratory episodes such as reintubation, upper airway obstruction requiring airway maneuvers, and severe desaturation under oxygen therapy [54]. Considering the discrepancy between 0.8% incidence of the residual muscle paralysis-induced postoperative critical respiratory events and 30–50% incidence of postoperative residual muscle paralysis defined as train of four less than 0.9 [55, 56], there must be a specific risk population vulnerable for residual muscle paralysis. Adult and pediatric patients with obstructive sleep apnea are considered to compensate pharyngeal anatomical abnormalities by increasing pharyngeal muscle activity during wakefulness [12, 13], and therefore, the residual muscle paralysis may be of greater impact on these patients.

Neostigmine has been used to reverse the effects of non-depolarizing muscle relaxants, it takes more than 20 min to normalize muscle contraction [57], and overdose of neostigmine results in recurarization [58, 59]. Eikermann et al. demonstrated increased postoperative respiratory complications in patients receiving an overdose of neostigmine [59]. Sugammadex rapidly reverses muscle paralysis by encapsulating rocuronium molecule, and it dramatically decreased the incidence of residual paralysis by less than 10% [60]. However, there are recurarization case reports even after the injection of the recommended dose of sugammadex [61].

Muramatsu et al. identified older age, renal dysfunction, and obesity as independent risk factors for recurarization [62].

3.4 *Influences of Opioids and Pain on the Neural Mechanisms in Postoperative Patients*

Opioids are known to have depressant effects on spontaneous breathing. Due to prolongation of expiratory phase, opioids dose-dependently decrease respiratory rate, and eventually cause central apnea while tidal volume is preserved [63]. The occurrence of irregular breathing patterns commonly seen in chronic opioid users such as ataxic breathing, biot respiration, and Cheyne–Stokes respiration has not been systematically investigated in patients receiving opioids for surgical and postoperative pain management. Opioids significantly depress hypoxic ventilatory response as well as hypercapnic response with increasing the apneic threshold in humans [64]. Absence or depression of both hypoxic and hypercapnic responses is a life-threatening condition. Both animal and human studies indicate that the effect of morphine on hypoxic ventilatory response is likely to be mediated centrally rather than peripherally [64, 65]. Furthermore, micro-dialysis of the hypoglossal nucleus with fentanyl significantly depressed the genioglossus activity in rat preparation indicating opioids as a possible cause of obstructive apnea as well as central apnea [66]. In fact, both obstructive and central apneas were observed in postoperative patients receiving opioids while central sleep apneas are more common than obstructive apneas [67].

Opioids are prescribed for surgical and postoperative pain which has respiratory stimulating effects, and opioid dose needs to be determined to balance respiratory depressant and stimulating effects. However, it should be noted that acute surgical pain does not normalize the hypoxic and hypercapnic responses depressed by opioids although pain increases resting ventilation. Lam et al. examined the effects of surgical stimulation on breathing in enflurane anesthetized humans [68]. They found a significant increase in ventilation with a reduction of $PaCO_2$ without changing both hypercapnic and hypoxic ventilatory responses. The reduction of the apneic threshold for carbon dioxide with surgical stimulation was found in sevoflurane anesthetized humans by Nishino and Kochi [69]. Sarton et al. further demonstrated that experimentally induced acute pain increased resting ventilation without affecting the hypoxic ventilatory response significantly depressed by low-dose sevoflurane [70]. Similarly, Borgbjerg et al. found that experimental pain attenuated the morphine-induced respiratory depression with increasing the intercept of the hypercapnic response curve without changing the slope of the curve [71].

Catley et al. described a high prevalence of obstructive apnea with marked desaturation (SaO_2<80%) within 16 h after surgery in patients receiving morphine for analgesia (456 episodes in 10 out of 16 patients) while the use of regional analgesia resulted in no such event (0 episode in 16 patients) [72]. All the respiratory events occurred during sleep. Notably, there are individual differences in optimal opioid dose for analgesia without respiratory depression [73]. It is unclear whether the variation of opioid sensitivity is explained by genetic polymorphism or developed in particular circumstances. Brown et al. documented a direct association between severity of hypoxemia and opioid dose for achieving equivalent analgesia in children with obstructive sleep apnea suggesting higher opioid sensitivity [74]. They reported a successful reduction of postoperative respiratory complications from 30 to 11% by decreasing the amount of opioids for surgical and postoperative analgesia [75]. In adults, administration or requirement of less opioids in patients with more prolonged or severe nocturnal desaturation also suggests higher opioid sensitivity in patients with obstructive sleep apnea [76, 77]. While awake patients with obstructive sleep apnea do not have increased sensitivity to opioids [78], this does not mean normal sensitivity to opioids in these patients during sleep or unconscious state. Combined use of opioids and sedatives doubles the risk of in-hospital resuscitation [79]. Possibilities of opioids sensitivity during sleep need to be investigated in future studies.

3.5 *Effects of Postoperative Residual Drugs on the Neural Mechanisms*

Residual inhalation anesthetics and muscle relaxants after surgery influence chemical control of breathing. Knill and other researchers consistently found that low concentration of inhalational anesthetics less than MAC-awake significantly depresses hypoxic ventilatory response in humans whereas hypercapnic ventilatory response is preserved in such conditions [80]. Depression of carotid chemosensitivity with small doses of halothane was confirmed in animal experiments as the mechanism of the reduced hypoxic ventilatory response [81]. Partial paralysis with vecuronium is also reported to depress hypoxic ventilatory response in human volunteers [82] and to inhibit neurotransmission of the rat carotid body [83]. Accordingly, carotid body function may be significantly impaired in patients during recovery from inhalational anesthesia combined with muscle relaxant. Sevoflurane and desflurane, two major inhalation anesthetics currently available, are also evidenced to have depressant effects on carotid body but are more easily eliminated from the body decreasing its risk. Although sugammadex can rapidly and completely reverse non-depolarizing muscle relaxants, the hypoxic ventilatory response remains depressed even after complete reversal of the muscle paralysis [84]. No doubt, the absence of increased ventilation during hypoxemia is of clinical significance. Hypoxemia due to central hypoventilation primarily caused by depression of hypercapnic ventilatory response may be reversed by the preservation of the "back-up" hypoxic ventilatory response. However, it should be noted that increased ventilatory drive itself does not usually reverse the hypoxemia resulting from upper airway obstruction and impaired lung mechanics. Knill compared the carotid body to a watchdog, sensing hypoxemia and alarming the condition [85]. Increased ventilation, dyspneic sensation, and arousal are considered to be significant clinical signs and survival responses produced by carotid body activation. Clinicians should recognize possible depression of these alarming functions of the carotid body in patients during recovery from anesthesia.

3.6 Influences of Reduced Arousal Threshold During Anesthesia and Sedation

Burst of UA muscle activity upon cortical arousal has been believed to be an important mechanism for recovery from the pharyngeal obstruction, at least, in sleeping adult patients with obstructive sleep apnea while small children with obstructive sleep apnea resolve obstructive breathing without cortical arousal [22]. Furthermore, ventilatory overshoot at the cortical arousal reduces arterial carbon dioxide levels leading to unstable periodic breathing patterns. Younes recently proposed that cortical arousals are coincidental events that occur just prior to arousal-independent airway opening, through analysis of the breathing and cortical activity in experimentally induced apneas or hypopneas [86–88]. In support of Younes's concept, the use of hypnotics has been reported to increase the arousal threshold and chemical stimuli to the genioglossus muscle [89], and stabilize breathing resulting in decrease of the apnea–hypopnea index (AHI) in OSA patients [90]. Furthermore, pentobarbital sedation has been reported to augment upper airway negative pressure reflex as well as to increase the arousal threshold [91].

According to recent accumulated knowledge of arousal and obstructive sleep apnea [88], interactions between the cortical arousal threshold and pharyngeal airway opening and closing thresholds can explain mechanisms of periodic obstructive breathing (Fig. 5.4). When the arousal threshold is below the opening threshold (upper panel), the cortical arousal occurs before airway opening with ventilatory overshoot and burst of dilating muscles leading to unstable large ventilatory oscillation.

In contrast, anesthetics and sedatives increase the arousal threshold over the opening threshold and decrease respiratory drives to the diaphragm and pharyngeal muscles (middle panel). It is possible that the increase of pharyngeal airway negative pressure reflex, caused by an increase of chemical stimuli during apnea or hypoxemia, may elicit the dilating muscle activity to first reach the opening threshold, which consequently restores ventilation. Ventilation is maintained until the dilating muscle activity decreases below the closing threshold which is generally below the opening threshold [92]. Due to reduced oscillation of the dilating muscle activity, stable breathing can be established despite a reduced dilating muscle activity. In support of this concept, Urahama et al. reported that no cortical arousal was observed during recovery from 77% of obstructive episodes in elderly patients under propofol sedation [93]. It should be noted that the periodicity of obstructive breathing is not altered in this circumstance while its frequency is reduced.

Deeper sedation (lower panel) would significantly decrease the dilating muscle activity below the opening threshold resulting in persistent pharyngeal closure and the possible development of critical hypoxemia. Without a doubt, unnecessarily deeper sedation totally abolishing cortical arousal response to critical respiratory events should be avoided.

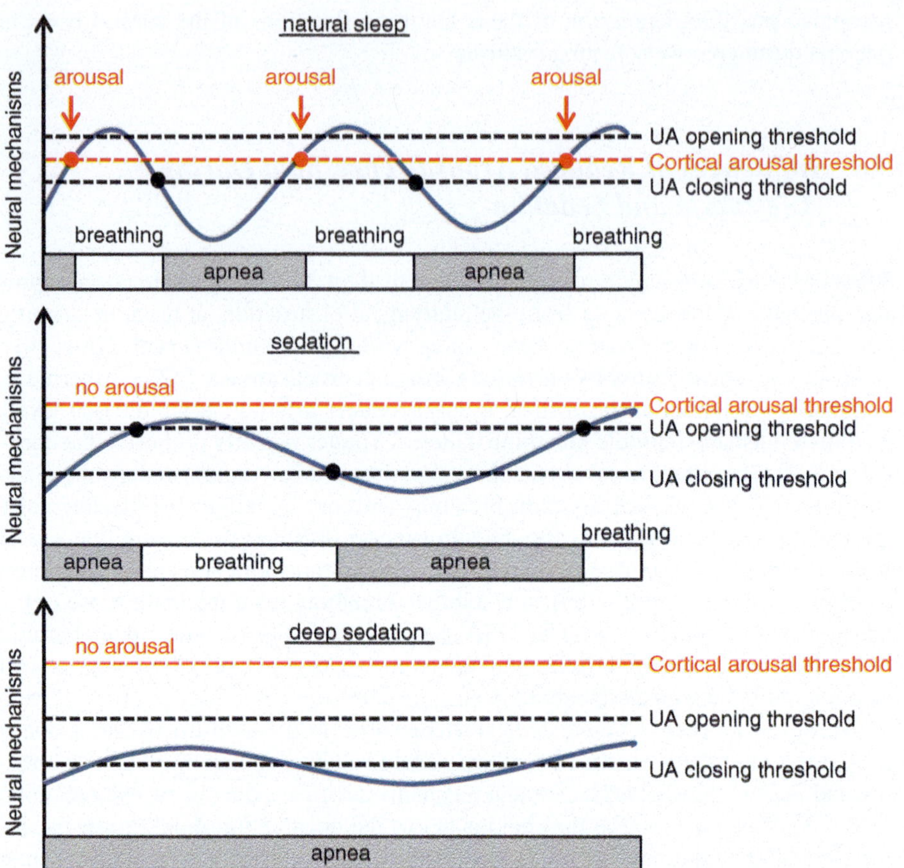

Fig. 5.4 Interactions between the cortical arousal threshold and upper airway (UA) opening and closing thresholds for explaining the mechanisms of periodic obstructive breathing during natural sleep (upper panel), sedation (middle panel), and deep sedation (lower panel). Sinusoidal curves represent changes of neural mechanisms regulating UA dilating muscle activity and respiratory efforts

3.7 *Effects of Surgical and Anesthetic Interventions on Postoperative Pharyngeal Obstruction*

Extensive surgical procedures on the pharyngeal airway were reported to result in the development of severe pharyngeal swelling followed by choking immediately after endotracheal extubation in obstructive sleep apnea patients [94, 95]. Similarly, a high incidence of severe respiratory compromises (6.1%: 19/311) including one death and six reintubations after anterior cervical spine surgery was attributable to pharyngeal edema [96]. Even modest pharyngeal swelling caused by laryngoscopy and excessive fluid infusion during surgery may possibly have significant influences on pharyngeal airway maintenance immediately after surgery, although no study has examined this possibility. Damage of the mucosal receptors responsible for the pharyngeal negative airway pressure reflex may impair the neural mechanisms, and pharyngeal swelling itself may shift the fulcrum of the balance model to the left, increasing pharyngeal collapsibility. Furthermore, placement of a gastric tube through the nasal passage increases nasal resistance, possibly creating more negative suction forces at the collapsible pharynx. In addition, retaining of secretion due to inability of swallowing during paralysis and increase in secretion due to administration of neostigmine for reversal of muscle relaxant may increase surface adhesive force, and thereby opening pressure of the pharyngeal airway, making the closed airway difficult to be re-opened [9, 97, 98]. All of these speculations need to be examined in future studies because of their potential clinical significance.

3.8 Advantageous Body Position and Head and Neck Position for Postoperative Airway Management

Body position is of great significance for pharyngeal airway maintenance in postoperative patients (Fig. 5.3). Due to the presence of a variety of drainage tubes or protection of the operation wound, postoperative patients are often lying on their back. If possible, the lateral and sitting positions are advantageous for pharyngeal airway maintenance because of less gravitational effects on the pharyngeal airway in these body positions [99, 100]. In particular, sitting posture increases functional residual volume which is advantageous for both pharyngeal airway maintenance and oxygenation [43]. In patients with full-stomach, sitting posture may decrease the chance of regurgitation of stomach contents.

Head elevation with a pillow produces neck flexion at the lower cervical spines and extension at the upper cervical spines as long as the face is maintained straight up. This so-called "sniffing position," which is a standard head position during induction of anesthesia, improves airway patency and may be advantageous during the postoperative period [101]. However, it is often difficult to maintain the face straight up with head elevation, and the neck is often flexed at the higher cervical spines, resulting in narrowing of the pharyngeal airway particularly when a high pillow is used [102]. Accordingly, it is recommended not to use a high pillow in the PACU when residual anesthetics possibly impair the neural compensatory mechanisms.

4 Laryngeal Airway Functions

4.1 Laryngeal Airway: Maintenance Mechanisms

Anesthesia and surgical procedures such as direct laryngoscopy for tracheal intubation, tracheal tube placement during surgery, and head and neck surgeries often impair two major laryngeal functions for stable breathing, e.g., laryngeal airway maintenance and lower airway protection, leading to various forms of postoperative

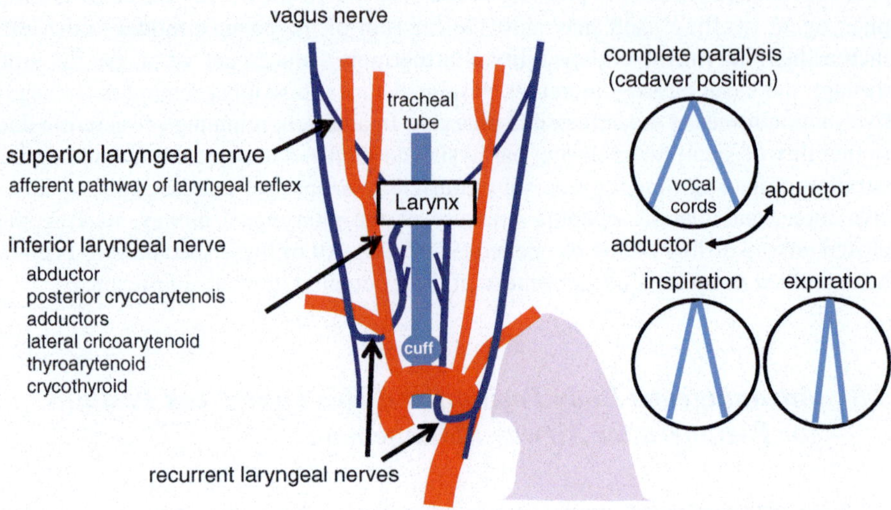

Fig. 5.5 Illustration of laryngeal sensory and motor nerves, and positions of vocal cords

complications such as laryngospasm, stridor, vocal cord paralysis, and postoperative sore throat and hoarseness. Chemical as well as reflexive control of vocal cord abductor and adductor muscles accomplishes these laryngeal functions. The adductors are numerous (lateral cricoarytenoid, cricothyroid, thyroarytenoid muscles), but the posterior cricoarytenoid muscle is the only laryngeal abductor while both are innervated by recurrent laryngeal nerves connecting to the nucleus ambiguous (Fig. 5.5.). The size of the laryngeal aperture is determined by a balance of forces between the adductors and abductor. Inspiratory activation of the abductor widens the vocal cords and expiratory activation of the adductors narrows the vocal cords braking the expiratory airflow preventing collapses of the alveoli [103]. Negative airway pressure augments the abductor and depresses diaphragm activity like the negative pressure reflex at the pharyngeal airway.

4.2 Laryngeal Reflex and Laryngospasm During Anesthesia and Sedation

Laryngeal reflexes such as the cough reflex, expiration reflexes, and laryngeal closure with apnea primarily operate to protect the airways from aspiration [104]. These reflexes are significantly modified by general anesthetics. The complete laryngeal closure (laryngospasm), in particular, often occurs during light anesthesia and emergence from anesthesia following series of cough and/or expiration reflexes leading to severe hypoxemia. Although several different types of laryngeal receptors have been reported, the laryngeal irritant receptors are considered to play the most essential role in the elicitation of defensive airway reflexes through the superior laryngeal nerve as an afferent pathway to the nucleus solitarius. Prevalence of laryngospasm during anesthesia is reported to be about 1% of 136,929 patients including infants to elderlies [105]. Half of the anesthesia-related laryngospasm cases requiring treatment occurs after tracheal extubation in children [106]. Light anesthesia level [107, 108], younger children [108], use of isoflurane [107], airway surgery [106, 108], and upper respiratory tract infection [107, 109] are considered to be risk factors. It should be noted that the latter three risk factors are related to the presence of irritants or irritative drugs on the airway mucosa.

Complete laryngeal closure can last for several minutes, and many case reports have documented the development of severe hypoxemia and "negative pressure pulmonary edema" after post-extubation laryngospasm [110]. Laryngospasm can be effectively prevented by intravenous administration of lidocaine while this is still controversial [111, 112]. Although the application of positive pressure is believed to be effective to resolve the laryngospasm, there is no systematic study evaluating the effectiveness. Administration of muscle relaxants such as succinylcholine is the most effective treatment of laryngospasm while intravenous administration of anesthetics such as propofol is effective in some cases [113].

4.3 Laryngeal Edema and Stridor After Tracheal Tube Placement

Interaction between a tracheal tube and the larynx during anesthesia and surgery significantly influences breathing after tracheal extubation. Tanaka et al. demonstrated that placement of a tracheal tube even for 3–4 h significantly increased laryngeal resistance in association with the development of asymptomatic but substantial laryngeal edema [114, 115]. An increase in laryngeal resistance and edema formation was less severe in patients with supraglottic airway placement than those with tracheal tube [115]. The normal glottic width of 6–8 mm in adults is much smaller than the standard tracheal tube size commonly used during anesthesia, suggesting that direct contact forces by the tracheal tube would cause trauma and edema formation at the vocal cords [116]. The contact pressure is reported to exceed over 200 mm Hg, far greater than the mucosal capillary perfusion pressure of 30 mm Hg [117]. Tracheal intubation with an oversized tube for 8 h in a dog preparation resulted in mucosal ischemia and necrosis with tissue edema, particularly at the arytenoids region [116]. The subglottic region is considered to be narrowest in neonates and small children while recent MRI analysis indicates the narrowest region to be at the vocal cords even in small children during wakefulness [118]. Even a mild level of edema formation at the subglottic region in small children results in a marked reduction of cross-sectional area and therefore increases laryngeal resistance. In fact, the use of a tight-fitting tracheal tube, younger children, trauma at tracheal intubation, duration of intubation, and coughing during intubation are the risk factors for the development of laryngeal edema after tracheal extubation [119]. The cuff-leak test, in which air leakage around the tube with the cuff deflated and with the tube occluded is detected, is recommended as a simple and useful method for diagnosis of severe laryngeal edema and for prediction of successful extubation [120, 121]. Intravenous administration of dexamethasone is concluded to decrease the need for reintubation in neonates after mechanical ventilation while the effectiveness of adrenaline nebulizer on the laryngeal edema is controversial [122, 123].

Regardless of the commonality of laryngeal edema after surgery, postoperative laryngeal stridor infrequently occurs immediately after emergence from anesthesia whereas post-extubation stridor occurs in 4–18% of patients after prolonged intubation in the intensive care unit [124, 125]. Stridor develops due to severe but incomplete laryngeal narrowing during spontaneous breathing even in awake subjects within 6 h after tracheal extubation [126]. Stridor should be treated, though the development of hypoxemia is easily prevented by supplementary oxygen administration. Increased inspiratory drive due to hypercapnia increases work of breathing and some cases require reintubation, tracheostomy, and prolonged intensive care [127]. Stridor results from the interaction between the gas traveling through the airway and narrowed laryngeal airway wall during inspiration, producing vibration of the vocal cords with a high-pitched characteristic sound different from the snoring sound [128]. The velocity of the air at the passively narrowed vocal cords increases during inspiration converting potential energy to kinetic energy. Due to the increased kinetic energy, lateral wall pressure at the vocal cords decreases, and therefore, narrows the vocal cords further during inspiration. Fiberoptic examination reveals paradoxical vocal cord movement [126]. While this passive mechanism is generally believed to be responsible for the postoperative stridor, no physiologic measurement has been systematically performed because of difficulty in predicting the occurrence of the postoperative stridor. Isono et al. found inspiratory activation of laryngeal adductor muscle in association with the development of stridor in patients with multiple system atrophy, and suggest an imbalance between forces of laryngeal adductors and abductor forces as a possible active mechanism for the generation of stridor [129]. Application of CPAP eliminated the stridor and inspiratory adductor muscle activation, strongly suggesting that the inspiratory adductor muscle activity is possibly induced by a negative pressure reflex and CPAP functions to depress the reflex in addition to an airway sprint effect. In this context, the effectiveness of benzodiazepines for the elimination of postoperative laryngeal stridor may be explained by reduction of negative inspiratory pressure [126]. No study has tested the active laryngeal narrowing hypothesis in the development of postoperative and/or post-extubation stridor and whether CPAP application is effective for the treatment of the stridor.

5 Conclusion

Interaction between upper airway structures and muscles achieves various physiological functions of the upper airway. Collapsibility of the pharyngeal airway is neutrally regulated by pharyngeal muscle contraction which significantly decreases during anesthesia and sedation. Obstructive sleep apnea patients with abnormally collapsible pharynx are at risk for unstable breathing leading to critical respiratory complications during anesthesia and sedation. In contrast, the laryngeal airway is well maintained even during anesthesia and sedation unless the laryngeal reflexes occur. Understandings of these fundamental differences are important for exploring the mechanisms of the pharyngeal and laryngeal obstruction and for determining their appropriate treatments to prevent life-threatening severe hypoxemia.

References

1. Remmers JE, DeGroot WJ, Sauerland EK, Anch AM. Pathogenesis of upper airway occlusion during sleep. J Appl Physiol. 1978;44:931–8.
2. Tangel DJ, Mezzanotte WS, White DP. Influence of sleep on tensor palatini EMG and upper airway resistance in normal men. J Appl Physiol. 1991;70:2574–81.
3. Parisi RA, Neubauer JA, Frank MM, et al. Correlation between genioglossal and diaphragmatic responses to hypercapnia during sleep. Am Rev Respir Dis. 1987;135:378–2.
4. Van Lunteren E, Van de Graaff WB, Parker DM, Mitra J, Haxhiu MA, Strohl KP, Cherniack NS. Nasal and laryngeal reflex responses to negative upper airway pressure. J Appl Physiol. 1984;56:746–52.
5. Horner RL, Innes JA, Holden HB, Guz A. Afferent pathway(s) for pharyngeal dilator reflex to negative pressure in man: a study using upper airway anesthesia. J Physiol Lond. 1991;436:31–44.
6. Nishino T, Shirahata M, Yonezawa T, Honda Y. Comparison of changes in the hypoglossal and the phrenic nerve activity in response to increasing depth of anesthesia in cats. Anesthesiology. 1984;60:19–24.
7. Ochiai R, Guthrie RD, Motoyama EK. Effects of varying concentrations of halothane on the activity of the genioglossus, intercostals, and diaphragm in cats: an electromyographic study. Anesthesiology. 1989;70:812–6.
8. Isono S. Developmental changes of pharyngeal airway patency: implications for pediatric anesthesia. Paediatr Anaesth. 2006;16:109–22.
9. Isono S. Upper airway muscle function during sleep. In: Loughlin GM, Marcus CL, Carroll JL, editors. Sleep and breathing in children: a developmental approach. New York: Marcel Dekker Inc.; 2000. p. 261–91.
10. Isono S. Obstructive sleep apnea of obese adults: pathophysiology and perioperative airway management. Anesthesiology. 2009;110:908–21.
11. Fenik VB, Davies RO, Kubin L. REM sleep-like atonia of hypoglossal (XII) motoneurons is caused by loss of noradrenergic and serotonergic inputs. Am J Respir Crit Care Med. 2005;172:1322–30.

12. Mezzanotte WS, Tanjel DJ, White DP. Waking genioglossal electromyogram in sleep apnea patients versus normal controls: a neuromuscular compensatory mechanism. J Clin Invest. 1992;89:1571–9.
13. Gozal D, Burnside MM. Increased upper airway collapsibility in children with obstructive sleep apnea during wakefulness. Am J Respir Crit Care Med. 2004;169:163–7.
14. Isono S, Remmers JE, Tanaka A, Sho Y, Sato J, Nishino T. Anatomy of pharynx in patients with obstructive sleep apnea and normal subjects. J Appl Physiol. 1997;82:1319–26.
15. Malhotra A, Pillar G, Fogel RB, Beauregard J, Edwards JK, Slamowitz DI, Shea SA, White DP. Genioglossal but not palatal muscle activity relates closely to pharyngeal pressure. Am J Respir Crit Care Med. 2000;162:1058–62.
16. Shea SA, Akahoshi T, Edwards JK, White DP. Influence of chemoreceptor stimuli on genioglossal response to negative pressure in humans. Am J Respir Crit Care Med. 2000;162:559–65.
17. Akahoshi T, White DP, Edwards JK, Beauregard J, Shea SA. Phasic mechanoreceptor stimuli can induce phasic activation of upper airway muscles in humans. J Physiol. 2001;531:677–91.
18. Fogel RB, Trinder J, Malhotra A, Stanchina M, Edwards JK, Schory KE, White DP. Within-breath control of genioglossal muscle activation in humans: effect of sleep-wake state. J Physiol. 2003;550:899–910.
19. Wilson SL, Thach BT, Brouillette RT, Abu-Osba YK. Upper airway patency in the human infant: influence of airway pressure and posture. J Appl Physiol. 1980;48:500–4.
20. Phillipson EA, Sullivan CE, Read DJ, Murphy E, Kozar LF. Ventilatory and waking responses to hypoxia in sleeping dogs. J Appl Physiol. 1978;44:512–20.
21. Phillipson EA, Kozar LF, Rebuck AS, Murphy E. Ventilatory and waking responses to CO2 in sleeping dogs. Am Rev Respir Dis. 1977;115:251–9.
22. McNamara F, Issa FG, Sullivan CE. Arousal pattern following central and obstructive breathing abnormalities in infants and children. J Appl Physiol (1985). 1996;81:2651–7.
23. Isono S, Tanaka A, Nishino T. Effects of tongue electrical stimulation on pharyngeal mechanics in anaesthetized patients with obstructive sleep apnoea. Eur Respir J. 1999;14:1258–65.
24. Fitzpatrick MF, Driver HS, Chatha N, Voduc N, Girard AM. Partitioning of inhaled ventilation between the nasal and oral routes during sleep in normal subjects. J Appl Physiol (1985). 2003;94:883–90.
25. Nishino T, Sugiyama A, Tanaka A, Ishikawa T. Effects of topical nasal anaesthesia on shift of breathing route in adults. Lancet. 1992;339(8808):1497–500.
26. Nishino T, Kochi T. Effects of sedation produced by thiopentone on responses to nasal occlusion in female adults. Br J Anaesth. 1993;71:388–92.
27. Kuna ST, Smickley J. Response of genioglossus muscle activity to nasal airway occlusion in normal sleeping adults. J Appl Physiol (1985). 1988;64:347–53.
28. van Lunteren E, Haxhiu MA, Cherniack NS. Mechanical function of hyoid muscles during spontaneous breathing in cats. J Appl Physiol (1985). 1987;62:582–90.
29. Kohno A, Kitamura Y, Kato S, Imai H, Masuda Y, Sato Y, Isono S. Displacement of the hyoid bone by muscle paralysis and lung volume increase: the effects of obesity and obstructive sleep apnea. Sleep. 2019;42(1):zsy198. https://doi.org/10.1093/sleep/zsy198.
30. Duara S, Rojas M, Claure N. Upper airway stability and respiratory muscle activity during inspiratory loading in full-term neonates. J Appl Physiol. 1994;77:37–42.
31. Roberts JL, Reed WR, Mathew OP, Thach BT. Control of respiratory activity of the genioglossus muscle in micrognathic infants. J Appl Physiol. 1986;61:1523–33.
32. Strohl KP, Hensley MJ, Hallett M, Saunders NA, Ingram RH Jr. Activation of upper airway muscles before onset of inspiration in normal humans. J Appl Physiol. 1980;49:638–42.
33. Kuna ST, Smickley JS. Superior pharyngeal constrictor activation in obstructive sleep apnea. Am J Respir Crit Care Med. 1997;156:874–80.
34. Strollo PJ Jr, Soose RJ, Maurer JT, de Vries N, Cornelius J, Froymovich O, Hanson RD, Padhya TA, Steward DL, Gillespie MB, Woodson BT, Van de Heyning PH, Goetting MG,

Vanderveken OM, Feldman N, Knaack L, Strohl KP, STAR Trial Group. Upper-airway stimulation for obstructive sleep apnea. N Engl J Med. 2014;370:139–49.
35. Miki H, Hida W, Chonan T, Kikuchi Y, Takishima T. Effects of submental electrical stimulation during sleep on upper airway patency in patients with obstructive sleep apnea. Am Rev Respir Dis. 1989;140:1285–9.
36. Langeron O, Masso E, Huraux C, Guggiari M, Bianchi A, Coriat P, Riou B. Prediction of difficult mask ventilation. Anesthesiology. 2000;92:1229–36.
37. Kheterpal S, Martin L, Shanks AM, Tremper KK. Prediction and outcomes of impossible mask ventilation: a review of 50,000 anesthetics. Anesthesiology. 2009;110:891–7.
38. Sato S, Hasegawa M, Okuyama M, Okazaki J, Kitamura Y, Sato Y, Ishikawa T, Sato Y, Isono S. Mask ventilation during induction of general anesthesia: influences of obstructive sleep apnea. Anesthesiology. 2017;126:28–38.
39. Sato Y, Ikeda A, Ishikawa T, Isono S. How can we improve mask ventilation in patients with obstructive sleep apnea during anesthesia induction? J Anesth. 2013;27:152–6.
40. Japanese Society of Anesthesiologists. JSA airway management guideline 2014: to improve the safety of induction of anesthesia. J Anesth. 2014;28:482–93.
41. Campbell IT, Beatty PC. Monitoring preoxygenation. Br J Anaesth. 1994;72:3–4.
42. Cressey DM, Berthoud MC, Reilly CS. Effectiveness of continuous positive airway pressure to enhance pre-oxygenation in morbidly obese women. Anaesthesia. 2001;56:680–4.
43. Tagaito Y, Isono S, Tanaka A, Ishikawa T, Nishino T. Sitting posture decreases collapsibility of the passive pharynx in anesthetized paralyzed patients with obstructive sleep apnea. Anesthesiology. 2010;113:812–8.
44. Isono S, Tanaka A, Ishikawa T, Tagaito Y, Nishino T. Sniffing position improves pharyngeal airway patency in anesthetized patients with obstructive sleep apnea. Anesthesiology. 2005;103:489–94.
45. Ikeda A, Isono S, Sato Y, Yogo H, Sato J, Ishikawa T, Nishino T. Effects of muscle relaxants on mask ventilation in anesthetized persons with normal upper airway anatomy. Anesthesiology. 2012;117:487–93.
46. Isono S. One hand, two hands, or no hands for maximizing airway maneuvers? Anesthesiology. 2008;109:576–7.
47. Izumi Y, Kochi T, Isono S, Ide T, Mizuguchi T. Breathing pattern and respiratory mechanics in sevoflurane-anesthetized humans. J Anesth. 1990;4:343–9.
48. Ochiai R, Guthrie RD, Motoyama EK. Differential sensitivity to halothane anesthesia of the genioglossus, intercostals, and diaphragm in kittens. Anesth Analg. 1992;74:338–44.
49. Wymore ML, Eisele JH. Differential effects of d-tubocurarine on inspiratory muscles and two peripheral muscle groups in anesthetized man. Anesthesiology. 1978;48:360–2.
50. Donati F, Antzaka C, Bevan DR. Potency of pancuronium at the diaphragm and the adductor pollicis muscle in humans. Anesthesiology. 1986;65:1–5.
51. Isono S, Kochi T, Ide T, Sugimori K, Mizuguchi T, Nishino T. Differential effects of vecuronium on diaphragm and geniohyoid muscle in anaesthetized dogs. Br J Anaesth. 1992;68:239–43.
52. D'Honneur G, Lofaso F, Drummond GB, Rimaniol JM, Aubineau JV, Harf A, Duvaldestin P. Susceptibility to upper airway obstruction during partial neuromuscular block. Anesthesiology. 1998;88:371–8.
53. Eikermann M, Vogt FM, Herbstreit F, Vahid-Dastgerdi M, Zenge MO, Ochterbeck C, de Greiff A, Peters J. The predisposition to inspiratory upper airway collapse during partial neuromuscular blockade. Am J Respir Crit Care Med. 2007;175:9–15.
54. Murphy GS, Szokol JW, Marymont JH, Greenberg SB, Avram MJ, Vender JS. Residual neuromuscular blockade and critical respiratory events in the postanesthesia care unit. Anesth Analg. 2008;107:130–7.
55. Debaene B, Plaud B, Dilly MP, Donati F. Residual paralysis in the PACU after a single intubating dose of nondepolarizing muscle relaxant with an intermediate duration of action. Anesthesiology. 2003;98:1042–8.

56. Murphy GS, Szokol JW, Franklin M, Marymont JH, Avram MJ, Vender JS. Postanesthesia care unit recovery times and neuromuscular blocking drugs: a prospective study of orthopedic surgical patients randomized to receive pancuronium or rocuronium. Anesth Analg. 2004;98:193–200.
57. Khuenl-Brady KS, Wattwil M, Vanacker BF, Lora-Tamayo JI, Rietbergen H, Alvarez-Gómez JA. Sugammadex provides faster reversal of vecuronium-induced neuromuscular blockade compared with neostigmine: a multicenter, randomized, controlled trial. Anesth Analg. 2010;110:64–73.
58. Payne JP, Hughes R, Al AS. Neuromuscular blockade by neostigmine in anaesthetized man. Br J Anaesth. 1980;52:69–76.
59. Eikermann M, Fassbender P, Malhotra A, Takahashi M, Kubo S, Jordan AS, Gautam S, White DP, Chamberlin NL. Unwarranted administration of acetylcholinesterase inhibitors can impair genioglossus and diaphragm muscle function. Anesthesiology. 2007;107:621–9.
60. Kotake Y, Ochiai R, Suzuki T, Ogawa S, Takagi S, Ozaki M, Nakatsuka I, Takeda J. Reversal with sugammadex in the absence of monitoring did not preclude residual neuromuscular block. Anesth Analg. 2013;117:345–51.
61. Iga S, Shimizu K, Kawade K, Kanazawa T, Nishitani K, Morimatsu H. Recurarization following adequate dose of sugammadex reversal in a patient monitored with aceleromyography. J Jpn Soc Clin Anesth. 2016;36:1–5.
62. Muramatsu T, Isono S, Ishikawa T, Nozaki-Taguchi N, Okazaki J, Kitamura Y, Murakami N, Sato Y. Differences of recovery from rocuronium-induced deep paralysis in response to small doses of Sugammadex between elderly and nonelderly patients. Anesthesiology. 2018;129:901–11.
63. Ferguson LM, Drummond GB. Acute effects of fentanyl on breathing pattern in anaesthetized subjects. Br J Anaesth. 2006;96:384–90.
64. Bailey PL, Lu JK, Pace NL, Orr JA, White JL, Hamber EA, Slawson MH, Crouch DJ, Rollins DE. Effects of intrathecal morphine on the ventilatory response to hypoxia. N Engl J Med. 2000;343:1228–34.
65. McQueen DS, Ribeiro JA. Inhibitory actions of methionine-enkephalin and morphine on the cat carotid chemoreceptors. Br J Pharmacol. 1980;71:297–305.
66. Hajiha M, DuBord MA, Liu H, Horner RL. Opioid receptor mechanisms at the hypoglossal motor pool and effects on tongue muscle activity in vivo. J Physiol. 2009;587:2677–92.
67. Van Ryswyk E, Antic NA. Opioids and sleep-disordered breathing. Chest. 2016;150:934–44.
68. Lam AM, Clement JL, Knill RL. Surgical stimulation does not enhance ventilatory chemoreflexes during enflurane anaesthesia in man. Can Anaesth Soc J. 1980;27:22–8.
69. Nishino T, Kochi T. Effects of surgical stimulation on the apnoeic thresholds for carbon dioxide during anaesthesia with sevoflurane. Br J Anaesth. 1994;73:583–6.
70. Sarton E, Dahan A, Teppema L, van den Elsen M, Olofsen E, Berkenbosch A, van Kleef J. Acute pain and central nervous system arousal do not restore impaired hypoxic ventilatory response during sevoflurane sedation. Anesthesiology. 1996;85:295–303.
71. Borgbjerg FM, Nielsen K, Franks J. Experimental pain stimulates respiration and attenuates morphine-induced respiratory depression: a controlled study in human volunteers. Pain. 1996;64:123–8.
72. Catley DM, Thornton C, Jordan C, Lehane JR, Royston D, Jones JG. Pronounced, episodic oxygen desaturation in the postoperative period: its association with ventilatory pattern and analgesic regimen. Anesthesiology. 1985;63:20–8.
73. Ray DC, Drummond GB. Continuous intravenous morphine for pain relief after abdominal surgery. Ann R Coll Surg Engl. 1988;70:317–21.
74. Brown KA, Laferrière A, Lakheeram I, Moss IR. Recurrent hypoxemia in children is associated with increased analgesic sensitivity to opiates. Anesthesiology. 2006;105:665–9.
75. Raghavendran S, Bagry H, Detheux G, Zhang X, Brouillette RT, Brown KA. An anesthetic management protocol to decrease respiratory complications after adenotonsillectomy in children with severe sleep apnea. Anesth Analg. 2010;110:1093–101.

76. Doufas AG, Tian L, Padrez KA, Suwanprathes P, Cardell JA, Maecker HT, Panousis P. Experimental pain and opioid analgesia in volunteers at high risk for obstructive sleep apnea. PLoS One. 2013;8(1):e54807.
77. Turan A, You J, Egan C, Fu A, Khanna A, Eshraghi Y, Ghosh R, Bose S, Qavi S, Arora L, Sessler DI, Doufas AG. Chronic intermittent hypoxia is independently associated with reduced postoperative opioid consumption in bariatric patients suffering from sleep-disordered breathing. PLoS One. 2015;10(5):e0127809.
78. Doufas AG, Shafer SL, Rashid NHA, Kushida CA, Capasso R. Non-steady state modeling of the ventilatory depressant effect of remifentanil in awake patients experiencing moderate-to-severe obstructive sleep apnea. Anesthesiology. 2019;130:213–26.
79. Overdyk FJ, Dowling O, Marino J, Qiu J, Chien HL, Erslon M, Morrison N, Harrison B, Dahan A, Gan TJ. Association of opioids and sedatives with increased risk of in-hospital cardiopulmonary arrest from an administrative database. PLoS One. 2016;11(2):e0150214.
80. Knill RL, Gleb AW. Ventilatory responses to hypoxia and hypercapnia during halothane sedation and anesthesia in man. Anesthesiology. 1978;49:244–51.
81. Ide T, Sakurai Y, Aono M, Nishino T. Contribution of peripheral chemoreception to the depression of the hypoxic ventilatory response during halothane anesthesia in cats. Anesthesiology. 1999;90:1084–91.
82. Eriksson LI, Sato M, Severinghaus JW. Effect of a vecuronium-induced partial neuromuscular block on hypoxic ventilatory response. Anesthesiology. 1993;78:693–9.
83. Igarashi A, Amagasa S, Horikawa H, Shirahata M. Vecuronium directly inhibits hypoxic neurotransmission of the rat carotid body. Anesth Analg. 2002;94:117–22.
84. Broens SJL, Boon M, Martini CH, Niesters M, van Velzen M, Aarts LPHJ, Dahan A. Reversal of partial neuromuscular block and the ventilatory response to hypoxia: a randomized controlled trial in healthy volunteers. Anesthesiology. 2019;131:467–76.
85. Knill RL, Gelb AW. Peripheral chemoreceptors during anesthesia: are the watchdogs sleeping? Anesthesiology. 1982;57:151–2.
86. Younes M. Role of arousals in the pathogenesis of obstructive sleep apnea. Am J Respir Crit Care Med. 2004;169:623–33.
87. Younes M, Loewen AH, Ostrowski M, Laprairie J, Maturino F, Hanly PJ. Genioglossus activity available via non-arousal mechanisms vs. that required for opening the airway in obstructive apnea patients. J Appl Physiol. 2012;112:249–58.
88. Eckert DJ, Younes MK. Arousal from sleep: implications for obstructive sleep apnea pathogenesis and treatment. J Appl Physiol (1985). 2014;116:302–13.
89. Younes M, Park E, Horner RL. Pentobarbital sedation increases genioglossus respiratory activity in sleeping rats. Sleep. 2007;30:478–88.
90. Eckert DJ, Owens RL, Kehlmann GB, Wellman A, Rahangdale S, Yim-Yeh S, White DP, Malhotra A. Eszopiclone increases the respiratory arousal threshold and lowers the apnoea/hypopnea index in obstructive sleep apnoea patients with a low arousal threshold. Clin Sci (Lond). 2011;120:505–14.
91. Eikermann M, Fassbender P, Zaremba S, Jordan AS, Rosow C, Malhotra A, Chamberlin NL. Pentobarbital dose-dependently increases respiratory genioglossus muscle activity while impairing diaphragmatic function in anesthetized rats. Anesthesiology. 2009;110:1327–34.
92. Wilson SL, Thach BT, Brouillette RT, Abu-Osba YK. Upper airway patency in the human infant: influence of airway pressure and posture. J Appl Physiol Respir Environ Exerc Physiol. 1980;48:500–4.
93. Urahama R, Uesato M, Aikawa M, Kunii R, Isono S, Matsubara H. Occurrence of cortical arousal at recovery from respiratory disturbances during deep propofol sedation. Int J Environ Res Public Health. 2019;16(18):3482. https://doi.org/10.3390/ijerph16183482.
94. Riley RW, Powell NB, Guilleminault C, Pelayo R, Troell RJ, Li KK. Obstructive sleep apnea surgery: risk management and complications. Otolaryngol Head Neck Surg. 1997;117:648–52.

95. Fairbanks DNF. Uvulopalatopharyngoplasty complications and avoidance strategies. Otolaryngol Head Neck Surg. 1990;102:239–45.
96. Sagi HC, Beutler W, Carroll E, Connolly PJ. Airway complications associated with surgery on the anterior cervical spine. Spine. 2002;27:949–53.
97. Van der Touw T, Crawford AB, Wheatley JR. Effects of a synthetic lung surfactant on pharyngeal patency in awake human subjects. J Appl Physiol. 1997;82:78–85.
98. Olson LG, Strohl KP. Airway secretions influence upper airway patency in the rabbit. Am Rev Respir Dis. 1988;137:1379–81.
99. Isono S, Tanaka A, Nishino T. Lateral position decreases collapsibility of the passive pharynx in patients with obstructive sleep apnea. Anesthesiology. 2002;97:780–5.
100. Boudewyns A, Punjabi N, Van de Heyning PH, De Backer WA, O'Donnell CP, Schneider H, Smith PL, Schwartz AR. Abbreviated method for assessing upper airway function in obstructive sleep apnea. Chest. 2000;118:1031–41.
101. Shorten GD, Armstrong DC, Roy WI, Brown L. Assessment of the effect of head and neck position on upper airway anatomy in sedated paediatric patients using magnetic resonance imaging. Paediatr Anaesth. 1995;5:243–8.
102. Sivarajan M, Joy JV. Effects of general anesthesia and paralysis on upper airway changes due to head position in humans. Anesthesiology. 1996;85:787–93.
103. Remmers JE, Bartlett D Jr. Reflex control of expiratory airflow and duration. J Appl Physiol Respir Environ Exerc Physiol. 1977;42:80–7.
104. Nishino T, Isono S, Tanaka A, Ishikawa T. Laryngeal inputs in defensive airway reflexes in humans. Pulm Pharmacol Ther. 2004;17:377–81.
105. Olsson GL, Hallen B. Laryngospasm during anaesthesia. A computer-aided incidence study in 136,929 patients. Acta Anaesthesiol Scand. 1984;28:567–75.
106. Tait AR, Malviya S, Voepel-Lewis T, Munro HM, Seiwert M, Pandit UA. Risk factors for perioperative adverse respiratory events in children with upper respiratory tract infections. Anesthesiology. 2001;95:299–306.
107. Pounder DR, Blackstock D, Steward DJ. Tracheal extubation in children: halothane versus isoflurane, anesthetized versus awake. Anesthesiology. 1991;74:653–5.
108. Schreiner MS, O'Hara I, Markakis DA, Politis GD. Do children who experience laryngospasm have an increased risk of upper respiratory tract infection? Anesthesiology. 1996;85:475–80.
109. Patel RI, Hannallah RS, Norden J, Casey WF, Verghese ST. Emergence airway complications in children: a comparison of tracheal extubation in awake and deeply anesthetized patients. Anesth Analg. 1991;73:266–70.
110. Halow KD, Ford EG. Pulmonary edema following post-operative laryngospasm: a case report and review of the literature. Am Surg. 1993;59:443–7.
111. Baraka A. Intravenous lidocaine controls extubation laryngospasm in children. Anesth Analg. 1978;57:506–7.
112. Leicht P, Wisborg T, Chraemmer-Jorgensen B. Does intravenous lidocaine prevent laryngospasm after extubation in children? Anesth Analg. 1985;64:1193–6.
113. Afshan G, Chohan U, Qamar-Ul-Hoda M, Kamal RS. Is there a role of a small dose of propofol in the treatment of laryngeal spasm? Paediatr Anaesth. 2002;12:625–8.
114. Tanaka A, Isono S, Sato J, Nishino T. Effects of minor surgery and endotracheal intubation on postoperative breathing patterns in patients anaesthetized with isoflurane or sevoflurane. Br J Anaesth. 2001;87:706–10.
115. Tanaka A, Isono S, Ishikawa T, Sato J, Nishino T. Laryngeal resistance before and after minor surgery: endotracheal tube versus laryngeal mask airway. Anesthesiology. 2003;99:252–8.
116. Bishop MJ, Weymuller EA Jr, Fink BR. Laryngeal effects of prolonged intubation. Anesth Analg. 1984;63:335–42.
117. Weymuller EA Jr, Bishop MJ, Fink BR, Hibbard AW, Spelman FA. Quantification of intralaryngeal pressure exerted by endotracheal tubes. Ann Otol Rhinol Laryngol. 1983;92:444–7.
118. Litman RS, Weissend EE, Shibata D, Westesson PL. Developmental changes of laryngeal dimensions in unparalyzed, sedated children. Anesthesiology. 2003;98:41–5.

119. Koka BV, Jeon IS, Andre JM, MacKay I, Smith RM. Postintubation croup in children. Anesth Analg. 1977;56:501–5.
120. Fisher MM, Raper RF. The 'cuff-leak' test for extubation. Anaesthesia. 1992;47:10–2.
121. Miller RL, Cole RP. Association between reduced cuff leak volume and postextubation stridor. Chest. 1996;110:1035–40.
122. Davis PG, Henderson-Smart DJ. Intravenous dexamethasone for extubation of newborn infants. Cochrane Database Syst Rev. 2001;4:CD000308.
123. Davies MW, Davis PG. Nebulized racemic epinephrine for extubation of newborn infants. Cochrane Database Syst Rev. 2002;1:CD000506.
124. Darmon JY, Rauss A, Dreyfuss D, Bleichner G, Elkharrat D, Schlemmer B, Tenaillon A, Brun-Buisson C, Huet Y. Evaluation of risk factors for laryngeal edema after tracheal extubation in adults and its prevention by dexamethasone. A placebo-controlled, double-blind, multicenter study. Anesthesiology. 1992;77:245–51.
125. Efferen LS, Elsakr A. Post-extubation stridor: risk factors and outcome. J Assoc Acad Minor Phys. 1998;9:65–8.
126. Roberts KW, Crnkovic A, Steiniger JR. Post-anesthesia paradoxical vocal cord motion successfully treated with midazolam. Anesthesiology. 1998;89:517–9.
127. Hammer G, Schwinn D, Wollman H. Postoperative complications due to paradoxical vocal cord motion. Anesthesiology. 1987;66:686–7.
128. Kakitsuba N, Sadaoka T, Kanai R, Fujiwara Y, Takahashi H. Peculiar snoring in patients with multiple system atrophy: its sound source, acoustic characteristics, and diagnostic significance. Ann Otol Rhinol Laryngol. 1997;106:380–4.
129. Isono S, Shiba K, Yamaguchi M, Tanaka A, Hattori T, Konno A, Nishino T. Pathogenesis of laryngeal narrowing in patients with multiple system atrophy. J Physiol. 2001;536:237–49.

Part III
Lower Airway

Chapter 6
Convective and Diffusive Mixing in Lower and Acinar Airways: Is Diffusive Mixing Effective in the Lung Periphery?

Kazuhiro Yamaguchi and Peter Scheid

Abstract To understand the details of primary mechanisms for gas transport and mixing in the lung periphery, one should analyze the interaction between convection and diffusion in the lung periphery by employing an acinar model that is supported by correct anatomical backgrounds. Before the most detailed morphometric information regarding the acinar structure became available from Haefeli-Bleuer and Weibel in 1988 (defined as the HB acinus in the present chapter), many investigators had analyzed a time-dependent gas concentration profile along the acinar airways during normal breathing. They introduced interdependence of convection and axial diffusion as the major mechanism for gas transport and mixing into the symmetrical model A of Weibel that has a simple geometry with trumpet or thumbtack shape. After 1988, a couple of investigators certified the importance of interdependence of convection and diffusion for describing acinar gas transport and mixing in the model with asymmetrical multi-branching geometry constructed from geometrical data of the HB acinus. Although the models used were radically different, earlier simulation studies before 1988 and recent studies after 1988 provided qualitatively similar results, i.e., that diffusion-related spatial differences in gas concentration (i.e., stratified heterogeneity) does exist during inspiration, but it is rapidly eliminated during expiration under normal breathing conditions, indicating that stratified heterogeneity has little impact on overall acinar gas transport and mixing during normal breathing. The basic difference between earlier and recent studies is the sensitivity to detecting the effect of interdependence of convection and diffusion. The simulation based on asymmetrical multi-branching models can more sensitively detect the heterogenous interdependence of convection and diffusion in acinar regions. Although

K. Yamaguchi (✉)
Department of Respiratory Medicine, Tokyo Medical University, Tokyo, Japan

Division of Respiratory Medicine, Tohto Clinic, Kenko-Igaku Association, Tokyo, Japan
e-mail: yamaguc@sirius.ocn.ne.jp

P. Scheid
Department of Physiology, Ruhr University of Bochum, Bochum, Germany

K. Yamaguchi (ed.), *Structure-Function Relationships in Various Respiratory Systems*, Respiratory Disease Series: Diagnostic Tools and Disease Managements, https://doi.org/10.1007/978-981-15-5596-1_6

stratified heterogeneity is thought to be insignificant for overall acinar gas transport and mixing during normal breathing, it exerts an appreciable impact on diffusing capacities measured at artificial breathing such as breath-holding or rebreathing.

Keywords Acinus · Symmetrical branching model · Asymmetrical multi-branching model · Convection · Axial diffusion · Stratified heterogeneity · Diffusing capacities for CO and NO

1 Introduction

The main objective of this book is to elucidate the structure–function relationships in the human lung in diverse directions. Among many issues regarding the structure–function relationships in the lung, the most difficult task is how to detect them in the microregion at the acinar level. This is because it is difficult to reliably analyze the morphological abnormalities in the microregion close to the acinus, even by making full use of high-precision diagnostic imaging technologies, e.g., high-resolution computed tomography (HRCT), magnetic resonance-assisted proton image (proton-MRI), positron emission tomography (PET), and synchrotron radiation CT. HRCT can detect the morphological abnormalities in upper and lower conductive airways up to the fifth generation of airway tree (central airways) but not those in peripheral and acinar airways. Furthermore, although proton-MRI, PET, and synchrotron radiation CT are promising tools for quantifying the microregional distribution of ventilation in the lung, they cannot directly diagnose the morphological changes in acinar airways. In this respect, the structure–function relationships in the acinar regions are very peculiar and different from other regions of the lung. In the regions larger than the acinus, morphological information and physiological information may be somehow acquired from radiological and physiological examinations, respectively. Therefore, the structure–function relationships there are straightforwardly obtained from either direction. However, in the acinar regions, the situation is different in that it is very difficult to clinically obtain the rigorous morphological information about structural abnormalities of the acinus. Thus, the structure–function relationships in the acinar regions are discerned in one direction, i.e., they can be identified only from the physiological parameters that are thought to be sensitive to the acinar structures, but there is no reverse direction. Therefore, to elucidate the structure–function relationships in the acinar regions, it is indispensable to have a profound knowledge of microscopic features of acinar airways with and without alveolar sleeve as well as basic physiological mechanisms for prescribing gas transport and mixing in the acinus. Based on these considerations, we tried to explain the complicated phenomena occurring in the pulmonary acinus as concisely as possible. For achieving this, we shed light on three issues in the present chapter: (1) anatomical design for understanding gas transport in the lung periphery, including symmetrical and asymmetrical models for lower conductive airways, acinar microstructures, and collateral pathways; (2) interaction of convection and diffusion for gas transport and mixing in the acinus, including a basic understanding of convection and diffusion and results of numerical analyses in the models with symmetrical and asymmetrical branching geometries; and (3) clinical importance of stratified heterogeneity in lungs with various pathophysiological conditions.

2 Anatomical Design for Understanding Gas Transport in the Lung Periphery

The lung is anatomically divided into two parts. The region with branching airways from the trachea to the terminal bronchiole is called the lower conducting airways, which is composed of central and peripheral airways. On the other hand, the region distal to the terminal or transitional bronchiole, in which a tremendous number of alveoli engaging in gas exchange are present, is defined as the acinus [1]. The branching pattern of acinar airways qualitatively differs from that of lower conductive airways. To address the complicated mechanism prescribing the gas transport dynamics along the airways, we should have a profound knowledge about the anatomical structures, including the lower conductive airways and acinar airways surrounded by alveoli.

2.1 Symmetrical and Asymmetrical Models for Lower Conductive Airways

To make physiological approaches useful, Weibel [2] abstracted the morphometric data of lower conductive airway trees into two types of models that extend from the trachea to the terminal alveolar sacs. Model A of Weibel considers the basic properties of airway branching by assuming a regular dichotomy and thus defining a symmetric typical-path model, while model B of Weibel focusses on the irregularities of tree architecture with asymmetric branching and defines the properties of a variable-path model.

In the symmetrical typical-path model (equivalent to the model A of Weibel [2]), the airway branches in a certain generation are assumed to have the same diameter and length. Thus, the diameter-to-length ratio is treated to be identical in all generations. The total cross-sectional area of all airways in each generation increases tremendously toward the periphery. The typical-path model has, thus, been called "trumpet" or "thumbtack" model [3]. In this case, it is necessary to note that the essential factor that defines the airway dimension is simply the generation number. Detailed criticisms on the typical-path model is found elsewhere [4, 5]. Briefly, (1) this model assumes that all anatomical gas exchange units (i.e., acini) are found at an equal distance from the trachea. This is an oversimplification if we appreciate the anatomical fact that some terminations of conductive airways end after a shorter distance than others. (2) While the diameter shows a relatively symmetric distribution in airways belonging to the same generation, the segment length varies significantly among generations. (3) The diameter of acinar airways, precisely examined by Haefeli-Bleuer and Weibel [1], does not follow the regression line constructed for estimating the diameter of conductive airways. In consideration of these facts, the typical-path model should be supplemented by a variable-path model that pays attention to the asymmetry of airway branching. The generation of airway branching in the typical-path model, including conductive and acinar airways, is numbered according to the system of Z (for all airways) and that of Z' (only for acinar airways), as proposed by Haefeli-Bleuer and Weibel [1]. This numbering system for airway generation differs somewhat from that classically proposed by Weibel [2] (Tables 6.1 and 6.2).

The asymmetrical model B of Weibel [2] is based on the anatomical finding that airways with a certain diameter occur in several generations and at different distances from the origin of airways. For instance, airways of 2 mm diameter are found in generations 4–14, with a maximum in generation 8. Alternatively, these bronchi are located at 18–31 cm from the origin of the trachea with a maximum at 24 cm. Note that in the symmetrical typical-path model, bronchi with 2 mm diameter are located maximally in generation 8 corresponding to 23.6 cm from the trachea,

Table 6.1 Microstructural traits of HB acinus

Structural parameters	Original values measured at TLC	Values scaled to FRC
Entrance	Transitional bronchiole (TrB) ($Z = 15$, $Z' = 0$)	
Airway branching pattern	Irregular, asymmetrical dichotomy (trichotomy: rare)	
Number of branches	Average: 9.1 (ranging from 6 to 12)	
Total longitudinal path-length	8.8 ± 1.4 mm (from TrB to terminal alveolar sacs)	7.0 mm
Inner airway diameter	Decrease from 500 μm ($Z' = 0$) to 270 μm ($Z' = 10$)	
Outer airway diameter	Approximately constant at 700 μm including alveolar sleeve independent of Z'	
Total surface area of alveoli	Average: 67 cm^2	42 cm^2
Total volume of an acinus	187 ± 79 mm^3	94 mm^3
Total number of acini in human lung	26,000–32,000	

Z: number of airway generations counted from trachea. Z at trachea is zero, while Z at transitional bronchiole is 15. Z': number of airway generations counted from transitional bronchiole. Z' at transitional bronchiole is zero. Values are measured under condition at full expansion of lung tissue specimen [1]. FRC is assumed to be half of TLC. The values at FRC are calculated assuming that volume change of acinus is proportional to that of whole lung volume. Change in total surface area of alveoli is taken to be in proportion to the two-thirds power of that in whole lung volume (i.e., 0.63·(original value)), while change in total path-length is taken to be in proportion to cube root of that in whole lung volume (i.e., 0.79·(original value))

Table 6.2 Dimensions of airways in HB acinus at TLC and FRC

Generation						
Z	Z'	Number of branches	L (mm)	d_{in} (mm)	s (mm^2)	S (mm^2)
15	0	1	0.80 (0.63)	0.50 (0.40)	0.20 (0.13)	2.50 (1.57)
16	1	2	1.33 (1.05)	0.50 (0.40)	0.39 (0.25)	16.7 (10.5)
17	2	4	1.12 (0.88)	0.49 (0.39)	0.75 (0.47)	41.4 (26.1)
18	3	8	0.93 (0.73)	0.40 (0.32)	1.00 (0.63)	74.8 (47.1)
19	4	16	0.83 (0.66)	0.38 (0.30)	1.81 (1.14)	159 (100)
20	5	32	0.70 (0.55)	0.36 (0.28)	3.26 (2.05)	253 (159)
21	6	64	0.70 (0.55)	0.34 (0.27)	5.81 (3.66)	479 (302)
22	7	128	0.70 (0.55)	0.31 (0.24)	9.11 (5.74)	873 (550)
23	8	256	0.67 (0.53)	0.29 (0.23)	16.9 (10.7)	1563 (985)
24	9	512	0.70 (0.55)	0.25 (0.20)	25.1 (15.8)	3260 (2054)
Total			8.48 (6.70)		64.3 (40.5)	6722 (4235)

Z: airway generations, the number of which begins at trachea (Z = zero) and transitional bronchiole corresponds to $Z = 15$. Z': airway generations, the number of which starts from transitional bronchiole (Z' = zero) down to terminal alveolar sacs ($Z' = 9$). L and d_{in} indicate path-length and inner diameter of a certain acinar airway, respectively. s is total cross-section of acinar airways without alveoli belonging to the same generation, which is calculated on the assumption that airway duct is approximated by cylinder. S indicates total surface area including alveoli of airways belonging to the same generation. Values without parentheses are original values estimated at TLC [1], while values in parentheses are those scaled to FRC

which is in close agreement with the distance that shows the maximum distribution of bronchi with 2 mm diameter in the asymmetrical model of B [5]. Based on these facts, one may conceive that, although the symmetrical typical-path model has various limitations, it works well, without any major problem, as a model for predicting airway branching, at least in lower conductive airways from the trachea to the terminal bronchioles.

Considering the airway tree as a convergent system in the manner of a river, an entirely different approach in comparison with Weibel's model B was introduced by Strahler [6, 7], Horsfield and Cuming [8], and Woldenberg et al. [9]. In the convergent system, the decisive factor for airway geometry is the flow there, i.e., the increased flow in a certain airway makes the airway diameter wider. Thus, the tube width is necessarily related to both morphology and function upstream to the flow. In this approach, the airway branches are counted from the terminal bronchioles toward the trachea and the branch numbers thus established were termed "order" as opposed to "generation" in the symmetrical model A of Weibel, resulting in that the higher the order, the greater the volume of lung supplied. The problem of the asymmetric model based on the convergent theory is that the different orders lack connectivity so that the pathway model for gas transport cannot be constructed properly from the asymmetrical, convergent system. To cope with this drawback of the asymmetrical model considered from the convergent theory, Fredberg and Moore [10, 11] contrived the airway branching model, which they called "self-consistent tree," modifying Horsfield's ordering model [8]. Such a tree has a "principal path" along which the orders follow in numerical sequence. The principal-path model with asymmetry does not exhibit a trumpet shape but shows a pear-like shape.

When attempting to precisely evaluate the convection-related gas transport in conductive airways toward the terminal bronchioles, a realistic model such as the variable-path model B of Weibel or the asymmetrical principal-path model will be needed. However, when considering the gas transport and mixing in acinar regions, the difference in models for conductive airways may exert little impact. This is because in acinar regions, the relative contribution to gas transport by diffusion predominates that by convection [3]. The gas transport by convection is largely prescribed by the airway branching pattern so that a conductive airway model certainly supported by the anatomical fact is needed. However, the gas transport by diffusion predominantly depends on the airway cross-section with and without alveoli in the acinus, which requires a correct acinar model independent of geometrical features of conductive airways.

2.2 *Microstructure of Acinus*

The extensive analysis of the microscopic structure of the acinus was made by Haefeli-Bleuer and Weibel [1], which is defined as the HB acinus in the present chapter. They considered the acinus as the region severed by the proximal part of a transitional bronchiole without identifiable alveoli (Z = 15th generation), which may be larger than the acinus of Aschoff that is defined as the region distal to the first-order respiratory bronchiole (Z = 16th generation) but is smaller than the acinus of Loeschcke that is defined as the region distal to the terminal bronchiole (Z = 14th generation) [4]. However, as it is extremely difficult to correctly separate the transitional bronchiole from the first-order respiratory bronchiole in a microscopic examination [1], there is a certain possibility that the acinus of HB is practically the same as that of Aschoff.

The airways in the HB acinus are characterized by asymmetrical, irregular dichotomy (rarely, trichotomy), which branch over nine generations (Tables 6.1 and 6.2). The inner airway diameter is reduced as the branching progresses, while the outer diameter, including the sleeve of alveoli, remains almost constant. The total path-length from the transitional bronchiole to the terminal alveolar sac averages 8.8 mm, while the total acinar volume averages 0.2 cm^3. The total alveolar surface area participating in gas exchange reaches about 70 cm^2. However, it should be noted that these values were measured at an almost full expansion of the lung tissue specimens, i.e., close to total lung capacity (TLC). Therefore, under physiological conditions, i.e., at functional residual capacity (FRC), the dimension of acinar structures, including acinar volume, alveolar surface area, and airway path-length should be modified accordingly [12].

Interestingly, Haefeli-Bleuer and Weibel [1] separated the HB acinus into eight small regions at the third-order respiratory bronchiole (Z = 18th generation). They named these small regions "1/8-subacini." They found that the volume of each

1/8-subacinus differed considerably, indicating that there are small and large subacini arranged in parallel in the HB acinus and these subacini may have different airway path-lengths and alveolar surface areas. Each of the 1/8-subacini is composed of two primary lobules of Miller, entrance of which is defined by the first-order alveolar duct ($Z = 19$th generation) [4].

2.3 Collateral Pathways in Acinus

There are three types of collateral pathways in the acinus, i.e., alveolar pores (pores of Kohn), channels of Lambert, and channels of Martin. Although the size of alveolar pores ranges from 2 to 30 μm, there are many studies certifying that the alveolar pores are not significantly open in air-filled lungs [13]. These studies indicate that the alveolar pores take no part in augmenting diffusive gas flux at the alveolar level.

Lambert [14] described accessory communications between bronchioles and alveoli in the normal human lung. The diameter of Lambert channels varies from zero to 30 μm, and they are considered to serve as a collateral pathway connected to adjacent alveoli when airway obstruction occurs at the level of terminal bronchioles [13]. However, the physiological importance of the Lambert channels for augmenting diffusive gas flux in the acinus was not certified.

Martin [15] reported intra-acinar communications at the level of respiratory bronchioles and/or alveolar ducts in the canine lung. The Martin channels were found in the human lung as well [16]. As the Martin channels have a sufficiently low resistance to gas flow, they act as the primary pathway for inducing collateral, diffusive flux in the acinus [13]. Therefore, one can conceive that only the Martin channels, but neither the alveolar pores nor the Lambert channels, serve to equalize the concentration difference of indicator gases in the acinus.

3 Interaction of Convection and Diffusion for Gas Transport in the Acinus

Since the direct measurement of gas transport and mixing dynamics in the acinar region is technically impossible at present, the details of the gas transport mechanism in this region were elucidated based on a theoretical analysis. In addition to the convective mixing elicited by ventilation, the diffusive mixing plays an important role in maintaining gas transport along acinar airways with large cross-sectional areas (Fig. 6.1). Numerous mathematical attempts have been made to estimate the effect of convection, diffusion, or interactions of both on gas transport in acinar airways using a model with symmetrical or asymmetrical branching of airways (*cf.* [3, 17]). Before discussing the specifics of the numerical analysis, we will first take a basic view on the nature of convection and diffusion.

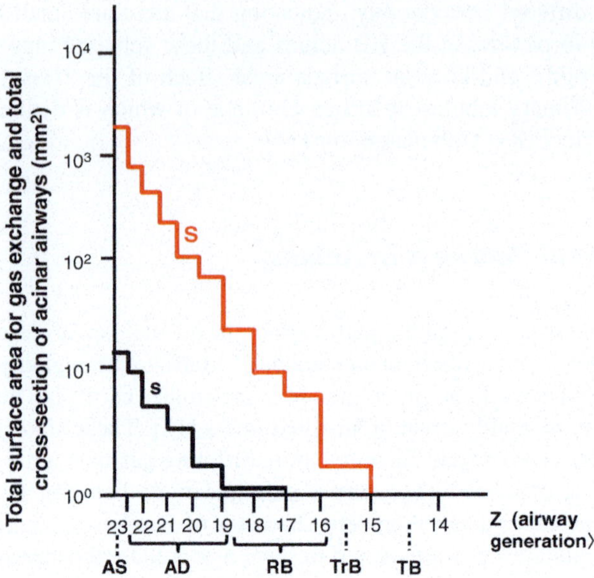

Fig. 6.1 Total surface area for gas exchange and total cross-sectional area of airways in HB acinus. *S*: total surface area for gas exchange (mm^2) of acinar airways with alveolar sleeves having the same generation of Z and *s*: total cross-section (mm^2) of acinar airways without alveoli having the same generation of Z. *S* and *s* are estimated based on the data reported by Haefeli-Bleuer and Weibel measured at TLC [1]. Values are scaled to those at FRC and expressed by the *n*th power of 10. *TB* terminal bronchiole, *TrB* transitional bronchiole, *RB* respiratory bronchiole, *AD* alveolar duct, *AS* alveolar sac

3.1 Basic Rationale of Convection and Diffusion

The term of convection is defined as a mass transport resulting from the unidirectional displacement of molecules included in a certain volume [3, 17]. Convective displacement is identical for all molecular species and thus, it is not different for gases with different diffusivity. This indicates that convection (laminar or turbulent flows) occurs along gradients of total pressure but not along those of partial pressure. On the other hand, the term of diffusion is defined as resulting from the random motion of molecules due to their thermal energies, depending on their concentration (partial pressure) gradients and diffusivity [3, 17]. Hence, a separation of two gases with different diffusivities occurs during gas transport when diffusion constitutes a main limiting process, whereas a separation of two gases does not occur when gas transport is predominantly limited by convection. Although the convective transport of a gas in the lung is provided by ventilation in general, there is a specific type of convection that occurs through the mechanical action of the heart (i.e., cardiogenic oscillation). Fukuchi et al. [18] found that the gas mixing in the canine lung is conspicuously enhanced by the cardiogenic oscillation. However, Drechsler and Ultman [19], using bolus dispersion of He and SF_6 in human subjects, did not find any enhancement of mixing when heart rate was increased up to 153% under a bicycle exercise condition. Based on these findings, Paiva and Engel [20] concluded that the cardiogenic oscillation plays no role in gas mixing in the lung periphery, particularly in the human lung, and the convective effect elicited by cardiogenic oscillation is indeed explained by the interaction of diffusion and ventilation-generated convection in the lung model with asymmetrical airway branching.

The important role of diffusion in the lung periphery was first described with gases of different diffusivities by Georg et al. in 1965 [21]. Since then, many groups of investigators have focused on the diffusion as the basic mechanism for explaining the gas transport in the lung periphery. Diffusive transport in airways and alveoli occurs in axial and radial directions. Radial diffusion into acinar airways and alveolar spaces can be assumed to be negligible. This is because the radial distance for diffusion, which is given by the alveolar diameter (~0.2 mm) plus the radius of acinar airway duct (~0.2 mm), is much smaller than the axial distance for diffusion, which is about 7 mm under physiological conditions (Table 6.1). Furthermore, the coupling of convection and diffusion occurs under conditions where the contribution of convection and diffusion to gas transport is of similar magnitude, which may occur predominantly in peripheral airways at generations between 8th and 12th [22]. The degree of convection–diffusion coupling under conditions with laminar or turbulent flow is estimated using the dispersion coefficient proposed by Taylor and others [23]. However, the Taylor dispersion has been identified to be insignificant for gas transport during normal breathing [24, 25]. This leads to the conclusion that it is enough to introduce only convection and axial diffusion, but neither the radial diffusion nor the Taylor dispersion, as the basic mechanisms when assessing gas transport along acinar airways under normal breathing.

In addition to the convective transport by ventilation, the contribution of diffusion to overall gas transport rises substantially in acinar regions. The respective importance of convection and diffusion for gas transport in a certain part of the lung can be appreciated from the Peclet number (P_e) as:

$$P_e = d \cdot (u / D) \tag{6.1}$$

where d is diameter of an airway tube, u is mean convective flow velocity, and D is binary diffusion coefficient of a gas. When the Pe number reaches 1.0, the diffusive transport is quantitatively the same as the convective transport [26]. Under resting conditions with air-breathing, the convective flow velocity at the trachea is about 100 cm/s while about 1 cm/s at the acinar entrance. This is because the total cross-section of the bronchioles forming the acinar entrance is about 100-times larger than that of the trachea. Therefore, the convective flow velocity falls 100-fold from the trachea toward the acinar airways and the contribution of diffusion to gas transport significantly dominates that of convection within the acinus [5, 26]. The position where the Pe number is equal to 1.0 is called the "critical zone," originally coined by Gomez [27].

3.2 Results of Numerical Analyses

3.2.1 Symmetrical Branching Geometry

Introducing both convection and axial diffusion as the gas transport mechanisms into the symmetrical model A of Weibel that has a simple geometry with trumpet or thumbtack shape, several groups of investigators analyzed the time-dependent change in gas concentration profile of an insoluble gas that is not absorbed into the blood (i.e., no gas exchange through alveolocapillary membrane) along peripheral airways and acinar airways during normal breathing [28–30]. Among them, the study by Paiva [29] is noteworthy. Paiva assumed (1) no dispersion in airways, (2) no radial diffusion in airways, (3) no axial diffusion in alveolar regions, (4) no convective flow in alveolar regions, and (5) no collateral pathway for diffusive flux in acinar regions. The theoretical calculation done by Paiva was made under a situation where tidal volume is set at 500 ml, inspiratory flow rate at 250 ml/s (the same expiratory flow rate was used but the direction is reversed), and one respiratory cycle at 4 s (2 s each for inspiratory and expiratory time). Paiva analyzed gas mixing kinetics of O_2 along airways distal to the 13th generation, which correspond closely to the secondary lobule of Miller. In his analysis, O_2 was taken to be a gas mimicking an insoluble, foreign inert gas that is not absorbed into capillary blood. The results during inspiration and expiration emphasized by Paiva are summarized as follows:

1. On inspiration, the front of a gas (this is alternatively called the diffusion front of a gas) lags appreciably behind the front of tidal volume. The front of tidal volume attains to the 22th generation of airways (i.e., the alveolar duct), whereas the front of a gas stays at the 17th generation (i.e., the second-order respiratory bronchiole). The gas front is practically stationary, suggesting that the convective advancement of a gas into the periphery is counterbalanced by the diffusion-elicited proximal retreat of a gas. The existence of a stationary gas front in the acinus was first described by Wilson and Lin [31] in the symmetrical model A of Weibel. Given the quasi-static nature and the sigmoid shape of the gas front, the location of the representative point of the gas front, which is given by the inflection point (i.e., the point at which the change in gas concentration is greatest), could be identified by solving the equation $\partial C/\partial t = 0$ and $\partial^2 C/\partial x^2 = 0$, in which C is gas concentration, t is time, and x is distance from the most terminal alveolar sacs:

$$ds/dx = (u \cdot S)/D \tag{6.2}$$

where S and s are the cross-sectional areas of an airway tube with and without surrounding alveoli, respectively. D is the binary diffusion coefficient of a gas and u is mean convective flow velocity over the cross-sectional area of S. $(u \cdot S)$ indicates the inspiratory flow rate which is generally set at 500–1000 ml/s in most studies. The ds/dx thus obtained predicts that the inflection point is located at the second-order respiratory bronchiole. Furthermore, Eq. (6.2) indicates that increasing u but decreasing D moves the gas front more peripherally in the acinus, resulting in the fact that the gas front of He (smaller molecular weight with higher gas diffusivity) is formed around the first-order respiratory bronchiole ($Z = 16$th generation), while that of SF_6 (larger molecular weight with lower gas diffusivity) is formed around the third-order respiratory bronchiole ($Z = 18$th generation), i.e., deeper penetration of SF_6 than He in the lung periphery. However, it should be noted that the concept of gas front should not be applied for explaining the gas transport dynamics elicited by high-frequency ventilation (HFV) or by high-frequency oscillation (HFO), in which the gas front has no time to be established.

2. On inspiration, a significant difference in the gas concentration along the acinar airways, i.e., the spatial or stratified heterogeneity, is found. The term "stratification" was first introduced by Rauwerda in his PhD thesis in 1946 [32].
3. On expiration, all acinar concentration gradients of a gas persisting during inspiration are rapidly abolished, which may be caused by the thumbtack-like shape of the acinus. The diffusive process in the acinar gas phase imposes an impact on gas transport in a very early stage, but not in a later stage, of expiration. The findings during one respiratory cycle indicate that the stratification (i.e., the spatial concentration gradient of a gas due to diffusion) is evident along acinar airways during inspiration but minimal during expiration.

The most serious problem identified in the analysis based on the symmetrical model A of Weibel by Paiva [29] is that it did not produce the sloping alveolar plateau (phase III) of the expirogram. To solve this problem, Paiva and Engel [33] elaborated on a model with asymmetrical branches of acinar airways. They coined the term "single branch point-two trumpet model," in which the two parallel, heterogenous trumpet-like units differing in total cross-sectional area and path-length were assumed. These two units in the acinus join at the branch point. The degree of convection–diffusion-elicited, parallel heterogeneity for gas concentration varies depending on the location of the branch point. In fact, if the branch point is situated at the more proximal part of the gas front (e.g., outside the acinus), there is no axial concentration gradient for diffusion to take effect. On the other hand, if the branch point is moved to the more peripheral end of the gas front (e.g., around the alveolar sacs), convective flows are negligibly small, resulting in the fact that any concentration difference is rapidly equilibrated by diffusion between the units. When the branch point is formed somewhere in the middle of the two extremes, the interaction of convection and diffusion acts as the primary factor deciding gas concentrations in the two units. A typical analysis performed by Paiva and Engel [33] demonstrated the following results in the asymmetrical model with two parallel units (small and large units in size) in which the branch point is located at the entrance of alveolar ducts and the convective flow/volume ratio of both units is assumed identical for the sake of simplicity. (1) At end-inspiration during pure O_2 inhalation, the higher O_2 concentration is formed in the small unit. This is due to diffusion aiding convective gas transport into the small unit, whose smaller axial dimensions favor diffusional flux and whose smaller volume yields a rapid increase in O_2 concentration in the small unit. (2) During expiration, diffusion promotes convection-dependent gas leaving from the small unit, thus O_2 concentration in the small unit decreases rapidly. On the other hand, the change in O_2 concentration in the large unit is relatively flat during expiration. This is because convective flow in the large unit is high enough to prevent back-diffusion of O_2 from the branch point. Consequently, the O_2 concentration at the branch point, which is the mixture exiting from both small and large units arranged in parallel, significantly varies as the expiration progresses, resulting in forming a sloping alveolar plateau of the expirogram. The analytical results of Paiva and Engel [33] convincingly refute the concept that the stratified heterogeneity generated during inspiration persists during expiration and contributes to the formation of the sloping alveolar plateau. Furthermore, their analysis suggests that neither parallel heterogeneity of convective flow nor flow asynchrony (i.e., sequential emptying) is needed for generating the sloping alveolar plateau. Many results concerning gas transport dynamics in the acinar region were derived from the single branch point-two trumpet model. Unfortunately, however, the anatomical backgrounds that decide the branch point in the acinus were not argued conclusively in the study of Paiva and Engle [33].

3.2.2 Asymmetrical Multi-branching Geometry

Before the most detailed morphometric information regarding acinar structures became available from Weibel's group in 1988 [1], several groups of investigators had already used the models with asymmetrical multi-branching geometry to simulate gas transport in the human acinus (*cf.* [26]). Although these early simulations used radically different acinar structures, they predicted several common conclusions, including the presence of very different fluctuations in alveolar gas concentration in different parts of the acinus and the pattern of expired gas concentration which does not resemble the pattern of alveolar gas concentration in any given portion of the lung.

The prominent study on gas transport dynamics in the acinus was conducted by Swan and Tawhai in 2011 [12]. They adopted the model with the asymmetrical multi-branching acinar airways, which was constructed based on the anatomical data of Haefeli-Bleuer and Weibel [1] for the human acinus (i.e., the HB acinus). Since the original data of the HB acinus were investigated at a full expansion (i.e., at a TLC level), Swan and Tawhai [12] changed them to those at an FRC level. Under resting breathing of ambient air (tidal volume: 300 ml, one respiratory cycle: 5 s (inspiration and expiration: 2.5 s each), and FIO_2: 21%), they solved the one-dimensional, partial differential equations governing the convective and diffusive O_2 transport process in cooperation with gas exchange of O_2 through the alveolo-capillary membrane in respective acinar airway generations (defined as the element in their paper). The perfusion of capillary network surrounding the alveolar sleeve of a certain element was assumed to be proportional to the surface area of gas exchange there. The PO_2 profile in the capillary blood approaching alveolar PO_2 from the mixed venous blood (P_VO_2 = 40 mmHg) in the element was estimated using the morphometric value of O_2 diffusing capacity by Weibel et al. [34]. Furthermore, the model assumed that during breathing, the volume of an element changes in proportion to the ratio of element volume to total volume of the HB acinus, while the path-length of an element changes in proportion to the cube root of volume change. The cross-sectional area of an airway tube without alveoli is taken to be constant, while the change in the cross-sectional area of an airway tube with alveolar sleeve is calculated to satisfy the volume change. The important messages from their analysis are as follows. (1) The PO_2 at the entrance of HB acinus (the transitional bronchiole) decreases slightly below inspired PO_2 just after inspiration begins, but this value is maintained over inspiration ($\approx$ 140 mmHg). (2) The spatial (stratified) heterogeneity for PO_2 distribution is evident during inspiration, that is, the element with a shorter path-length has a higher PO_2 than that with a longer path-length. (3) The PO_2 gradients from the entrance to the peripheral elements formed during inspiration (i.e., stratified heterogeneity) are almost abolished during expiration and the spatial distribution of PO_2 is relatively uniform at the end of expiration (Fig. 6.2). The coefficient of variation of PO_2 in HB acinar elements is 3% of the mean PO_2 value averaged by the element volume. (4) The position of PO_2 profiles along the acinar airways estimated from the asymmetrical multi-branching model is always located much lower than that from the symmetrical branching model, irrespective of the acinar airway generations. This is most likely because the heterogenous interdependence of convection and diffusion in the acinar regions cannot be sensitively captured by the symmetrical model. However, the disadvantage of the analysis performed by Swan and Tawhai [12] is that, since they adopted a simple differential equation describing the interaction of convection and diffusion in the

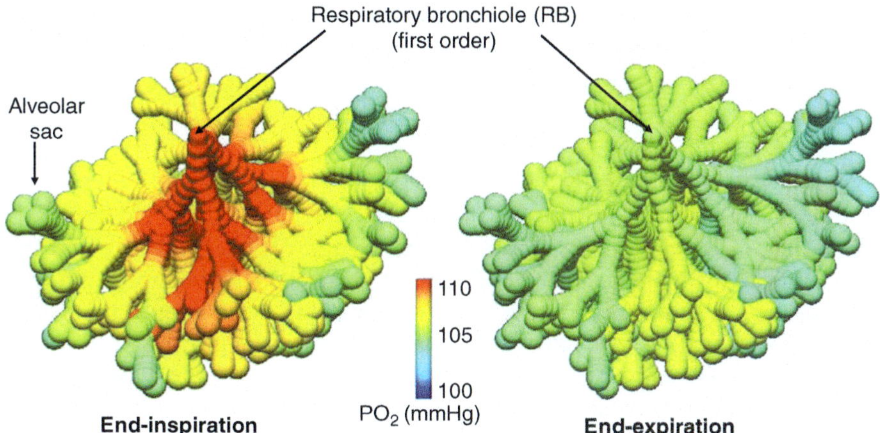

Fig. 6.2 Spatial distribution of PO_2 in the HB acinus. Left side: end-inspiration, right side: end-expiration. Transitional bronchiole (entrance of HB acinus) is excluded. The spatial difference (i.e., stratified heterogeneity) in PO_2 is evident during inspiration but is minimal during expiration under conditions of air breathing. Adopted from Swan and Tawhai [12] with permission of the publisher

acinus, the term correcting the influence of geometrical change along the axial direction upon diffusive gas transport was not included in their analysis. Furthermore, they did not consider the effect of changing volume of the airway tube caused by lung expansion and deflation during breathing. Therefore, we attempted to confirm the analytical results obtained by Swan and Tawhai [12], applying the asymmetrical HB acinar model, in which the geometrical dimensions were scaled to those at the FRC (Fig. 6.1, Tables 6.1 and 6.2). The calculation of PO_2 profiles in the HB acinus was conducted under resting conditions with air-breathing (inspired $PO_2 = 150$ mmHg), using the full differential equation describing the interaction of convection and diffusion proposed by Verbanck and Paiva [26]. The TLC was set at 5000 ml, while the lung volume at end-inspiration and end-expiration was set to be 60% and 50% of the TLC, respectively. This results in a lung volume at the FRC and a tidal volume of 2500 ml and 500 ml, respectively. The time for one respiratory cycle was set at 4 s, which was divided into two equal time periods for inspiration and expiration (i.e., 2 s for each). The initial PO_2 in the HB acinus was set at 100 mmHg. The volume change of an acinus during breathing was assumed to be proportional to that of the whole lung volume from end-inspiration to end-expiration. The length and diameter of an acinar airway were considered to change in proportion to the cube root of the volume change. The cross-sectional area of an acinar airway without alveoli (s) was calculated on the assumption that the shape of the airway tube was cylindrical. Therefore, the volume of an acinar airway (V) was deduced by multiplying the inner cross-sectional area by the length. The surface area for gas exchange of an acinar airway with alveolar sleeve (S) was considered to change in proportion to the two-thirds power of the volume change. The gas exchange of O_2 across the alveolocapillary membrane was disregarded in our calculation. This is because the O_2 flux into or from the capillary blood is small in comparison with the total flow in the airways, and the O_2 equilibration by diffusion within alveolar spaces is fast [35]. The equation used for our calculation is:

$$\begin{aligned}\partial C/\partial t = {} & D_{\mathrm{eff}}\left(\partial^2 C/\partial x^2\right)+\left(D/S\right)\left(\partial s/\partial x\right)\left(\partial C/\partial x\right)\\ & -u\left(\partial C/\partial x\right)-\left(C/V\right)\left(\partial V/\partial t\right)\end{aligned} \tag{6.3}$$

where C, V, t, and x are concentration of a gas, volume of an airway, time, and axial distance from the most distal alveolar sacs, respectively. u is the mean convective flow velocity at x (a positive value of u corresponds to inspiration and a negative value to expiration). S and s are surface area for gas exchange of an airway with alveolar sleeve and cross-sectional area of an airway without alveoli, respectively. D is binary diffusion coefficient of a gas, while D_{eff} means effective D, which is empirically defined as $D\cdot(s/S)$ and indicates the impairment of axial diffusion by the presence of alveolar septa in an alveolated airway tube [36]. The first term of Eq. (6.3) corresponds to the modified diffusion equation, describing the axial spread of a gas in an alveolated airway tube in consideration of impediment by the alveolar septa (i.e., D is replaced by D_{eff}). The second term indicates the influence of the

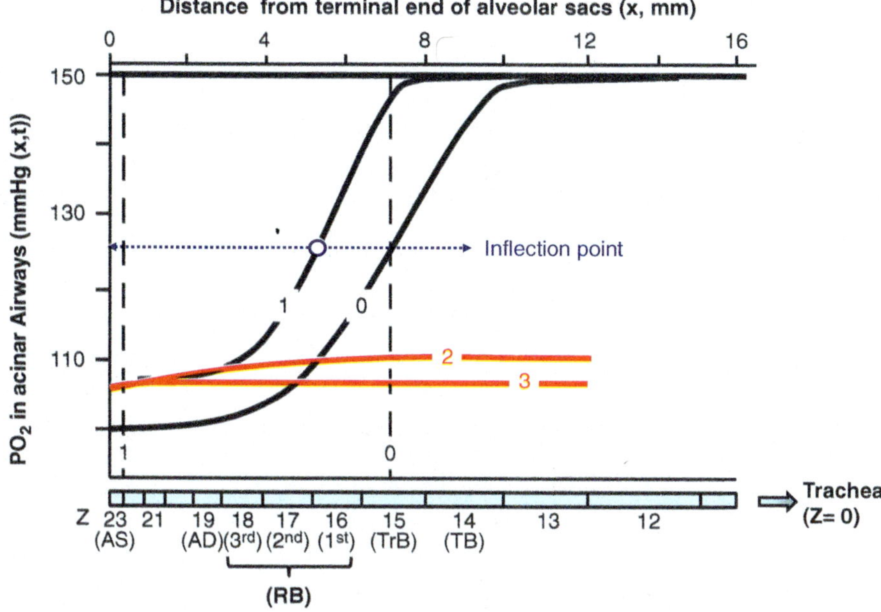

Fig. 6.3 Time- and distance-dependent PO_2 profiles during one respiratory cycle under normal breathing of air. Upper abscissa: axial distance from terminal end of alveolar sacs, lower abscissa: airway generation number of *Z* when the trachea is taken to be zero. *TB* terminal bronchiole, *TrB* transitional bronchiole, *RB* respiratory bronchiole, *AD* alveolar duct, *AS* alveolar sac. Dashed lines: tidal ventilation fronts just after beginning of inspiration (0) and at end-inspiration (1). Tidal ventilation front just after beginning of inspiration is assumed to be in transitional bronchiole ($Z = 15$th generation, i.e., entrance of acinus), while that at end-inspiration is considered to reach terminal alveolar sac ($Z = 23$th generation). Black solid line (0): gas front of PO_2 (PO_2 profile) just after beginning of inspiration. Black solid line (1): gas front of PO_2 at end-inspiration. Blue circle: inflection point (representative point of gas front) that is calculated from Eq. (6.2). Red solid line (2): gas front of PO_2 at 0.2 s after beginning of expiration. Red solid line (3): gas front of PO_2 during expiration excluding beginning of expiration. Various assumptions and conditions necessary for computation are provided in the text

structure of airway branching, which generates the diffusive flux toward the mouth [26]. The third term represents the convective transport of a gas with the velocity of *u*. The fourth term accounts for the effect of volume change in an airway during breathing. Using this equation system, we found qualitatively the same results (Fig. 6.3) as those demonstrated by Swan and Tawhai [12] (Fig. 6.2), i.e., the spatial, stratified heterogeneity of PO_2 distribution exists along the acinar airways during inspiration, but it is almost, but not fully, eliminated during expiration. The PO_2 in the most distal portion is slightly lower than that in the proximal portion of the acinus. These findings provide an insight into gas exchange mechanisms under conditions of normal breathing; that is, since most of the acinar regions have a PO_2 close to the mean value, the assumption of an HB acinus serving as a well-mixed gas exchange unit is therefore reasonable at least during expiration in a first approximation.

However, the sloping alveolar plateau was not generated in our calculation (Fig. 6.3). This is because we did not consider the different units arranged in parallel in the HB acinus such as the 1/8-subacini (each of them is consisting of two primary lobules of Miller) proposed by Haefeli-Bleuer and Weibel [1]. These subacini are expected to have different dimensions, including volume, alveolar surface area, and path-length (i.e., smaller and larger units arranged in parallel). Since these subacini are separated at the third respiratory bronchioles ($Z = 18$th generation) located at a portion distal to the gas front, the consideration of these parallel subacini may yield the sloping alveolar plateau through qualitatively the same mechanism considered by Paiva and Engel [33] (see Sect. 3.2.1). Although Paiva and Engel [33] postulated that the branch point separating the parallel units is located at alveolar duct regions ($Z = 19$th–22th generations), the precise anatomical structure of the HB acinus indicates that it may be anatomically more realistic to assume that the branch point is located at the third respiratory bronchiole ($Z = 18$th generation). The reason why we did not consider the parallel units in the HB acinus is that the detailed dimensions of 1/8-subacni were not described in the original paper of Haefeli-Bleuer and Weibel [1].

4 Clinical Importance of Stratified Heterogeneity in Lungs with Various Pathophysiological Conditions

The important role of stratified heterogeneity to describe gas exchange has been identified particularly in birds and fishes, both having very special structural features as the gas exchanger (*cf.* [17]). However, as discussed in detail in previous sections, the theoretical analyses have consistently shown that the diffusion-dependent, spatial, stratified heterogeneity for an indicator gas under normal breathing obviously exists within the human acinus during inspiration, but is rapidly abolished during expiration (Figs. 6.2 and 6.3). These findings indicate that the diffusion-dependent stratification plays no role in prescribing overall gas mixing within human acini at least under conditions with normal breathing. The exception is the transport of NO that is generated endogenously. In this case, the serial concentration difference in NO along acinar airways, i.e., the stratified heterogeneity, persists during expiration [37]. This is explained by the back-diffusion of NO produced endogenously in the bronchial tree into the alveolar space against the expiratory convective flow.

Many studies attempting to quantify stratification in canine and human lungs were performed between the 1960s and the 1980s. Among them, the study by Adaro [38] deserves special attention. In experiments on anesthetized dogs, Adaro simultaneously administered acetylene and freon 22, inert gases with considerably different diffusivity in the gas phase but almost identical solubility (blood/gas partition coefficient (λ)) in the blood phase. Based on the classical ventilation–perfusion (V_A/Q) theory, the inert gases with identical solubility should display the same clearance values independent of V_A/Q heterogeneity in the lung [4]. In contrast to this expectation, Adaro found less clearance for freon 22 with less diffusivity in the gas phase, which suggests the gas-phase diffusion-dependent stratification to impose a higher resistance to a gas with less diffusivity in the normal lung. However, since the λ values of acetylene and freon 22 are not identical (Fig. 6.4), the study of Adaro was reevaluated by Yamaguchi et al. [39], who infused three inert gases, including ethylene, acetylene, and freon 22 through the peripheral vein under steady-state conditions. Their basic rationale for using these three gases is that ethylene and acetylene

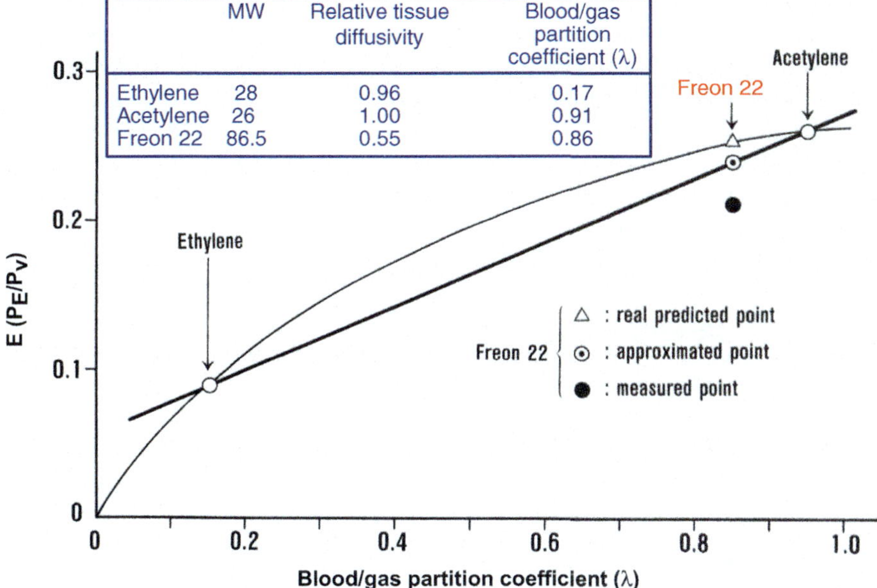

Fig. 6.4 Excretion (E)-solubility (λ) diagram for three inert gases. E indicates P_E/P_v, where P_E is inert gas concentration in expired gas, while P_v is that in mixed venous blood. λ is blood/gas partition coefficient of each inert gas. Ethylene and acetylene have almost the same diffusivity in lung tissue and gas phase so that behaviors of these two gases are simply predicted from classical V_A/Q theory, which is denoted by the curved line drawn between E values for ethylene and acetylene. The straight line shows the approximated effect of V_A/Q heterogeneity estimated from ethylene and acetylene. The point for freon 22 located on the curved line (triangle) indicates the real predicted point for freon 22, which explained purely by V_A/Q heterogeneity (denoted as "real predicted point" in the figure). Unfortunately, however, this point cannot be decided correctly. On the other hand, it is easy to decide the E value of freon 22 located on the straight line (middle black dot). The point corresponding to this E value is defined as the point of freon 22 approximately (but mainly) decided by V_A/Q heterogeneity (denoted as "approximated point" in the figure). Since the contour of E–λ diagram is convex against λ, the approximated point of freon 22 is always located below the real predicted point of freon 22. Therefore, in case that the measured E value of freon 22 (whole black dot) lies beneath the approximated point of E value of freon 22, one can conclude that acinar gas exchange of freon 22 is significantly aggravated by diffusion (gas-phase or aqueous-phase) in addition to V_A/Q heterogeneity. Adopted from Yamaguchi et al. [39] with permission of the publisher

have identical diffusivity in the gas phase but considerably different λ in the blood so that they can predict the correct effect of V_A/Q heterogeneity on the inert gas exchange (Fig. 6.4). Yamaguchi et al. [39] demonstrated that the measured excretion ($E = P_E/P_v$, where P_E is gas concentration in the expired gas, while P_v is that in the mixed venous blood) of freon 22 did not differ significantly from E predicted from the ethylene and acetylene behaviors in anesthetized, normal dogs (Fig. 6.5). Their findings suggest that gas-phase diffusion plays no role in impairing the gas transport in the lung periphery under steady-state conditions with normal breathing. However, Yamaguchi et al. [39] also found that the measured E value of freon 22 was conspicuously lower than that predicted from ethylene and acetylene in a canine model with ARDS (Fig. 6.5), indicating that the diffusion-elicited stratified heterogeneity in the alveolar space acts as an important factor for impeding inert gas exchange in lungs with ARDS. However, this is not caused by the gas-phase diffusion, it is rather caused by the aqueous-phase diffusion that is induced by the edematous fluid flooding into the alveolar space. These experimental findings indicate that the results obtained from the theoretical analyses are valid, i.e., the stratified heterogeneity generated during inspiration does not significantly affect overall gas transport and mixing under steady-state conditions with normal breathing.

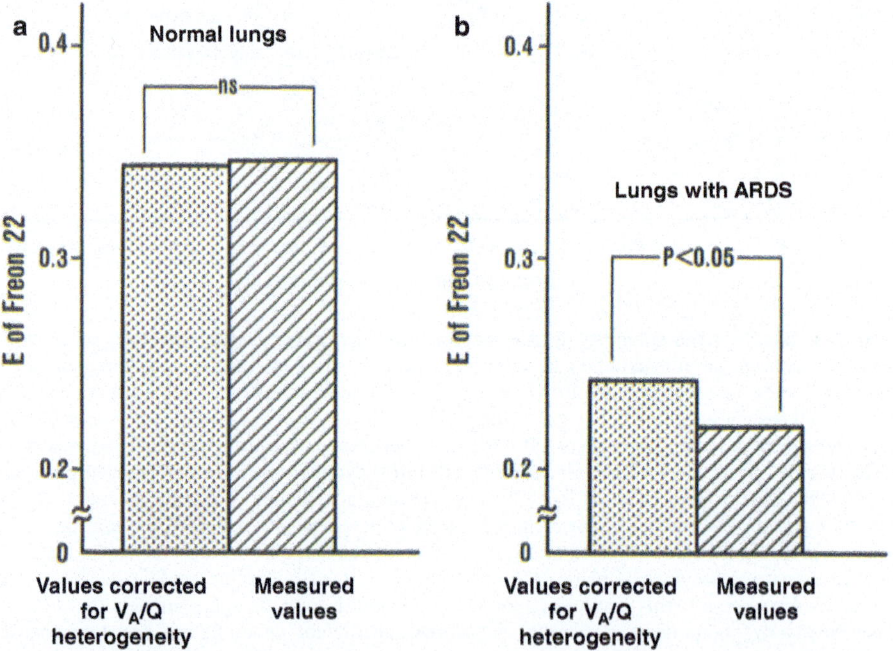

Fig. 6.5 Excretion of freon 22 in normal and ARDS lungs (canine model). There is no difference between the approximated value of freon 22 corrected for V_A/Q heterogeneity and measured values of freon 22 in normal lungs (**a**). However, there is a considerable difference between them in lungs with ARDS (**b**), suggesting that in addition to V_A/Q heterogeneity, aqueous-phase diffusion plays an important role in impairing gas exchange in lungs with ARDS. Modified from Yamaguchi et al. [39] with permission of the publisher

However, the inspiration-related, diffusion-elicited stratified heterogeneity has the effect that is not ignored under an artificial breathing such as breath-holding or rebreathing, both of which are the main maneuvers for measuring diffusing capacities for carbon monoxide (D_{LCO}) and nitric oxide (D_{LNO}). Magnussen et al. [40] measured single-breath D_{LCO} at different breath-holding times in patients with bronchial asthma and found increasing D_{LCO} with decreasing breath-holding time. This is expected from the classical theory that the parallel heterogeneity of inspired gas distribution is not diminished when breath-holding time is substantially short. In COPD patients, however, D_{LCO} decreased with shortening of breath-holding time, sometimes reaching a negative value when breath-holding time is very short. This is thought to be explained from the incomplete gas mixing between inspired and resident gas, which is caused by the augmented inspiration-related stratified heterogeneity in patients with COPD suffering from destruction of acinar airways and airspaces [3]. Recently, the important role of inspiration-related stratified heterogeneity on rebreathing D_{LNO} was confirmed by Hsia et al. [41] in an immature canine model with elongated airways. In immature animals several months after pneumonectomy, the remaining lung rapidly expands into the empty thorax, followed by an accelerated increase in number of airway segments (especially in the acinus), length and diameter of conducting airways, and length of intra-acinar airways. Using this special canine model, Hsia et al. [41] found that the D_{LNO} value measured in the background gas with He and O_2 (lower density but higher kinematic viscosity than N_2-O_2 gas, thus gas-phase NO diffusivity is increased) was high, but that in the background gas with SF_6 and O_2 (higher density but lower kinematic viscosity than N_2-O_2 gas, thus gas-phase NO diffusivity is lowered) was low, in comparison with that observed in the background gas with N_2 and O_2. These findings call for a couple of explanations. The gas-phase resistance to diffusive transport of NO is higher in the SF_6-O_2 background gas due to a decreased diffusivity of NO, leading to the decreased D_{LNO}. However, the SF_6-O_2 background gas has a lower kinematic viscosity, leading to the generation of greater turbulence at airway branches despite a low Reynold's number. This will improve the gas mixing of NO and acts as the factor for increasing D_{LNO}. The net effect of the background gas on D_{LNO} depends on which mechanism dominates. In the case of D_{LNO}, the former prevails over the latter. More recently, Linnarsson et al. [42] examined the effect of atmospheric pressures on breath-holding D_{LNO}. They found that the D_{LNO} measured at 4.0 atm was significantly lower than that at 1.0 atm, indicating that diffusive transport of NO in the lung periphery is inversely related to the gas density. Based on these experimental findings, we concluded that although the diffusion-dependent stratification plays no role in prescribing overall gas mixing within the acini under conditions with normal breathing, it exerts a detrimental impact on acinar gas mixing, i.e., incomplete gas mixing between inspired and resident gas, under artificial breathing conditions such as breath-holding or rebreathing, both being clinically important while measuring diffusing capacities for CO and NO in patients with a variety of lung diseases.

5 Conclusion Remarks

The results obtained from many theoretical analyses based on symmetrical and asymmetrical geometries of the acinus consistently confirm that the importance of molecular diffusion outweighs the importance of ventilation-induced convection for prescribing gas transport and mixing in the acinar region. However, the reality is that the interdependence of diffusion and convection works as the most decisive factor there. The diffusion-elicited spatially different gas concentration in the acinus, i.e., the stratified heterogeneity, is evident during inspiration but it rapidly fades out during expiration under conditions of normal breathing. This shows that stratification takes little effect on the overall gas mixing efficiency during normal breathing. However, the stratification has an impact that cannot be ignored on impeding acinar gas mixing under conditions with artificial breathing such as breath-holding or rebreathing, both of which are the breathing maneuvers employed when measuring the diffusing capacities. Therefore, the values of diffusing capacities measured for patients with different lung diseases should be interpreted with caution, because they are sometimes underestimated by the stratification-dependent, incomplete gas mixing in the acinar region.

References

1. Haefeli-Bleuer B, Weibel ER. Morphometry of the human pulmonary acinus. Anat Rec. 1988;220:401–14. https://doi.org/10.1002/ar.1092200410.
2. Weibel ER. Morphometry of human lung. Berlin: Springer; 1963.
3. Piiper J, Scheid P. Diffusion and convection in intrapulmonary gas mixing. In: Fishman AP, Farhi LE, Tenney SM, Geiger SR, editors. Handbook of physiology, Section 3: The respiratory system, Vol. IV: Gas exchange. Bethesda, MD: American Physiological Society; 1987. p. 51–69.
4. Yamaguchi K, Tsuji T, Aoshiba K, Nakamura H, Abe S. Anatomical backgrounds on gas exchange parameters in the lung. World J Respirol. 2019;9(2):8–28. https://doi.org/10.5320/wjr.v9.i2.8.
5. Hsia CCW, Hyde DM, Weibel ER. Lung structure and the intrinsic challenges of gas exchange. Compr Physiol. 2016;6:827–95. https://doi.org/10.1002/cphy.c150028.
6. Strahler AN. Revision of Horton's quantitative factors in erosional terrain. Am Geophys Union. 1953;34:345.
7. Strahler AN. Quantitative analysis of watershed geomorphology. Trans Am Geophys Union. 1957;38:913–20. https://doi.org/10.1029/TR038i006p00913.
8. Horsfield K, Cumming G. Morphology of the bronchial tree in man. J Appl Physiol. 1968;24:373–83. https://doi.org/10.1152/jappl.1968.24.3.373.
9. Woldenberg MJ, Cumming J, Harding K, Horsfield K, Prowse K, Singhal S. Law and order in the human lung. Official Naval Research Report (AD 709602). Springfield, VA: NTIS, US Department of commerce; 1970.
10. Fredberg JJ, Moore JA. The distributed response of complex branching duct networks. J Acoust Soc Am. 1978;63:954–61. https://doi.org/10.1121/1.381775.

11. Fredberg JJ. Airway dynamics: recursiveness, randomness, and reciprocity in linear system simulation and parameter estimation. In: Chang HK, Paiva M, editors. Respiratory physiology: an analytical approach, vol. 40. New York: Marcel Dekker; 1989. p. 167–94.
12. Swan AJ, Tawhai MH. Evidence for minimal oxygen heterogeneity in the healthy human pulmonary acinus. J Appl Physiol. 2011;110:528–37. https://doi.org/10.1152/japplphysiol.00888.2010.
13. Mitzner W. Collateral ventilation. In: Crystal RG, West JB, Weibel ER, Barnes PJ, editors. The LUNG scientific foundations. Philadelphia, PA: Lippincott-Raven; 1997. p. 1425–35.
14. Lambert MW. Accessory bronchiole-alveolar communications. J Pathol Bacteriol. 1955;70:311–4. https://doi.org/10.1002/path.1700700206.
15. Martin H. Respiratory bronchioles as the pathway for collateral ventilation. J Appl Physiol. 1966;21:1443–7. https://doi.org/10.1152/jappl.1966.21.5.1443.
16. Raskin SP, Herman PG. Interacinar pathways in the human lung. Am Rev Respir Dis. 1975;111:489–95.
17. Scheid P, Piiper J. Intrapulmonary gas mixing and stratification. In: West JB, editor. Pulmonary gas exchange. Vol. I. Ventilation, blood flow, and diffusion. New York: Academic Press; 1980. p. 88–130.
18. Fukuchi Y, Cosio M, Kelly S, Engel LA. Influence of pericardial fluid on cardiogenic gas mixing in the lung. J Appl Physiol. 1977;42:5–12. https://doi.org/10.1152/jappl.1977.42.1.5.
19. Drechsler DM, Ultman JS. Cardiogenic mixing in the pulmonary conducting airways of man? Respir Physiol. 1984;56:37–44. https://doi.org/10.1016/0034-5687(84)90127-0.
20. Paiva M, Engel LA. Influence of bronchial asymmetry on cardiogenic gas mixing in the lung. Respir Physiol. 1982;49:325–38. https://doi.org/10.1016/0034-5687(82)90120-7.
21. Georg J, Lassen NA, Mellemgaard K, Vinther A. Diffusion in the gas phase of the lungs in normal and emphysematous subjects. Clin Sci. 1965;29:525–32.
22. Al Mukahal FHH, Duffy BR, Wilson SK. Advection and Taylor-Aris dispersion in rivulet flow. Proc Math Phys Eng Sci. 2017;473:20170524. https://doi.org/10.1098/rspa.2017.0524.
23. Berezhkovskii AM, Skvortsov AT. Aris-Taylor dispersion with drift and diffusion of particles on the tube wall. J Chem Phys. 2013;139:084101. https://doi.org/10.1063/1.4818733.
24. Chang DB, Lewis SM, Young AC. A theoretical discussion of diffusion and convection in the lung. Math Biosci. 1976;29:331–49. https://doi.org/10.1016/0025-5564(76)90110-3.
25. Nixon W, Pack A. Effect of altered gas diffusivity on alveolar gas exchange—a theoretical study. J Appl Physiol Respir Environ Exerc Physiol. 1980;48:147–53. https://doi.org/10.1152/jappl.1980.48.1.147.
26. Verbanck S, Paiva M. Gas mixing in the airways and airspaces. Compr Physiol. 2011;1:809–34. https://doi.org/10.1002/cphy.c100018.
27. Gomez DM. A physico-mathematical study of lung function in normal subjects and in patients with obstructive pulmonary diseases. Med Thorac. 1965;22:275–94. https://doi.org/10.1159/000192413.
28. Scherer PW, Shendalman LH, Greene NM. Simultaneous diffusion and convection in single breath lung washout. Bull Math Biophys. 1972;34:393–412. https://doi.org/10.1007/BF02476450.
29. Paiva M. Gas transport in the human lung. J Appl Physiol. 1973;35:401–10. https://doi.org/10.1152/jappl.1973.35.3.401.
30. Davidson MR, Fitz-Gerald JM. Transport of O_2 along a model pathway through the respiratory region of the lung. Bull Math Biol. 1974;36:275–303. https://doi.org/10.1007/BF02461329.
31. Wilson A, Lin K-H. Convection and diffusion in the airways and the design of the bronchial tree. In: Bouhuys A, editor. Airway dynamics. Springfield, IL: Thomas; 1970. p. 5–19.
32. Rauwerda PE. Unequal ventilation of different parts of the lung and the determination of cardiac output. PhD Thesis. University of Groningen, The Netherlands; 1946.
33. Paiva M, Engel LA. The anatomical basis for the sloping N_2 plateau. Respir Physiol. 1981;44:325–37. https://doi.org/10.1016/0034-5687(81)90027-X

34. Weibel ER, Federspiel WJ, Fryder-Doffey F, Hsia CCW, König M, Stalder-Navarro V, et al. Morphometric model for pulmonary diffusing capacity I. Membrane diffusing capacity. Respir Physiol. 1993;93:125–49. https://doi.org/10.1016/0034-5687(93)90001-Q.
35. Paiva M, Engel LA. Theoretical studies of gas mixing and ventilation distribution in the lung. Physiol Rev. 1987;67:750–96. https://doi.org/10.1152/physrev.1987.67.3.750.
36. Paiva M. Gaseous diffusion in an alveolar duct simulated by a digital computer. Comput Biomed Res. 1974;7:533–43. https://doi.org/10.1016/0010-4809(74)90030-5.
37. Van Muylem A, Noël C, Paiva M. Modeling of impact of gas molecular diffusion on nitric oxide expired profile. J Appl Physiol. 2003; 94:119–27. https://doi.org/10.3917/deba.127.0094
38. Adaro F. Separation of stratification effects from unequal distribution effects by study of inert gas exchange. Bull Physio Pathol Respir. 1972;8:164–5.
39. Yamaguchi K, Mori M, Kawai A, Takasugi T, Asano K, Oyamada Y, et al. Ventilation-perfusion inequality and diffusion impairment in acutely injured lungs. Respir Physiol. 1994;98:165–77. https://doi.org/10.1016/0034-5687(94)00061-1.
40. Magnussen H, Holle JP, Hartmann V, Schoenen JD. D_{CO} at various breath-holding times: comparison in patients with chronic bronchial asthma emphysema. Respiration. 1979;37:177–84. https://doi.org/10.1159/000194024.
41. Hsia CCW, Yan X, Dane DM, Johnson RL Jr. Density-dependent reduction of nitric oxide diffusing capacity after pneumonectomy. J Appl Physiol. 2003;94:1926–32. https://doi.org/10.1152/japplphysiol.00525.2002.
42. Linnarsson D, Hemmingsson TE, Frostell C, Van Muylem A, Kerckx Y, Gustafsson LE. Lung diffusing capacity for nitric oxide at lowered and raised ambient pressures. Respir Physiol Neurobiol. 2013;189:552–7. https://doi.org/10.1016/j.resp.2013.08.008.

Chapter 7
Radiological Evaluation of Lower Airway Dimensions Deciding Ventilatory Dynamics: Can Radiologically Determined, Static Airway Structures Precisely Predict Ventilatory Dysfunction?

Susumu Sato and Toyohiro Hirai

Abstract Airway narrowing causes airflow limitation because of the increase in airway resistance during expiration. Radiological evaluation of airway narrowing during breath-hold has been widely performed for various respiratory diseases. In particular, bronchial asthma and chronic obstructive pulmonary disease (COPD) are major respiratory diseases that present with airflow limitations. Quantitative assessment using computed tomography (CT) images is key to evaluating airway lesions and airway narrowing in both clinical and research fields.

On CT images, airway wall thickening, and airway narrowing are evaluated with designated software; however, there are serious concerns about and limitations to the resolution of CT images. Ultra-high resolution CT (U-HRCT) is a promising variation of CT that could produce new findings to help deepen our understanding of respiratory diseases.

Moreover, for acquiring precise and meaningful measurements of airway dimensions in vivo, various new parameters, such as airway tree shape irregularity, airway visibility, and airway volume to lung volume ratio, have been proposed.

Keywords Bronchial asthma · Chronic obstructive pulmonary disease · High-resolution CT · Ultra-high resolution CT · Airflow limitation

---

S. Sato (✉) · T. Hirai
Graduate School of Medicine and Faculty of Medicine, Kyoto University, Sakyo-ku, Kyoto, Japan
e-mail: ssato@kuhp.kyoto-u.ac.jp

K. Yamaguchi (ed.), *Structure-Function Relationships in Various Respiratory Systems*, Respiratory Disease Series: Diagnostic Tools and Disease Managements, https://doi.org/10.1007/978-981-15-5596-1_7

1 Introduction

1.1 Anatomy of the Lower Airways

As described in the previous chapter, human airways consist of tracheobronchial trees and upper airways (Fig. 7.1). Since inspiratory and expiratory air must go through these airways, airway structure, pathology, and anatomical geometry are important determinants of the physiological properties of the respiratory system. Therefore, in respiratory diseases, the anatomical properties of the airways are key components in determining and characterizing ventilatory dysfunction.

In particular, airway narrowing causes airflow limitation because of the increase in airway resistance during expiration and inspiration. Exhaled air must go through the airway pathways from the lungs, and airway narrowing results in decreasing airflows. Decreasing airflows are defined as "obstructive ventilatory dysfunction." Various diseases, including bronchial asthma, chronic obstructive pulmonary disease (COPD), and bronchiectasis or other diseases, may present such a clinical manifestation. In particular, patients with bronchial asthma may have chronic airway inflammation and consequently present airway wall thickening (Fig. 7.1b).

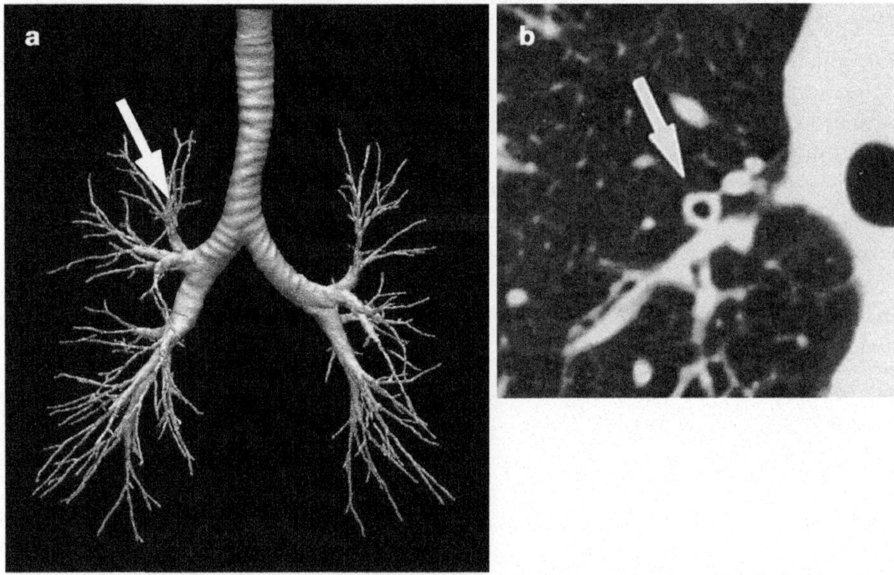

Fig. 7.1 (**a**) Airway tree. The arrow indicates the right apical bronchus (RB1). (**b**) Cross-sectional CT image of the RB1. The arrow indicates the RB1

1.2 Principle of Radiological Evaluation of Airways

Quantitative evaluation of the morphological changes of the respiratory tract using computed tomography (CT) has been performed since the 1990s. Initially, studies were conducted using experimental animals, followed by studies in which an evaluation of the airways of near-fatal asthma patients was attempted. A number of subsequent reports evaluated dynamic airway contraction in healthy subjects and asthmatic patients with stimulation using methacholine and other chemicals.

For these analyses, airways were typically measured in cross-section, and measurements were originally performed manually [1, 2]. It soon became obvious that there were measurement errors associated with visualizing features of CT images as well as inter- and intra-observer variations.

In addition, both cross-sectional and longitudinal studies require a comparison of scans at identical airway levels/generations. Especially in a longitudinal study, both preintervention and postintervention evaluations should be performed for each patient to compare the effect of these interventions on airways.

To minimize measurement errors, quantitative analysis using automated computer-aided software is required. Thanks to developments in CT scanning technology, such as helical scans and HRCT scans, image analysis has become popular in clinical and research fields. From the 2000s to the 2010s, numerous studies had been performed using quantitative image analysis of airway dimensions. For example, Nakano et al. proposed a new computer-aided algorithm to evaluate airway dimensions using helical CT scan images [3]. They found that the severity of emphysematous lesion and changes in airway dimensions (wall thickness and percentage of wall area) independently correlated with measures of airflow obstruction. Likewise, various studies revealed that airway measurements using chest CT is poised to make a major contribution to the understanding of obstructive airway diseases.

However, radiological evaluations of airway generations are still conducted manually and with visual confirmation almost exclusively; therefore, these measurements are limited to at least the visible airways. Indeed, it had become obvious that the visibility of even the fifth- and sixth-generation airways was impaired in COPD [4], and thinner airway walls are almost invisible, making spatially matched comparisons of these airway dimensions extremely challenging.

Recently, a great step forward in CT scanner development was achieved. A more precise image acquisition technique was implemented in a human study, which yielded an ultra-high resolution with up to a 1024 × 1024 (or a virtual 2048 × 2048 after additional processing) pixel matrix and a 0.25 mm slice thickness.

Quantitative airway analysis using CT may take us into another world.

2 Morphometry of Airways

The desire to acquire precise measurements of airway dimensions in vivo is driven by the goal to make meaningful measurements of airway wall remodeling in obstructive pulmonary diseases, such as asthma and COPD.

2.1 Indexes of Airway Geometry

Indexes of airway geometry are listed in Table 7.1. In particular, airway narrowing is represented by a decrease in the airway luminal area (Ai), and airway remodeling could be represented by an increase in the airway wall area (WA). Consequently, an increase in the percentage of the airway wall area to the total airway area (Ao) is presented as WA%.

2.2 Identification of Inner and Outer Borders

To perform the morphometry of airways, identification of the inner area and outer field is key. As illustrated in Fig. 7.2 [5], airways look like tubes in the lung parenchyma, which may also present with neighboring structures, such as vessels.

CT indices, such as the ratio of the airway wall area to the total airway wall area (WA%) and the luminal area (Ai), have been used for the quantitative analysis of

Table 7.1 Airway parameters measured and derived using quantitative computed tomography

Measured parameters	Confounding factors	Derived parameter	Reference
Luminal area (Ai)	Spatial resolution, image noise, inspiration level	Airway wall area% (WA%)	[1–3]
Airway wall area (WA)		AWT Pi10 = $\sqrt{WA}$ Pi of 10 mm calculated using regression of measured Pi vs. $\sqrt{WA}$	
Total airway area (Ao)			[10]
Airway wall thickness (WT)			
Internal diameter (ID)			
Internal perimeter (Pi)			
Total airway count (TAC)			[4]
Airway tree volume		Airway volume ratio (AWV%)	[30]
Branch angle	Continuous images, spatial resolution, image noise		
Segment length			
SD_Ri			[25]

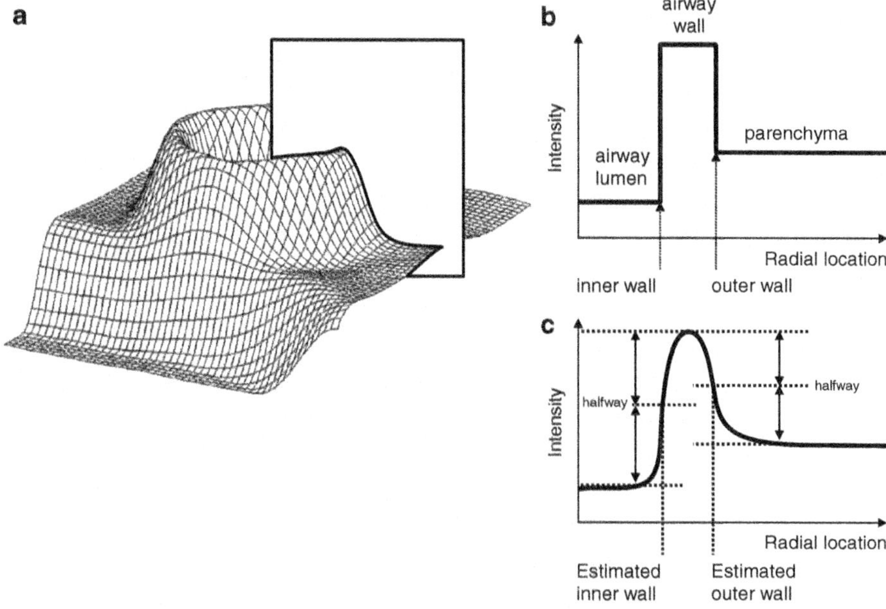

Fig. 7.2 (**a**) Representation of the X-ray density of a tube with sharp edges on a single CT slice as a topographical relief. This tube was surrounded by two different materials (adopted from Weinheimer et al. [5]) Diagram (**b**) shows a line profile across an ideal and a blurred airway wall and (**c**) illustrates how the FWHM technique is applied. *FWHM* full-width-at-half-maximum

airway thickening and narrowing. In addition, airway size may present as the equivalent diameter of the airway (inner diameter, ID).

2.3 *Artifacts Resulting from Oblique Airway Directions* [6]

Since most previous studies used axial cross-sections of the body and, therefore, of the bronchi, control of the airway orientation in the Z-axis during scanning would have been difficult; therefore, the right apical segmental bronchus (RB1) and right basal segmental bronchi (RB8-10) were typically chosen, because these airways are usually sliced in axial cross-section. However, such airways may also be affected by oblique directionality with respect to the Z-axis [7].

Recently, 3-D image analysis has become more widely used, including methods such as MPR image acquisition, and techniques to create perpendicular slices to the airway directions [6] could then be applied to airways acquired with these imaging modalities. In phantom studies, this technique can resolve artifacts from undetected oblique airway directions; however, such complicated computer processing is difficult to apply in clinical settings.

2.4 *Visual Evaluation*

A number of visual scoring systems have been developed to assess the extent and distribution of airway abnormalities [8]. These scoring systems have been applied to various diseases. Each scoring system consists of grading airway wall thickening, bronchiectasis, mosaic perfusion, and/or gas trapping. However, qualitative studies are sensitive to the display settings (e.g., window width and level) of the images and are prone to poor reproducibility.

2.5 *Quantitative Evaluation*

2.5.1 Manual Analysis

In 1992, McNamara, Okazawa and colleagues proposed methods to quantify airway geometry in animal and human lungs [1, 9]. CT images were enlarged using image analysis software, and regions of interest (ROI) were traced manually along the internal border (the luminal border of the airway) and the external border (the parenchymal border). In airways where a vessel abutted the external border, an extrapolated line was traced on the assumption that the airway wall thickness was constant throughout the areas of vascular contact. After the luminal area and total airway area were determined, the wall area, airway wall thickness, and WA% could be calculated.

2.5.2 Computer-Aided Analysis [3, 10]

Since manual tracing is extremely time-consuming and prone to error, computer-aided and automated techniques have been developed. The first methods used a Hounsfield unit (HU) threshold cut-off value, such as −500 HU, and most common methods reported employ "full-width-at-half-maximum (FWHM)" techniques (Fig. 7.3). FWHM methods can provide standardized and unbiased measurements.

In 2000, Nakano, Mishima and their colleagues tested their hypothesis that the measurement of airway dimensions would provide additive value to the measurement of emphysema in the prediction of lung function abnormalities. They used modified FMHM methods in their study and employed thin-section helical (spiral) CT to quantify airway dimensions. Unlike later and very recent studies, only one airway, the trunk of the apical bronchus of the right upper lobe, was evaluated. As mentioned above, the right apical bronchus is typically sliced in cross-section and easily identified on CT scans; in addition, it was large enough to be accurately measured. With the identified bronchus, the luminal area (Ai), airway wall area, and airway wall thickness (WT) were calculated automatically.

Their computer-aided procedure consisted of various steps to identify adjacent structures, such as vessels (Fig. 7.3b–e). Each step is subject to modifications of original optimization methods to adjust computer processing, image quality, and acquisition settings [3]:

1. The lumen of the bronchus was identified using a threshold of −500 HU. The area of the lumen was considered Ai (Fig. 7.4b).

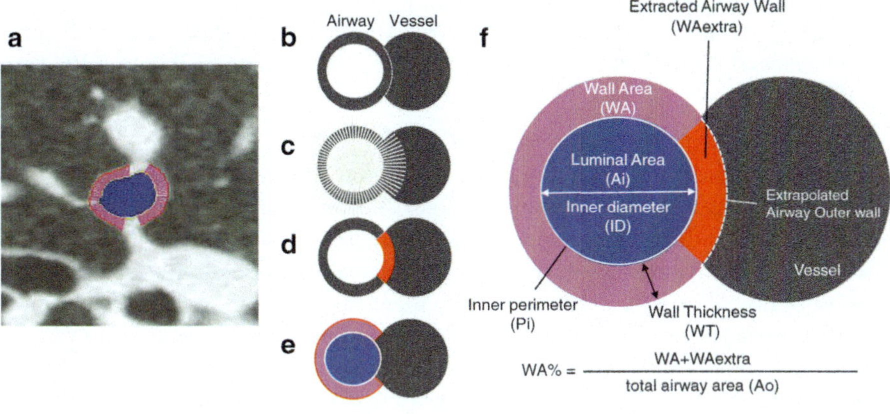

Fig. 7.3 Measurement of airway dimensions. (**a**) An example of the measurement of airway dimensions using in-house software. (**b**–**e**) Step-by-step schematic of airway measurements. (**f**) Indexes of airway dimensions

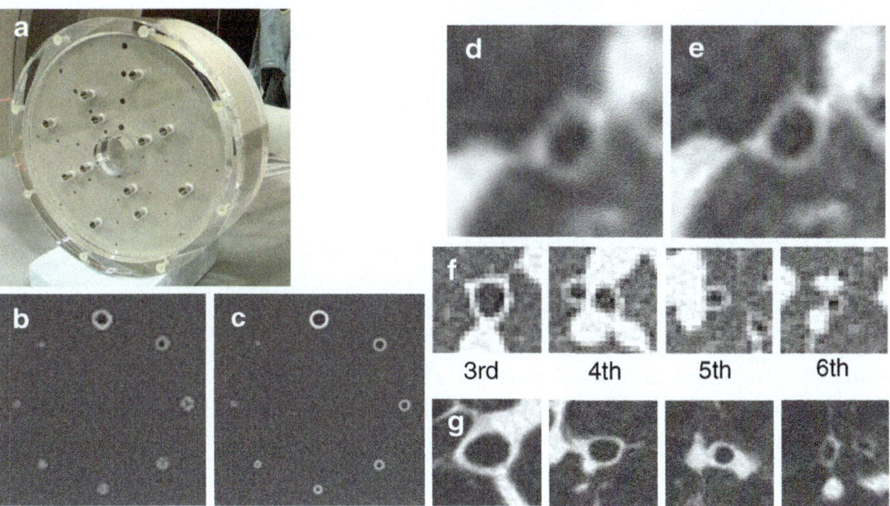

Fig. 7.4 Ultra-high resolution CT analysis. (**a**) Phantom with small and thin tubes. (**b**, **c**) Axial CT images of phantoms. (**d**, **e**) Examples of a chest CT scan obtained using conventional HRCT and U-HRCT, respectively. (**f**, **g**) Representative cross-sectional images in each airway generation from 3rd to 6th

2. The inner diameter was calculated as the equivalent diameter, which is equal to the diameter of a circle with the same Ai (Fig. 7.4f).
3. From the centroid of the lumen, 128 (or 64 [10]) rays fanning out over 360° were examined to determine WT along each ray using the "full-width-at-half-maximum" principle (Fig. 7.3c).
4. The rays that projected onto the adjacent vessel were excluded if their WTs exceeded the threshold (mean +1 SD of all WTs). WT was then calculated from those of the included rays (Fig. 7.3d, e). The excluded airway wall was determined as WAextra. The total airway area (Ao) consisted of Ai, WA, and WAextra (Fig. 7.3f).

This multiple step-by-step procedure can eliminate overestimation of the outer airway wall due to including adjacent vessels or other connective tissues in the lungs. Unfortunately, FWHM methods are still capable of over- and underestimating airway walls due to factors including the spatial resolution of the CT scanner, the angle of airway orientation, the edge detection ability of the CT scanner, and the reconstruction algorithm used. Therefore, these methods are subject to the limitations of airway sizes, which deserves the attention of researchers. Nevertheless, similar and modified methods and techniques are included by many manufacturers that provide commercially available software packages nowadays. These software packages can also create multiplanar reformation (MPR) images and cross-sections vertically to the airway direction, then automatically calculate airway dimensions at various sites (with or without manual selection of airways/pathways). Using these packages, attempts have been made to quantitatively measure the third- (segmental) and fourth- (subsegmental) generation airways [3, 11–13] and even the fifth- or sixth-generation airways in some cases [14, 15]; these studies showed that luminal narrowing in these airways is significantly associated with airflow limitation. From these results, it is evident that the evaluation of the airway dimension yields certain information for predicting ventilatory dysfunction in patients with obstructive pulmonary diseases.

2.6 Conventional Versus Ultra-High Resolution CT

2.6.1 Standard HRCT (Conventional HRCT)

Currently, standard high-resolution CT (HRCT) can provide images with a 512 pixel × 512 pixel size and ≥0.5 mm slice thickness. Standard HRCT can be used to measure airway dimensions from axial and MPR images in both central and proximal airways. Various studies have already presented a significant association of airway narrowing and airway wall thickening with airflow limitation; however, these findings might be subject to potential bias from the limited image resolution.

2.6.2 Phantom Study for Standard HRCT

As previously mentioned, CT analysis is subject to systematic errors due to limitations in image resolution. Therefore, phantom studies are necessary, especially in small airways or for thinner airway walls. Several phantom studies have revealed serious systematic errors in the analysis of smaller airways. For example, Oguma and colleagues conducted a phantom study using four different CT scanners [6]. The results from an Aquilion 64 (Toshiba, Tokyo, Japan), the primary CT unit, were then compared to those of the other three CT scanners (Light Speed VCT, GE Healthcare UK, Buckinghamshire, UK; Brilliance 64, Philips, Eindhoven, Netherlands; and SOMATOM Definition, Siemens, Munich, Germany). In this study, they tested variability and measurement errors in airway dimensions, luminal area (Ai), and wall area percentage (WA%). The authors found that for small airways with a wall thickness of <1.0 mm, the measurement accuracy according to quantitative CT parameters can decrease as the walls become thinner. Therefore, they concluded that airway wall thickness had the most profound effect on the accuracy of measurements with all CT scanners under all scanning conditions, with the magnitude of the errors for WA% and Ai varying depending on the wall thickness of airways with wall thicknesses <1.0 mm.

As previously mentioned, this study confirmed that the parameters of the measured airway dimensions are affected by airway size, the reconstruction algorithm, the composition of the airway phantom, and the type of CT scanner. Moreover, the results from numerous previous studies confirmed that thinner and small airway dimensions had more profound effects than those of larger airways, which could result in inevitable and serious systematic errors in the measurement of airway dimensions. We must be more careful in extrapolating these results to clinical applications.

2.6.3 Ultra-High Resolution CT (U-HRCT)

A recently introduced ultra-high resolution computed tomography (U-HRCT) appears to be a promising technique for overcoming these resolution limitations because of its ability to provide images of living persons at 1024 × 1024 pixel resolution and a 0.25 mm slice thickness without an increase in radiation dose. As a result, U-HRCT allows a clearer visualization of lung structures compared to conventional CT scans [16]. Therefore, even in smaller airways, precise measurement of airway dimensions can be achieved due to the finer image resolution, and therefore, U-HRCT has the potential to allow the direct evaluation of small airways 1–2 mm in diameter and to provide unbiased information regarding their clinical and physiological impacts.

2.6.4 Phantom Study for U-HRCT

Phantom tube analysis was conducted using both conventional/standard (512 × 512 pixel) and ultra-high resolution (U-HRCT; 1024 × 1024 pixel) scans [17]. Reconstructions were performed using the forward-projected model-based iterative reconstruction solution (FIRST) algorithm for the 512 × 512 and 1024 × 1024 pixel resolution scans and the adaptive iterative dose reduction 3-D (AIDR-3D) algorithm for all scans.

Compared to the conventional 512 × 512 pixel scan, variations in the CT values for air were increased in the U-HRCT scans, except in the 1024 × 1024 pixel scan reconstructed through FIRST. The measurement error in the lumen area of tubes with a 2-mm diameter and a 0.5-mm wall thickness was minimal in the U-HRCT scans but not in the conventional 512 × 512 pixel scan. In contrast to the conventional scans, the U-HRCT scans resolved phantom tubes with a ≥0.6-mm WT at an error rate of <11%. Airways with a diameter of 1–2 mm could be accurately quantified by U-HRCT but not by conventional CT. Therefore, we can conclude that the U-HRCT scan may allow more accurate measurement of bronchioles with smaller dimensions compared with conventional scans [17] (Fig. 7.4b, c).

The lower limit of quantitation for the lumen size of smaller airways on U-HRCT remains to be established; therefore, a second airway phantom with a smaller and thinner tube wall was developed (Fig. 7.4a) and the performance between a U-HRCT scanner (Aquilion Precision, Cannon Medical, Tokyo, Japan) and a conventional HRCT scanner (Aquilion One, Cannon Medical, Tokyo, Japan) was compared [18]. Figure 7.4a shows images illustrating the lumen of the phantom tubes. The lumen of tubes 2 mm down to 1 mm in internal diameter was clearly visualized on U-HRCT (Fig. 7.4c) but not on conventional CT (Fig. 7.4b).

As a result, errors noted in airway dimension measurement were greater in the tubes of small inner radius made of a material with a high CT density and on images reconstructed with a body algorithm ($p < 0.001$), and there were some variations in error between the CT scanners under different fields of view.

The errors for tubes ≥1.3 mm and 1.0 mm in diameter on U-HRCT were <10% and –24%, respectively [18].

2.6.5 Clinical Study for U-HRCT

It should be noted that artificial tubes are not identical to in vivo airways; however, the results of the present comparative study can still be used to confirm the image quality and accuracy of U-HRCT scans compared to standard/conventional HRCT. Additional clinical study in which patients with COPD were scanned with the Aquilion Precision scanner using the same scanning parameters and autoexposure control was conducted [18]. Figure 7.4 shows representative images obtained by U-HRCT (Fig. 7.4e) and conventional HRCT scanning (Fig. 7.4d). As the images show, airway visualization is much better in the U-HRCT images from the third- to the sixth-generation airways. We will discuss data from this clinical study later in this chapter (see Sect. 3.2).

Moreover, it is obvious that the visualization of the lumen of the smaller phantom tubes on conventional CT was quite poor. This indicates that CT radiologists should be aware of the possibility that a small nodule on conventional CT might reflect small airways (Fig. 7.4b, c).

3 Studies and Applications of Airway Measurements in Respiratory Diseases

3.1 Asthma

Among respiratory diseases, bronchial asthma was the first target of in vivo airway dimension quantification. This is because asthma lesions exist only in the "airways," and assessments of dynamically changing lung functions could be used to evaluate airflow obstructions but not to quantify underlying airway lesions. Asthma is characterized by chronic airway eosinophilic inflammation, and recently "airway remodeling" and airway wall thickening have attracted much attention. Therefore, in evaluating remodeled airways, quantitative and less invasive image analysis is required, and such information cannot be detected by usual pulmonary function tests.

In human asthma, visual evaluation of CT images was first performed in the 1990s. However, as CT resolution and spatial resolution have improved, the accuracy of such evaluations has improved, thus opening up the path to precise quantitative evaluation.

3.1.1 Airway Remodeling

Airway wall thickening has been considered a sign of airway remodeling [2]. Niimi et al. conducted a quantitative analysis in asthmatic patients and found that airway wall thickening was significantly more observed in severe persistent asthma patients and also associated with the severity of asthma control, duration of the disease, and airflow limitation.

Additionally, wall thickening was assumed to cause airway hyperresponsiveness and to be related to asthma severity [19]. In fact, a protective effect against airway narrowing has also been suggested. To date, thickened airway walls in COPD patients have been considered a notable sign of asthma-COPD overlap (ACO).

3.1.2 Ventilatory Dysfunction

Considering the association of airway dimensions and ventilatory dysfunction in asthmatic patients, airway wall thickening is significantly but weakly correlated with the severity of airflow limitation (%predicted FEV_1) [2]. This is nevertheless undesirable because the condition of asthmatic airways is quite reversible and may show dynamic day-to-day variation and treatment response. Moreover, the site of airway obstruction may vary within patients. To address these concerns, a thorough analysis of multiple airways must be performed.

3.2 Chronic Obstructive Pulmonary Disease (COPD)

COPD is a major respiratory disease and a leading cause of death worldwide. COPD is characterized by the presence of airflow obstruction caused by parenchymal destruction (emphysema), airway narrowing, or both. Since coexisting emphysema has been easily identified by chest CT images and is correlated with various respiratory dysfunctions and symptoms, such as impaired diffusing capacity, hyperinflation, hypoxemia, and hypercapnia, much attention has been paid to quantifying the severity of emphysematous lesions, and many studies have elucidated the clinically important associations among emphysematous lesions, airflow limitations, and respiratory symptoms. In particular, low attenuation areas/voxels (LAAs or LAVs) on computed tomography (CT) have been shown to represent macroscopic or microscopic emphysema, or both.

3.2.1 Clinical Phenotypes

Despite the attention given to emphysematous lesions on chest CT, Nakano et al. presented a significant association between airway wall thickening (WA%) of the right apical bronchus and airflow limitation, and airway wall thickening was found to contribute independently to airflow limitation in addition to parenchymal destruction (emphysema) [8]. They also proposed "CT-derived phenotypes," such as emphysema-dominant, airway remodeling-dominant, and mixed [20]. In fact, there are apparent clinical phenotypes, known as "pink puffers" and "blue bloaters," among patients with COPD. In the literature, blue bloaters may represent the airway lesion-dominant phenotype [21].

Unfortunately, blue bloaters may present with many bronchitis symptoms (chronic bronchitis (CB) symptoms); these CB symptoms are less associated with airway wall thickening. In 2010, Grydeland et al. reported a significant association between airway wall thickening and CB symptoms in more than 400 patients with COPD [22].

3.2.2 Site of Airflow Obstruction

Because the major site of airflow obstruction in obstructive pulmonary diseases is unknown, much of the concern regarding quantitative image analysis of the airways is determining the location of the key lesion of airflow obstruction. Various studies have evaluated the different airway generations, such as the third- (segmental) and fourth- (subsegmental) generation airways [3, 11–13], and even the deeper fifth- and sixth-generation airways [14, 15] in patients with COPD.

To investigate the site of obstruction, most studies evaluate luminal area (Ai) rather than wall thickness (WA%, WT, or WA/BSA, etc.) because smaller airways may have thinner airway walls, making it impossible to accurately quantify their wall thickness.

Reports of quantitative analysis using 3-D MPR image acquisition techniques indicate that airflow limitation in COPD is more closely related to the dimensions of the small airways, such as the fifth to sixth generations, than those of the proximal (third) airways. These observations are quite consistent in reporting that small airways more strongly contribute to airflow limitation in COPD than large airways; however, in recent study using U-HRCT [17], more than half of the airways (70 and 62%) located at the sixth generation of the right apical bronchus (RB1) and the right basal bronchus (RB10) path, respectively, were <2 mm in diameter, which indicated that conventional CT is not appropriate to quantitatively measure the sixth-generation airways.

Furthermore, the finding that U-HRCT failed to visualize the lumen of a phantom tube 0.5 mm in diameter (Fig. 7.4c) suggests that the decrease in the number of sixth-generation airways in COPD might be because the sixth generation may include extremely small airways <0.5 mm in diameter or because those airways do not actually exist.

Consequently, among the correlations between FEV_1 and the mean and sum of the Ai of all measurable bronchi, the correlation with the sum of the Ai of the sixth-generation airways had the greatest correlation coefficients.

Moreover, multivariable linear regression analysis showed that the sum of the Ai in the sixth-generation airways of both the RB1 and RB10 segments was associated with $\%FEV_1$ independent of the severity of emphysema (LAV%). These findings suggest that the sum of the Ai of the sixth-generation airways can be a promising CT biomarker that sensitively reflects airflow in COPD.

Furthermore, the finding that the association between the sum of the Ai and FEV_1 was stronger in the sixth generation than in the fourth and fifth generations supports the hypothesis that small airways more strongly contribute to airflow limitation in COPD than large airways. Because of the limits of the resolution of U-HRCT, it is still possible that more distal and smaller airways, such as the seventh- or eighth-generation airways, have a stronger contribution to ventilatory dysfunction. We must wait until the development of a CT scanner to procide images with higher spatial resolution in the future.

3.2.3 Heterogeneity of Airway Lesions

Even with the use of automated airway detection, it is only possible to make a limited number of measurements in any individual at any time, and thus, the issue of heterogeneity in airway dimensions is important for both between- and within-subject comparisons. Moreover, distal airways in the left lungs are prone to motion artifacts caused by cardiac beating rhythms.

Very recently, 320-row multidetector CT (MDCT) scanners were introduced. This kind of scanner can continuously scan the thorax under free-breathing conditions, first visualizing the respiratory motion of the lung and bronchi directly, which can be analyzed quantitatively using designated software or workstations [23]. It is possible to scan with ECG gating methods to eliminate artifacts from cardiac beating; however, it is still difficult to manipulate 320-row MDCT scanners to perform the typical quantification of airway diseases.

In fact, these scanners have already presented several interesting findings, such as the asynchrony of respiratory movements in the lung lobes and impaired synchrony between airway Ai and mean lung density [24]. Further development of designated software is keenly awaited.

4 Further Analyses of Airway Dimensions

4.1 Shape Irregularity [25]

Bronchial asthma and COPD may have significant features, such as "airway remodeling," which may occur heterogeneously along the longitudinal pathway. Matsuoka and colleagues reported variations in various measures of the airway lumen and wall dimensions in normal subjects within individual scans and as a function of distance from the hilum to the periphery.

The evaluation of longitudinal airway lumen shape irregularities may be a new biomarker for characterizing ventilatory dysfunction in these patients, because remodeled airway structure changes may vary in the longitudinal plane.

Oguma and colleagues developed a new technique to investigate longitudinal airway lumen shape irregularities and demonstrated that the patients with COPD showed significantly greater shape irregularity than patients with bronchial asthma and healthy controls [25].

In addition, airway wall thickening was significantly greater and the luminal area was smaller in patients with bronchial asthma than in those with COPD and healthy controls. These results were consistent even among the bronchial asthma and COPD subgroups with similar airflow limitations.

The combination of cross-sectional and longitudinal airway structure analyses using CT images may suggest differences in the characteristics of airway remodeling between COPD and asthma.

4.2 *Prediction of Small Airway Dimensions Using Histological Sections* [10]

In many airway diseases, the most important site of airflow obstruction is the small airways [26]; however, as mentioned previously, there are large errors associated with the measurements of smaller airways. To overcome this limitation, Nakano et al. compared airway measurements from CT scans and histological examinations of excised lungs from smokers who had various degrees of airway obstruction. They found that the wall areas of small airways approximately 1 mm in diameter measured histologically and the wall area percentage of larger airways with a mean internal diameter of more than 3.0 mm were significantly related ($R^2 = 0.57$).

Therefore, at least for COPD, measuring larger airway dimensions can provide an estimate of small airway remodeling. From these associations, these authors proposed a new index of airway dimensions that represents small airway dimensions, namely, AWT-Pi10. Moreover, these associations suggest that the same pathophysiological processes that cause small airway obstruction also occur in larger airways.

4.3 *Visibility of Airways and Airway Counts*

Recent studies have revealed that the visibility of airways on chest CT images decreases as the severity of COPD increases [4, 27]. This can be seen in the 3-D rendered images of airway trees (Fig. 7.1a).

Tanabe et al. also reported that the airway counts for the sixth-generation airways of the RB1 and RB10 on U-HRCT decreased as FEV_1 was decreased. However, the correlation coefficients for these relationships were relatively small ($r = 0.28$ and 0.30, respectively). Airway counting is not an actual analysis of airway dimensions; however, a decrease of the airway inner lumen corresponds to a decrease in airway counts. Thus, the airway counts could be a parameter of airway dimension.

Moreover, this result can extend previous findings showing that total airway count on conventional CT and the number of terminal bronchioles on micro-CT was decreased in COPD compared to controls [28, 29].

4.4 *Airway Volume Percent (AWV%)* [30]

A decrease in airway lumen size and in airway counts results in a decrease in the total volume of the airway tree. Increased lung volume (hyperinflation) in the parenchyma is an important and major structural change in COPD (Fig. 7.5a, b). It is unclear whether structural changes in both the airway tree and the parenchyma occur simultaneously, and since the outer wall of the airways is connected with the lung parenchyma, the inspiration level also affects airway dimensions [31]. Therefore, the mechanical properties and dynamic conditions of the lung parenchyma affect the behavior of the airways, and it can be speculated that these interactions may play important roles in COPD pathophysiology.

We examined these effects by establishing a novel computed tomography (CT) index, namely, airway volume percent (AWV%), which is defined as the percentage ratio of the volume of the airway tree to lung volume. The whole airway tree was automatically segmented, and the percentage ratio of the airway tree volume was calculated as the AWV% using SYNAPSE VINCENT software (FUJIFILM Medical; Tokyo, Japan).

Interestingly, AWV% decreased as the Global Initiative for Chronic Obstructive Lung Disease (GOLD) spirometric grade increased (Fig. 7.5c, $p < 0.0001$). AWV% was more closely correlated with FEV1 (Fig. 7.5d) and other pulmonary functions than established CT indexes. Moreover, multivariate analyses showed that a lower AWV% was associated with a lower FEV_1 independent of the severity of emphysema and airway wall thickness or total airway count (TAC). AWV% is an easily measured CT biomarker and may elucidate the clinical impacts of the airway–lung interaction in COPD [30].

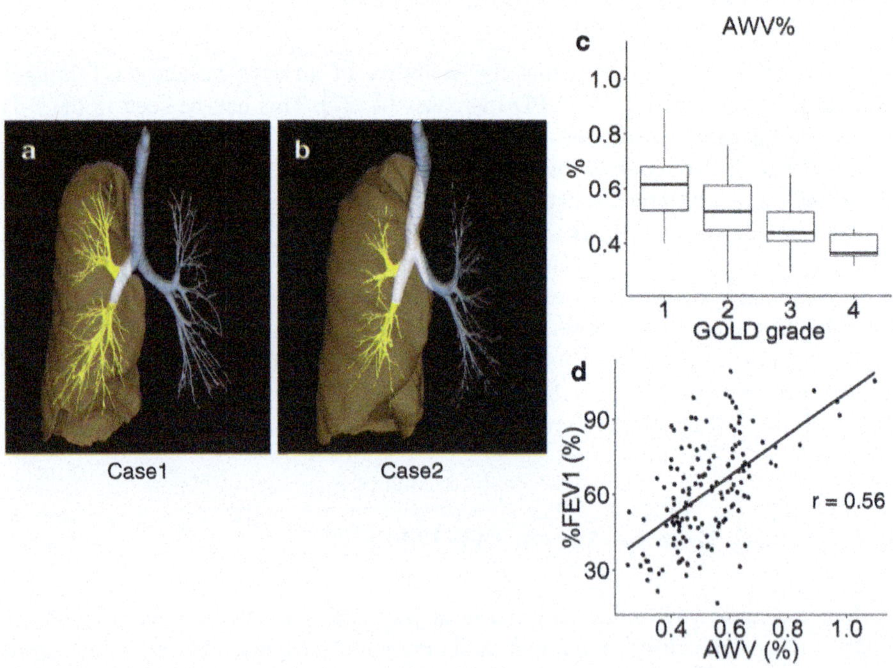

Fig. 7.5 Analysis of airway volume ratio (AWV%) in COPD [25]. (**a, b**) Representative 3-D renderings of the airway tree and right lung in 2 COPD cases. Case 1 shows mild airflow obstruction (**a**), and case 2 shows severe obstruction (**b**). (**c**) Distribution of AWV% among different GOLD spirometric grades. (**d**) Correlation between AWV% and $\%FEV_1$

5 Conclusion

While various studies have attempted to evaluate single or multiple airways at single or multiple sites, the most important issue is the accuracy in estimating the airway walls, identifying both internal and external borders. To achieve this, a more accurate image acquisition technique with a relatively lower radiation dose is required. Therefore, ultra-high resolution CT is quite promising; however, even with automated processing by commercially available software, we can speculate that the total airway tree volume and its ratio to lung volume (AWV%) may provide more accurate predictions of ventilatory abnormalities.

Ultra-high resolution CT evaluation of the small airway dimensions may be able to reveal more precise information on these airways; combined with a direct evaluation of the small airways, these analyses should be incorporated in future studies in asthma and COPD. Neverthless, "static" image aquisition technique has inevitable gap to estimate "dynamic" function of respiratory system precisely.

References

1. McNamara AE, Müller NL, Okazawa M, Arntorp J, Wiggs BR, Paré PD. Airway narrowing in excised canine lungs measured by high-resolution computed tomography. J Appl Physiol. 1992;73:307–16.
2. Niimi A, Matsumoto H, Amitani R, Nakano Y, Mishima M, Minakuchi M, Nishimura K, Itoh H, Izumi T. Airway wall thickness in asthma assessed by computed tomography. Relation to clinical indices. Am J Respir Crit Care Med. 2000;162:1518–23.
3. Nakano Y, Muro S, Sakai H, Hirai T, Chin K, Tsukino M, Nishimura K, Itoh H, Paré PD, Hogg JC, Mishima M. Computed tomographic measurements of airway dimensions and emphysema in smokers. Correlation with lung function. Am J Respir Crit Care Med. 2000;162:1102–8.
4. Kirby M, Tanabe N, Tan WC, Zhou G, Obeidat M, Hague CJ, Leipsic J, Bourbeau J, Sin DD, Hogg JC, Coxson HO, CanCOLD Collaborative Research Group, Canadian Respiratory Research Network. Total airway count on computed tomography and the risk of chronic obstructive pulmonary disease progression. Findings from a population-based study. Am J Respir Crit Care Med. 2018;197:56–65.
5. Weinheimer O, Achenbach T, Bletz C, Duber C, Kauczor H-U, Heussel CP. About objective 3-d analysis of airway geometry in computerized tomography. IEEE Trans Med Imaging. 2008;27:64–74. IEEE
6. Oguma T, Hirai T, Niimi A, Matsumoto H, Muro S, Shigematsu M, Nishimura T, Kubo Y, Mishima M. Limitations of airway dimension measurement on images obtained using multi-detector row computed tomography. Muñoz-Barrutia A, editor. PLoS One. 2013; 8: e76381. Public Library of Science
7. Dame Carroll JR, Chandra A, Jones AS, Berend N, Magnussen JS, King GG. Airway dimensions measured from micro-computed tomography and high-resolution computed tomography. Eur Respir J. 2006;28:712–20.
8. de Jong PA, Müller NL, Paré PD, Coxson HO. Computed tomographic imaging of the airways: relationship to structure and function. Eur Respir J. 2005;26(1):140–52.
9. Okazawa M, Müller N, McNamara AE, Child S, Verburgt L, Paré PD. Human airway narrowing measured using high resolution computed tomography. Am J Respir Crit Care Med. 1996;154:1557–62. American Public Health Association

10. Nakano Y, Wong JC, de Jong PA, Buzatu L, Nagao T, Coxson HO, Elliott WM, Hogg JC, Paré PD. The prediction of small airway dimensions using computed tomography. Am J Respir Crit Care Med. 2005;171:142–6.
11. Han MK, Kazerooni EA, Lynch DA, Liu LX, Murray S, Curtis JL, Criner GJ, Kim V, Bowler RP, Hanania NA, Anzueto AR, Make BJ, Hokanson JE, Crapo JD, Silverman EK, Martinez FJ, Washko GR, COPDGene Investigators. Chronic obstructive pulmonary disease exacerbations in the COPDGene study: associated radiologic phenotypes. Radiology. 2011;261:274–82.
12. Smith BM, Hoffman EA, Rabinowitz D, Bleecker E, Christenson S, Couper D, Donohue KM, Han MK, Hansel NN, Kanner RE, Kleerup E, Rennard S, Barr RG. Comparison of spatially matched airways reveals thinner airway walls in COPD. The Multi-Ethnic Study of Atherosclerosis (MESA) COPD Study and the Subpopulations and Intermediate Outcomes in COPD Study (SPIROMICS). Thorax. 2014;69:987–96.
13. Washko GR, Diaz AA, Kim V, Barr RG, Dransfield MT, Schroeder J, Reilly JJ, Ramsdell JW, McKenzie A, van Beek EJR, Lynch DA, Butler JP, Han MK. Computed tomographic measures of airway morphology in smokers and never-smoking normals. J Appl Physiol. 2014;116:668–73. American Physiological Society Bethesda, MD
14. Hasegawa M, Nasuhara Y, Onodera Y, Makita H, Nagai K, Fuke S, Ito Y, Betsuyaku T, Nishimura M. Airflow limitation and airway dimensions in chronic obstructive pulmonary disease. Am J Respir Crit Care Med. 2006;173:1309–15.
15. Shimizu K, Seto R, Makita H, Suzuki M, Konno S, Ito YM, Kanda R, Ogawa E, Nakano Y, Nishimura M. Computed tomography (CT)-assessed bronchodilation induced by inhaled indacaterol and glycopyrronium/indacaterol in COPD. Respir Med. 2016;119:70–7.
16. Kuyper LM, Paré PD, Hogg JC, Lambert RK, Ionescu D, Woods R, Bai TR. Characterization of airway plugging in fatal asthma. Am J Med. 2003;115:6–11.
17. Tanabe N, Oguma T, Sato S, Kubo T, Kozawa S, Shima H, Koizumi K, Sato A, Muro S, Togashi K, Hirai T. Quantitative measurement of airway dimensions using ultra-high resolution computed tomography. Respir Investig. 2018;56:489–96.
18. Tanabe N, Shima H, Sato S, Oguma T, Kubo T, Kozawa S, Koizumi K, Sato A, Togashi K, Hirai T. Direct evaluation of peripheral airways using ultra-high-resolution CT in chronic obstructive pulmonary disease. Eur J Radiol. 2019;120:108687.
19. Niimi A, Matsumoto H, Takemura M, Ueda T, Chin K, Mishima M. Relationship of airway wall thickness to airway sensitivity and airway reactivity in asthma. Am J Respir Crit Care Med. 2003;168:983–8.
20. Nakano Y, Müller NL, King GG, Niimi A, Kalloger SE, Mishima M, Paré PD. Quantitative assessment of airway remodeling using high-resolution CT. Chest. 2002;122:271S–5S.
21. Petty TL. The history of COPD. Int J Chron Obstruct Pulmon Dis. 2006;1:3–14. Dove Press
22. Grydeland TB, Dirksen A, Coxson HO, Eagan TML, Thorsen E, Pillai SG, Sharma S, Eide GE, Gulsvik A, Bakke PS. Quantitative computed tomography measures of emphysema and airway wall thickness are related to respiratory symptoms. Am J Respir Crit Care Med. 2010;181:353–9.
23. Yamashiro T, Moriya H, Matsuoka S, Nagatani Y, Tsubakimoto M, Tsuchiya N, Murayama S. Asynchrony in respiratory movements between the pulmonary lobes in patients with COPD: continuous measurement of lung density by 4-dimensional dynamic-ventilation CT. Int J Chron Obstruct Pulmon Dis. 2017;12:2101–9. Dove Press
24. Yamashiro T, Moriya H, Tsubakimoto M, Matsuoka S, Murayama S. Continuous quantitative measurement of the proximal airway dimensions and lung density on four-dimensional dynamic-ventilation CT in smokers. Int J Chron Obstruct Pulmon Dis. 2016;11:755–64. Dove Press
25. Oguma T, Hirai T, Fukui M, Tanabe N, Marumo S, Nakamura H, Ito H, Sato S, Niimi A, Ito I, Matsumoto H, Muro S, Mishima M. Longitudinal shape irregularity of airway lumen assessed by CT in patients with bronchial asthma and COPD. Thorax. 2015;70:719–24.

26. Hogg JC, Chu F, Utokaparch S, Woods R, Elliott WM, Buzatu L, Cherniack RM, Rogers RM, Sciurba FC, Coxson HO, Paré PD. The nature of small-airway obstruction in chronic obstructive pulmonary disease. N Engl J Med. 2004;350:2645–53.
27. McDonough JE, Yuan R, Suzuki M, Seyednejad N, Elliott WM, Sanchez PG, Wright AC, Gefter WB, Litzky L, Coxson HO, Paré PD, Sin DD, Pierce RA, Woods JC, McWilliams AM, Mayo JR, Lam SC, Cooper JD, Hogg JC. Small-airway obstruction and emphysema in chronic obstructive pulmonary disease. N Engl J Med. 2011;365:1567–75.
28. Koo H-K, Vasilescu DM, Booth S, Hsieh A, Katsamenis OL, Fishbane N, Elliott WM, Kirby M, Lackie P, Sinclair I, Warner JA, Cooper JD, Coxson HO, Paré PD, Hogg JC, Hackett T-L. Small airways disease in mild and moderate chronic obstructive pulmonary disease: a cross-sectional study. Lancet Respir Med. 2018;6:591–602.
29. Tanabe N, Vasilescu DM, Kirby M, Coxson HO, Verleden SE, Vanaudenaerde BM, Kinose D, Nakano Y, Paré PD, Hogg JC. Analysis of airway pathology in COPD using a combination of computed tomography, micro-computed tomography and histology. Eur Respir J. 2018;51:1701245.
30. Tanabe N, Sato S, Oguma T, Shima H, Sato A, Muro S, Hirai T. Associations of airway tree to lung volume ratio on computed tomography with lung function and symptoms in chronic obstructive pulmonary disease. Respir Res. 2019; 20:77. BioMed Central.
31. Kambara K, Shimizu K, Makita H, Hasegawa M, Nagai K, Konno S, Nishimura M. Effect of lung volume on airway luminal area assessed by computed tomography in chronic obstructive pulmonary disease. Arjomandi M, editor. PLoS One. 2014; 9:e90040. Public Library of Science.

Chapter 8
Functional Properties of Lower Airway Estimated by Oscillometry: Is Oscillometry Useful for Detecting Lower-Airway Abnormalities?

Hajime Kurosawa

Abstract Oscillometry (forced oscillation technique, FOT) is a noninvasive method which enables to measure lung mechanics such as respiratory system impedance (Zrs), resistance (Rrs), and reactance (Xrs) during quiet breathing. Obstruction of airways can be implied from an increase in Rrs and more negative shifting in Xrs, in other words, high resonant frequency (Fres). Because oscillometry and spirometry is not identical modality, correlation between Rrs and FEV1 is generally weak. Frequency dependence of Rrs, which comes from several conditions such as shunt effects, mechanical inhomogeneity, turbulence, and peripheral airway lesion, can be a marker for smokers, COPD, asthma, infants, and some other conditions such as space occupied lesion of large central airways. Respiratory cycle dependence of Rrs or Xrs would be useful to understand the difference of ventilatory mechanics between inspiratory and expiratory phase, which can be observed as dynamic airway narrowing in patients with COPD. In conclusion, oscillometry is useful to assess the physiology of lower airways.

Keywords Forced oscillation technique · Frequency dependence · Ventilatory inhomogeneity · Airway resistance · Resonant frequency

---

H. Kurosawa (✉)
Center for Environmental Conservation and Research Safety, Tohoku University, Sendai, Japan

Department of Occupational Health, Tohoku University School of Medicine, Sendai, Japan
e-mail: kurosawa-thk@med.tohoku.ac.jp

K. Yamaguchi (ed.), *Structure-Function Relationships in Various Respiratory Systems*, Respiratory Disease Series: Diagnostic Tools and Disease Managements, https://doi.org/10.1007/978-981-15-5596-1_8

1 Introduction

Oscillometry (also known as the forced oscillation technique, FOT) is a noninvasive method which enables to measure lung mechanics [1, 2]. In actual measurement, during quiet breathing, engineered vibration of air is transmitted to the airway through the mouthpiece as input signals of oscillation (Fig. 8.1). At this time, the airflow rate and the mouse pressure that are dynamically changing with breathing are continuously monitored to obtain respiratory system impedance (Zrs). The real part of the impedance (in-phase component) is called respiratory system resistance (Rrs), whereas the imaginary part (out-of-phase component) is called respiratory system reactance (Xrs). Since the first method using monofrequency oscillometry in the 1950s [3], several types of equipment have been evolved. Recently, with the widespread clinical application of broadband frequency oscillometry, a growing number of studies have been investigating their usefulness in assessing or managing obstructive pulmonary disease, including asthma and COPD.

2 Parameters of Oscillometry

When the oscillatory signals are added through the mouth, Zrs is defined as the complex ratio of pressure to airflow rate at the mouth. Zrs consists of its real part, Rrs, and imaginary part, Xrs, having the following relationship: $(\text{Zrs})^2 = (\text{Rrs})^2 + (\text{Xrs})^2$. Rrs reflects the dissipative mechanical property (such as frictional pressure loss) of the lung. Xrs is supposed to reflect elastic property of respiratory system and inertial properties of oscillating gas and respiratory system.

The parameters used for oscillometry are generally expressed as the mean values during a respiratory cycle (whole-breath). Also, the mean values during inspiratory and expiratory phases and differences between inspiratory and expiratory phases are used, if necessary.

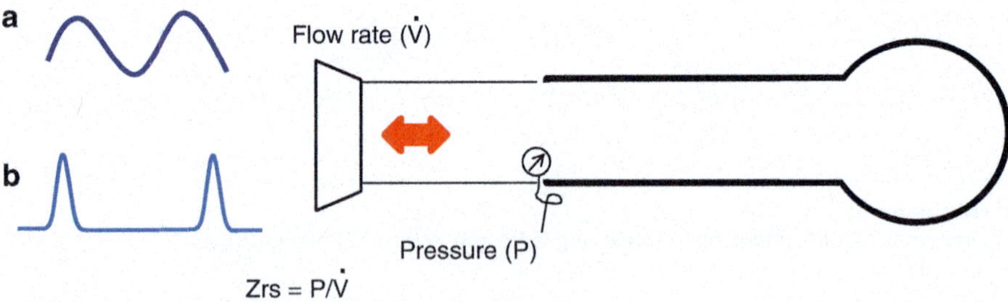

Fig. 8.1 Schema of oscillometry. Engineered sinusoidal (A) or impulse (B) vibration transmitted to the airway during quiet breathing, and the flow rate and pressure are monitored to obtain respiratory system impedance (Zrs)

Rrs indicates the total resistance of the respiratory system, including airway resistance (Raw), respiratory tissue resistance (Rti), and chest wall resistance (Rcw) [1, 2]. Although Rrs is not identical to Raw, Raw is a major component of Rrs. Hence, Rrs is a very sensitive indicator for Raw, and therefore, Rrs is frequently interpreted as an index of the airway caliber. In broad frequency oscillometry, Rrs at 5 Hz (R5) and 20 Hz (R20) (or sometimes R19 as an equivalent) are routinely used. Although both the increase in Rrs and the decrease in forced expiratory volume in 1 s (FEV1) reflects airway obstruction [1, 2], the oscillometry and spirometry are not the identical modality. Therefore, in general, Rrs and FEV1 are sometimes correlated with each other [4–6], but sometimes not.

Xrs is presumed to reflect the summation of elastance and inertia. In practical measurements in humans, Xrs shows frequency dependency [1, 2] (Fig. 8.2). In general, Xrs is negative at a low frequency whereas positive at a high frequency. The frequency at which Xrs = 0 is referred to as a resonant frequency (Fres) (Fig. 8.2).

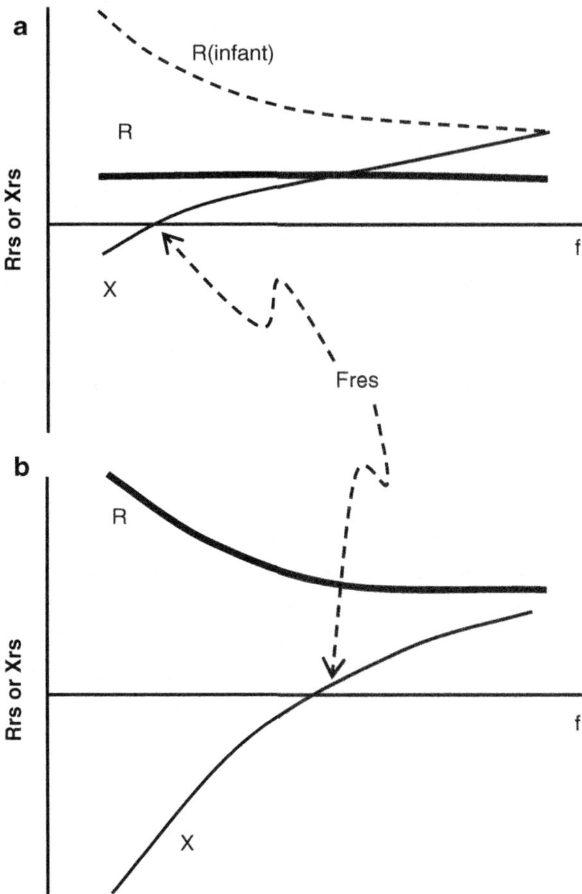

Fig. 8.2 Frequency properties of Rrs (R) and Xrs (X) in healthy subjects (**a**) and in patients with obstructive dysfunction such as asthma and COPD (**b**). Frequencies where Xrs = 0 are called as resonant frequency (Fres). Healthy infants have physiological frequency dependence of Rrs (a: dotted line)

Generally, more elastic properties result in more negative whereas more inertial properties result in more positive Xrs. Therefore, elasticity and inertia balance each other at Fres. Xrs at 5 Hz (X5) and Fres is frequently used as a representative marker of Xrs. The low-frequency reactance area (Ax, or ALX), which is the integral of X5 to the Fres, is also common.

3 Frequency Dependence of Rrs

In healthy adults, Rrs remains nearly constant over the oscillation frequencies [4, 7]. On the other hand, the predominant increase in Rrs in the lower frequency, so-called frequency dependence of Rrs, can be observed under certain circumstances [8]. Frequency dependence of Rrs in patients with COPD was the most well-known phenomenon that was firstly reported by Grimby and Takishima [9]. The difference from R5 to R20 (R5-R20) or an equivalent has been frequently used as a representative marker of the frequency dependence of Rrs in oscillometry. The physiological interpretation of frequency dependence of Rrs remains uncertain [1]. There are not published findings correlating pathology to frequency dependence so far. Further researches to pursue it are obviously needed. In this text, therefore, several speculations concerning those will be briefly reviewed.

3.1 Shunt Effects

Modeling studies concerning oscillometry often have replaced the respiratory system with electrical circuits and performed various simulations. When a circuit has a capacitor with a resistor, the higher the frequency, the easier it is for electricity to pass through the capacitor, and depending on how the circuit is assembled, the current may escape like a shunt and the resistance of the circuit may appear to be low, resulting in frequency dependence of Rrs [10]. Again, considering analogy to the airway, enough softness of the airway corresponds to the capacitor. In fact, this was experimentally confirmed as a causal factor of frequency dependence of Rrs in healthy infants (Fig. 8.2a), which has also confirmed to be minimized or disappeared with growing to adults [11, 12]. Besides, it is quite possible that these shunt effects also occur in the lower airways near various pathological structures such as bullous lesions both in adults and in infants.

3.2 *Mechanical Inhomogeneity*

As reported by Grimby and Takishima [9], frequency dependence of Rrs can be observed in COPD and smokers [5, 6, 9, 13]. The phenomenon was originally speculated as "non-uniform distribution of mechanical properties of the lungs to their overall mechanical behavior" [9]. It is theorized that the lung parenchyma affects each other locally which was called as interdependence [14]. Any lesions in the lung may break local uniformity, resulting in mechanical inhomogeneity. In this condition, ventilatory unevenness itself [15] and/or shunt effects in local lung parenchyma or airways as described above could be candidates to cause the frequency dependence of Rrs [1].

3.3 *Turbulence*

Involvement of turbulence in the central airways with Rrs was suggested by Wouters EF [16]. They assessed Rrs in normal subjects breathing air and helium–oxygen mixture, and found that breathing helium–oxygen reduced Rrs and frequency dependence. They interpreted the reduction of large airway turbulence as one of the main factors. Recently, a case of disappeared increased Rrs and its frequency dependency after resection of tracheal tumors was reported [17]. This could be attributed not only to some shunt effects discussed above, but also to airflow turbulence resulting from airflow passing through the space-occupied lesions such as tumor on the large airway.

3.4 *Peripheral Airway Resistance*

Obviously, the peripheral airway is a very important place in various conditions such as COPD and asthma. For researchers to study those fields using oscillometry, an assumption that the frequency dependence of Rrs expressed as R5-R20 or an equivalent reflects lesions in the peripheral airways may be rather common. This concept is based on the hypothesis that R20 reflects the resistance of central airway and R5 reflects the resistance of overall airway, which was simulated using various electrical circuit models [18]. Then inevitably, R5-R20 resulted to reflect peripheral airway resistance. Although this concept itself has not been accompanied by the basis of established physiological or pathological studies as seen in the shunt effects as described above so far, many researchers have published various studies in line with this central-peripheral hypothesis. In fact, some of them have reported correlations with other peripheral airway biomarkers or with image analysis [19, 20].

However, even if the association is true to some extent, it is unclear how much it contributes to the dependence. Before identifying them, there may be some contradictions to overcome. First, from the point of view of total cross-sectional areas of each generation of airway tree, the total area is exponentially larger at the periphery as taught by the basic respiratory physiology [21]. The lesion of peripheral airway alone should be apparently indistinguishable until it has progressed to some extent in terms of airway resistance, which has been known as the reason why the peripheral airways are called "silent zone" [22]. This basic concept is, however, inconsistent with the fact that even 60% of smokers with normal FEV1/FVC that could be implied as holders of very mild lesion have significant findings [13]. Second, to date, there has been no clear indication of where the border between the peripheral and central airways lies. Although diameter less than 2 mm historically defines small airways, there has been no evidence that oscillometry alone distinguishes any such anatomical locations. Given these things, special care is necessary to interpret the frequency dependence of Rrs as peripheral resistance or as a kind of biomarkers of peripheral lesions.

4 Assessment of Lower Airways in Disease

Assessment of lower airway mechanics has been variously performed in clinical settings. In this text, some of respiratory diseases will be briefly reviewed.

4.1 Asthma

In patients with asthma, increased Rrs is observed as a result of bronchoconstriction airway remodeling, or both. Frequency dependence of Rrs, in general, is observed in moderate to severe cases whereas it is not in mild ones [23]. Compared to healthy subjects, Xrs shifts to more negative and has higher Fres. As reported in studies performed in patients with asthma [5, 24], there is a close relationship between Fres and FEV1. There can be a significant correlation between FEV1 and R5 (or R20), but it is not stronger than between FEV1 and Fres. Although fractional exhaled nitric oxide (FeNO) is known to increase in patients with asthma, there is no correlation between FeNO and Rrs or Xrs [25, 26].

Pharmacological effects can be easily assessed by oscillometry as a decrease in R5, R5-R20, ALX, and Fres or shifting to less negative side in X5. Bronchial reversibility is assessed by change in FEV1 before and after bronchodilator administration.

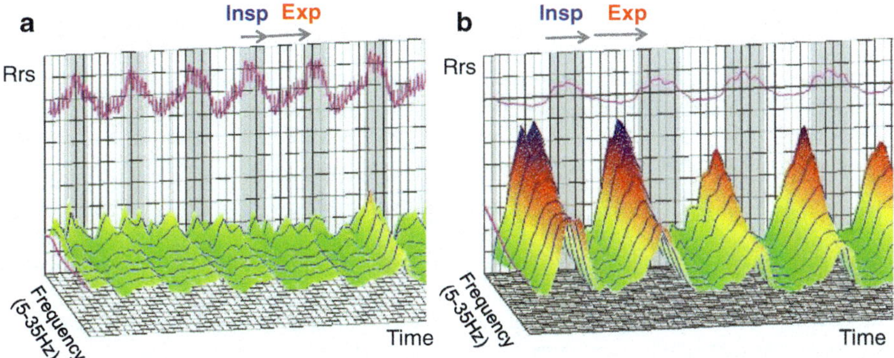

Fig. 8.3 Examples of Rrs profile in a healthy subject (**a**: male 47y) and in a patient with COPD (**b**: male 65y, GOLD stage 1) expressed by 3D colored graphics. Case A showed no frequency dependence and low Rrs. There was not respiratory cycle dependence. Case B showed remarkable frequency dependence and high Rrs at the expiratory phase though low Rrs at the inspiratory phase. Respiratory cycle dependence was obvious in this case. Measurements were performed using MostGraph (Chest MI, Tokyo, Japan)

4.2 COPD

In patients with COPD, an increase in Rrs and frequency dependence of Rrs, and high Fres are generally observed with severity dependent manner [4, 6, 8]. Greater variability of FOT parameters during expiration than during inspiration is reported [4, 6] (Fig. 8.3), which possibly reflects dynamic airway narrowing [27].

Early detection of changes in lung function in smokers is important for the prevention of COPD. Approximately 60% of smokers showed higher Rrs and more negative Xrs, mainly in the expiratory phase, as observed in COPD patients [13]. A case–control study of lung function in World Trade Center Health registry area residents and workers revealed that cases with persistent respiratory symptoms were more likely than control subjects to have elevated R5 and R5-R20 despite normal spirometry [28]. These studies suggest the potentiality of FOT.

5 Conclusion

Clinical application of oscillometry is useful to assess lower airway mechanics. It has progressed with the spread of commercially available devices, and increasing numbers of reported studies have examined the usefulness of oscillometry in the evaluation of various respiratory diseases such as asthma and COPD. Data interpretation, R5-R20 in particular, is needed to care.

References

1. King GG, Bates J, Berger KI, Calverley P, de Melo PL, Dellacà RL, Farré R, Hall GL, Ioan I, Irvin CG, Kaczka DW, Kaminsky DA, Kurosawa H, Lombardi E, Maksym GN, Marchal F, Oppenheimer BW, Simpson SJ, Thamrin C, van den Berge M, Oostveen E. Technical standards for respiratory oscillometry. Eur Respir J. 2019;55:1900753. https://doi.org/10.1183/13993003.00753-2019.
2. Shirai T, Kurosawa H. Clinical application of the forced oscillation technique. Intern Med. 2016;55:559–66.
3. Dubois AB, Brody AW, Lewis DH, Burgess BF Jr. Oscillation mechanics of lung and chest in man. J Appl Physiol. 1956;8:587–94.
4. Mori K, Shirai T, Mikamo M, et al. Colored 3-dimensional analyses of respiratory resistance and reactance in COPD and asthma. COPD. 2011;8:456–63.
5. Tsuburai T, Suzuki S, Tsurikisawa N, et al. Use of forced oscillation technique to detect airflow limitations in adult Japanese asthmatics. Arerugi. 2012;61:184–93. (in Japanese and abstract in English)
6. Ohishi J, Kurosawa H, Ogawa H, Irokawa T, Hida W, Kohzuki M. Application of impulse oscillometry for within-breath analysis in patients with chronic obstructive pulmonary disease: pilot study. BMJ Open. 2011;1:e000184. https://doi.org/10.1136/bmjopen-2011-000184.
7. Nakaji H, Petrova G, Matsumoto H, et al. Effects of 24-week add-on treatment with ciclesonide and montelukast on small airways inflammation in asthma. Ann Allergy Asthma Immunol. 2013;110:198–203.
8. Pride NB. Forced oscillation techniques for measuring mechanical properties of the respiratory system. Thorax. 1992;47:317–20.
9. Grimby G, Takishima T, Graham W, Macklem P, Mead J. Frequency dependence of flow resistance in patients with obstructive lung disease. J Clin Invest. 1968;47:1455–65.
10. Pimmel RL, Winter DC, Bromberg PA. Estimating central and peripheral respiratory resistance. J Appl Physiol. 1978;45:375–80.
11. Nguyen YT, Demoulin B, Schweitzer C, Bonabel-Chone C, Marchal F. Identification of bronchodilator responsiveness by forced oscillation admittance in children. Pediatr Res. 2007;62:348–52.
12. Beydon N, Stephanie DD, Lombardi E, et al. An official American Thoracic Society/European Respiratory Society statement: pulmonary function testing in preschool children. Am J Respir Crit Care Med. 2007;175:1304–45.
13. Shinke H, Yamamoto M, Hazeki N, Kotani Y, Kobayashi K, Nishimura Y. Visualized changes in respiratory resistance and reactance along a time axis in smokers: a cross-sectional study. Respir Investig. 2013;51:166–74.
14. Mead J, Takishima T, Leith D. Stress distribution in lungs: a model of pulmonary elasticity. J Appl Physiol. 1970;28:596–608.
15. Lutchen KR, Hantos Z, Petak F, Adamicza A, Suki B. Airway inhomogeneities contribute to apparent lung tissue mechanics during constriction. J Appl Physiol. 1996;80:1841–9.
16. Wouters EF, Lándsér FJ, Polko AH, Visser BF. Density dependence of respiratory input impedance in normal subjects. Clin Exp Pharmacol Physiol. 1990;17:477–84.
17. Hagiwara E, Gon Y, Hayashi K, Takahashi M, Iida Y, Hiranuma H, Nakagawa Y, Hataoka T, Mizumura K, Maruoka S, Shimizu T, Takahashi N, Hashimoto S. Evaluation of airway resistance in primary small cell carcinoma of the trachea by MostGraph: a case study. J Thorac Dis. 2016;8:E702–6. https://doi.org/10.21037/jtd.2016.07.86.
18. Goldman MD, Nazeran H, Ramos C, et al. Electrical circuit models of the human respiratory system reflect small airway impairment measured by impulse oscillation (IOS). Conf Proc IEEE Eng Med Biol Soc. 2010;2010:2467–72. https://doi.org/10.1109/IEMBS.2010.5626611.
19. Postma DS, Brightling C, Baldi S, et al. Exploring the relevance and extent of small airways dysfunction in asthma (ATLANTIS): baseline data from a prospective cohort study. Lancet Respir Med. 2019;7:402–16. https://doi.org/10.1016/S2213-2600(19)30049-9.

20. Cottini M, Licini A, Lombardi C, Berti A. Clinical characterization and predictors of IOS-defined small-airway dysfunction in asthma. J Allergy Clin Immunol Pract. 2020;8(3):997–1004.e2. https://doi.org/10.1016/j.jaip.2019.10.040.
21. West JB, Luks AM. West's pulmonary physiology: the essentials. 10th ed. Philadelphia: Wolters Kluwer; 2015.
22. Woolcock AJ, Vincent NJ, Macklem PT. Frequency dependence of compliance as a test for obstruction in the small airways. J Clin Invest. 1969;48:1097–106.
23. Cavalcanti JV, Lopes AJ, Jansen JM, Melo PL. Detection of changes in respiratory mechanics due to increasing degrees of airway obstruction in asthma by the forced oscillation technique. Respir Med. 2006;100:2207–19.
24. Shibasaki A, Kurosawa H, Tamura G. Evaluation of airway narrowing by MostGraph and spirometry. Examination using a reversibility test. Arerugi. 2013;62:566–73. (in Japanese and abstract in English)
25. Inoue H, Niimi A, Takeda T, et al. Pathophysiological characteristics of asthma in the elderly: a comprehensive study. Ann Allergy Asthma Immunol. 2014;113:527–33.
26. Shirai T, Mori K, Mikamo M, et al. Respiratory mechanics and peripheral airway inflammation and dysfunction in asthma. Clin Exp Allergy. 2013;43:521–6.
27. Kurosawa H, Kohzuki M. Images in clinical medicine. Dynamic airway narrowing. New Engl J Med. 2004;350:1036.
28. Friedman SM, Maslow CB, Reibman J, et al. Case-control study of lung function in world trade center health registry area residents and workers. Am J Respir Crit Care Med. 2011;184:582–9.

Part IV
Acinus

Chapter 9
Inhomogeneous Distribution of Ventilation–Perfusion (V_A/Q) and Diffusing Capacity–Perfusion (D/Q) in the Lung: What Abnormal V_A/Q and D/Q Distributions Are Detected in Diseased Lungs?

Kazuhiro Yamaguchi

Abstract Anatomical gas exchange unit is given by the acinus that is defined as the microregion served by the transitional bronchiole. The functional gas exchange unit conforms anatomical gas exchange unit during expiration but not during inspiration. Series arrangement of alveoli causing serial distribution of alveolar ventilation, which is termed "diffusional screening," is not necessary to be considered while estimating acinar gas exchange. Thus, distributions of alveolar ventilation (V_A) and perfusion (Q) in the acinus are taken to be arranged in parallel and connected through the alveolocapillary tissue barrier that configures a part of diffusing capacity (D). Continuous but representative distribution of V_A/Q in the lung periphery is decided in terms of multiple inert gas elimination technique (MIGET) that was elaborated by Wagner and colleagues. MIGET uses six foreign inert gases as indicators and has been shown to be clinically useful for elucidating structure–function relationships at acinar levels in various lung diseases. Distribution of D/Q in addition to that of V_A/Q can be estimated based on V_A/Q–D/Q concept first proposed by Piiper. This analysis is possible while making use of nine indicator gases, including O_2, CO_2, and CO besides six inert gases enforced for MIGET. V_A/Q–D/Q analysis demonstrated that D/Q heterogeneity-related aqueous-phase diffusion limitation impairs O_2 gas exchange, which leads to the emergence of hypoxemia in patients with fibrotic lung diseases but not in those with COPD.

K. Yamaguchi (✉)
Department of Respiratory Medicine, Tokyo Medical University, Tokyo, Japan

Division of Respiratory Medicine, Tohto Clinic, Kenko-Igaku Association, Tokyo, Japan
e-mail: yamaguc@sirius.ocn.ne.jp

K. Yamaguchi (ed.), *Structure-Function Relationships in Various Respiratory Systems*, Respiratory Disease Series: Diagnostic Tools and Disease Managements, https://doi.org/10.1007/978-981-15-5596-1_9

Keywords Anatomical gas exchange unit · Functional gas exchange unit Diffusional screening · Multiple inert gas elimination technique (MIGET) · V_A/Q D/Q · Aqueous-phase diffusion limitation · Reaction limitation

1 Introduction

From a historical point of view, the first suggestion that inspired gas is not homogeneously distributed to all parts of the lung was made by Keith in 1909 [1]. Based on the anatomical study using the lung dissection specimens, Rohrer [2] reached a similar conclusion that the ventilation in the lung may be heterogeneously distributed. The first recognition that the gas exchange which takes place in any lung region is decided not only by the ventilation (V_A) or by the blood flow (Q), but by the ratio of one to another (V_A/Q), was made by Krogh and Lindhard in 1917 [3]. In 1922, Haldane [4] recognized that V_A/Q heterogeneity could cause hypoxemia. In the late 1940s, Fenn et al. [5] and Riley and Cournand [6] actively addressed the relationships between V_A, Q, and gas exchange in the lung. These advances in respiratory physiology stimulated the San Diego group, including West, Wagner, and others [7, 8], to contrive the multiple inert gas elimination technique (MIGET) that allows one to predict the continuous V_A/Q distribution in patients with various types of lung disease. The intelligent point of the MIGET is that this method uses six inert gases that are simply perfusion-limited but neither diffusion-limited nor reaction-limited in the blood so that it can precisely detect the V_A/Q distribution in the acinar regions where gas exchange takes place. Differing from the paradigm of the story as mentioned above, Piiper [9] proposed, in 1961, an intriguing concept that in addition to the heterogeneous distribution of V_A/Q, the basic mechanism giving rise to hypoxemia should include the heterogeneous distribution of diffusing capacity for O_2 (D) to perfusion (i.e., D/Q) in the acinar regions, as well. His concept was named "V_A/Q–D/Q field," the basis of which was that O_2 is not purely perfusion-limited gas but it is limited also by diffusion in blood to some extent [10]. The V_A/Q–D/Q concept of Piiper was challenged by Yamaguchi et al. [11–13], who attempted to determine the V_A/Q–D/Q distributions under various pathophysiological circumstances. As such, the physiological aspects that are necessary for profoundly understanding gas exchange in the acinar regions made a great deal of progress in the late twentieth century. However, some crucial problems regarding functional parameters such as V_A/Q and/or D/Q describing the gas exchange kinetics in the acinar regions have remained unsolved. Among them, the most important question is what anatomical structure serves for stipulating the functional gas exchange unit, which provides the essential basis on expanding the concept of V_A/Q and/or D/Q distributions into the clinical practice; that is, if the microstructural aspects forming the functional gas exchange unit are surely recognized, it can provide an insight into the morphological abnormalities in the acinar regions from the functional abnormalities such as heterogeneous distributions of V_A/Q and/or D/Q, i.e., the assessment of structure–function relationships in the acinar regions judged primarily from the functional parameters. This is very important when physicians should do pathophysiological decision-making against a patient with a certain lung disease. This is because it is not possible to clinically acquire the rigorous morphological information about the acinus even when making full use of high-precision diagnostic imaging technologies. In view of these facts and clinical requirements, the present review highlights the following issues. (1) What is the most appropriate organization serving as the functional gas exchange unit, which is certainly backed by the structural design of the lung periphery? (2) Basic rationales describing gas exchange dynamics in functional gas exchange unit with V_A/Q or with V_A/Q and D/Q. (3) Based on the distributional measurements of V_A/Q alone and those of V_A/Q and D/Q in patients with pathologically and/or radiologically confirmed lung diseases, the attempt was made to certify the structure–function relationships in the lungs with various types of specific disease.

2 What Is the Most Appropriate Organization Serving as Functional Gas Exchange Unit?

2.1 Anatomical Gas Exchange Unit

As extensively argued in our previous papers [14, 15], the anatomical gas exchange unit is defined by the pulmonary acinus that is consisting of two anatomical units, including ventilation and perfusion units. Although there are many definitions concerning the acinus, including the acinus of Loeschcke, that of Haefeli-Bleuer (HB), that of Aschoff, the primary lobule of Miller, and so forth, we selected the HB acinus as the most appropriate anatomically defined organization meeting the definition of the acinus [15]. This is owing to the fact that in comparison with other acini, the structure of HB acinus was studied in detail by making full use of the optical microscopic means such as the scanning electron microscope [16]. The transitional bronchiole (15th generation of the bronchial tree) forms the entrance of HB acinus. The acinar airways are characterized by asymmetrical irregular dichotomy, which branch over 9 generations. The total path-length from the transitional bronchiole to the terminal alveolar sac averages 8.8 mm. The total acinar volume averages 0.2 cm^3 and the acinar size ranges between 7 and 10 mm. The total alveolar surface area participating in gas exchange reaches about 70 cm^2 at a full expansion of the lung tissue specimens, resulting in that about 30,000 acini are contained in one lung. However, it should be noted that these values were estimated at a full expansion of the lung tissue specimens, i.e., close to a total lung capacity (TLC). Therefore, under a physiological condition, i.e., at a functional residual capacity (FRC), the dimension of acinar structures, including acinar volume, alveolar surface area, and airway path-length should be modified accordingly (see Chap. 6). There is no doubt about the fact that the HB acinus participates in gas exchange as the ventilation unit. However, there is a problem regarding the perfusion unit formed by the pulmonary microcirculation. The dense intertwined capillary network embedded in the alveolar wall forms a continuum of blood flow. The perfusion of the capillary network is supplied by the adjacent arteriole and is drained into the adjacent venule. The distance between these two microvessels is of an order between 0.5 and 1.0 mm [17], resulting in that there are about 12,000 anatomically defined perfusion units in the acinus. This fact certainly indicates that the perfusion unit is not congruent with the ventilation unit. However, a couple of investigators [18, 19] identified that the entire region of acinus is perfused almost homogeneously, indicating that all capillary networks residing in the acinus are closely connected to each other, which leads to the conclusion that the acinus can be assigned, in a first approximation, to the anatomical gas exchange unit in which ventilation and perfusion units are virtually matched.

2.2 *Functional Gas Exchange Unit*

The absolute condition that should be satisfied while defining the functional gas exchange unit is the homogeneous distribution of a gas over the region that forms the unit. Regarding this issue, the important messages were derived from the studies of Paiva [20] and Swan and Tawhai [21]. A series, stratified heterogeneity of a gas concentration along acinar airways from the entrance to terminal alveolar sacs does exist during inspiration, resulting in that the anatomically defined acinus does not satisfy the absolute condition indispensable for the functional gas exchange unit during inspiration. However, the situation differs during expiration. A couple of groups of investigators [20, 21] demonstrated that the inspiration-elicited stratified inhomogeneity for a gas diminishes rapidly during expiration. Furthermore, the Martin channels, the communications locating at respiratory bronchioles or alveolar ducts, may further promote the homogenization of spatial gas difference in the HB acinus [14, 15]. Hence, we can conclude that the HB acinus serves as the functional gas exchange unit during expiration. However, we should appreciate the fact that the HB acinus does not rigorously satisfy the definition of the functional gas exchange unit during inspiration. These facts lead to an important conclusion that the gas exchange parameters estimated from the gas samples harvested over expiration can be interpreted on the basis of the functional gas exchange unit, which implies that the expiration-associated gas exchange parameters, such as the continuous distribution of V_A/Q and/or that of $V_A/Q–D/Q$, reflect the structural abnormalities of bronchioles, airspaces, microvessels, and/or alveolocapillary tissue barriers in the lung periphery represented by the HB acinus.

2.3 *Series Arrangement of Alveoli: Diffusional Screening*

One of the crucial problems while considering the gas exchange in the acinus is the distributional pattern of alveolar ventilation. Although the classical respiratory physiology commonly assumes that ventilation and perfusion are arranged in parallel in the lung periphery, the detailed microscopic examination on acinar structures revealed that alveoli are distributed in series (but perfusion is in parallel), i.e., alveoli are arranged laterally along the acinar airway from the entrance to the last generation. The indicator gas, e.g., O_2 molecules, entering the acinus from the conductive airways will diffuse both along the axis of acinar airways and radially into the alveoli and toward the alveolocapillary membrane (Fig. 9.1). O_2 molecules first encounter the surface of the alveolocapillary membrane in the proximal acinar airways where O_2 molecules are preferentially absorbed so that the distal generations receive less O_2; that is, O_2 molecules are screened at the proximal acinar airways. This phenomenon is first discovered by Sapoval et al. [22] and named the "diffusional screening." Based on the concept of diffusional screening, Felici et al. [23]

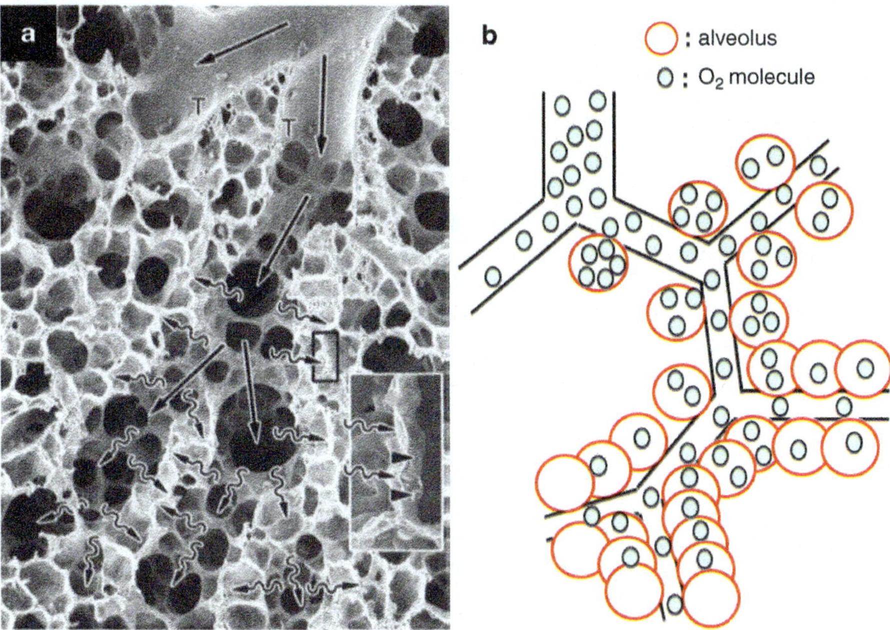

Fig. 9.1 Diffusional screening. (**a**) Scanning electron microscopic image of proximal portion of HB acinus. T: transitional bronchiole. Black straight arrows: axial diffusion of O_2 molecules along acinar airways. Black winding arrows: radial diffusion of O_2 molecules into alveoli encompassing acinar airways. Personal provision from Weibel. (**b**) Schematic presentation of left-side figure. O_2 molecules are preferentially absorbed into alveoli attached to proximal airways, where alveoli are serially arranged along an airway duct. Therefore, distal airways receive less O_2, indicating O_2 being screened at proximal acinar airways. However, the significance of diffusional screening for acinar gas exchange is rebutted by Swan and Tawhai [21]

estimated how much alveolar surface area is used for gas exchange, arriving at the conclusion that only 30–40% of the alveolar surface is required for gas exchange. This problematic finding observed by Felici et al. was reevaluated by Swan and Tawhai [21], who constructed the acinar model using the anatomical data obtained for the HB acinus and showed that the diffusional screening does exist during inspiration in the sense that O_2 molecules are absorbed by the alveoli located in proximal acinar regions before they reach the more peripheral acinar regions, i.e., the stratified PO_2 gradient from the entrance toward the terminal alveolar sacs during inspiration. However, they [21] identified that the stratified PO_2 gradient in the acinus is conclusively caused by the remarkable increase in surface area of acinar airways with alveolar sleeve as the generation of acinar airways is advanced but it does not need the diffusional screening-related mechanism. This indicates that the assumption of series distribution of ventilation in the acinus is not required when considering the acinar gas exchange kinetics. Hence, we conclude that the common hypothesis in the field of respiratory physiology, i.e., the parallel arrangements of ventilation and perfusion within and between the acini (i.e., the functional gas exchange units) are reasonable (Fig. 9.2).

2.4 Physical Properties of Gases Used for Estimating Acinar Gas Exchange

The classical gas exchange theory postulates that any gas existing in the acinus is equally prescribed by an effective ventilation-elicited convention that is decided by converting the effect of gas-phase diffusion occurring in the acinar region to the equivalent effect of convection [14, 15]. In the capillary blood, however, the gas transport mechanism differs largely depending on the physical properties of the gases concerned (Fig. 9.3), which is defined as the limitation by perfusion, aqueous-phase diffusion, or reaction with hemoglobin (Hb). As extensively argued in Sect. 4.1, the relative contribution of limitation by perfusion and that by aqueous-phase diffusion to overall gas transport in the capillary blood is decided by $D/(\beta Q)$ [10–13], i.e., when $D/(\beta Q)$ is large, limitation by perfusion is predominant over that by diffusion. D is diffusing capacity of an indicator gas, which is formed by alveolocapillary tissue barrier (i.e., alveolocapillary membrane and plasma layer) and erythrocytes, while Q is capillary blood flow. β denotes capacitance coefficient of a gas. β values of inert gases equal their Bunsen solubility coefficients (α) corrected for the difference in total pressure, while β value of a reactive gas combined with erythrocyte Hb (O_2, CO_2, carbon monoxide (CO), or nitric oxide (NO)), which is defined as the effective solubility, equals the slope of each dissociation curve. Thus, the effective solubility of a reactive gas is not constant and appreciably changes depending on its partial pressure in the blood. Although physical solubility of a

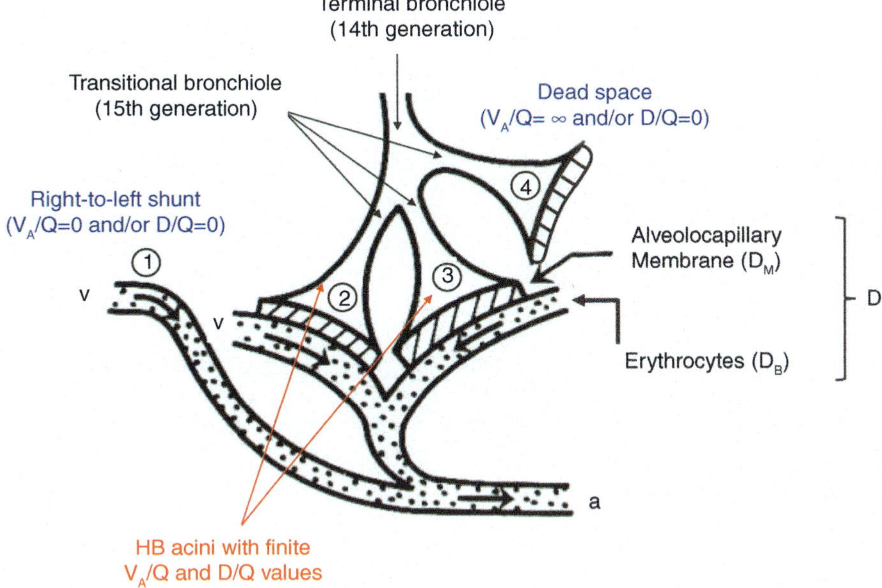

Fig. 9.2 Schematic diagram of the functional gas exchange unit. HB acinus assumes a functional gas exchange unit in which alveolar ventilation (V_A) and capillary perfusion (Q) are arranged in parallel. V_A and Q are separated by resistive tissue barrier, total gas conductance of which is defined as diffusing capacity (D). D is formed by alveolocapillary tissue barrier including plasm layer (membrane component of D (D_M)) and erythrocytes (aqueous-phase diffusion and reaction with Hb and defined as blood component of D (D_B)). This provides the basic concept of V_A/Q-D/Q field (Fig. 9.7). When D/Q=infinite, the functional gas exchange unit is prescribed solely by V_A/Q. Right-to-left shunt is given by units with $V_A/Q = 0$, while dead space by units with V_A/Q = infinite. However, units with $D/Q = 0$ provide a shunt effect for capillary blood and dead space effect for alveolar gas in a simultaneous manner. Models of parallelly arranged functional gas exchange units are reasonable while considering V_A/Q or V_A/Q-D/Q distribution in the lung

reactive gas in the blood is of the same order of magnitude as that of an inert gas, the effective solubility of a reactive gas is very large in comparison with that of an inert gas because of its combination with Hb. Thereby, the $D/(\beta Q)$ value of an inert gas is much larger than that of a reactive gas, leading to the conclusion that the inert gas transport is not limited by aqueous-phase diffusion in general but predominantly prescribed by perfusion. The pathological condition where inert gas transport is significantly limited by aqueous-phase diffusion will be discussed in Sect. 4.1. On the other hand, the gas transport of a reactive gas is limited by aqueous-phase diffusion besides perfusion. For the profound comprehension of the underlying gas transfer mechanism in the blood, we will first consider the equilibration kinetics of various gases, including inert gases, O_2, CO_2, CO, and NO, between capillary blood and gas phase in the acinus.

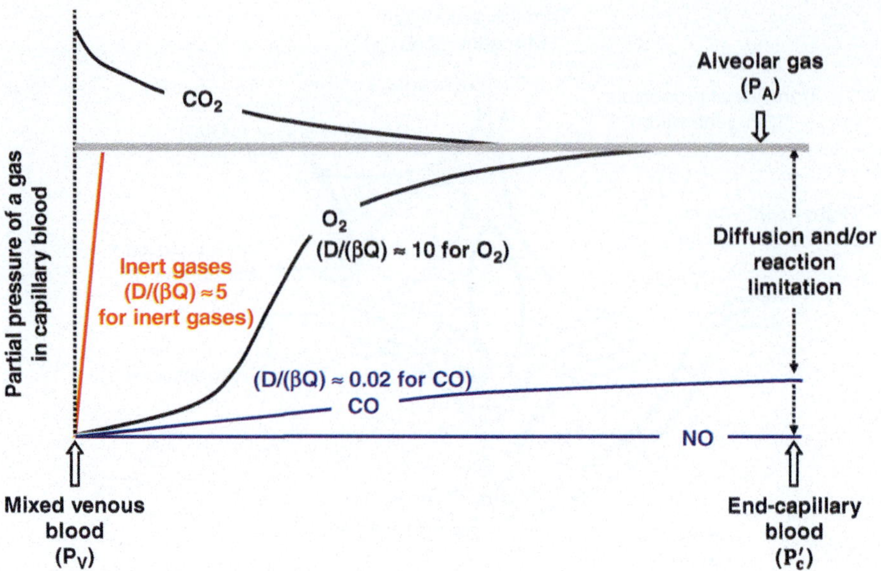

Fig. 9.3 Equilibration kinetics of gases with different physical properties between capillary blood and alveolar gas. Relative contribution of limitation by perfusion and that by aqueous-phase diffusion is decided by $D/(\beta Q)$, in which D is diffusing capacity of an indicator gas. Q is capillary blood flow. β denotes the capacitance coefficient of a gas. $D/(\beta Q)$ of an inert gas is much larger than that of a reactive gas (O_2, CO_2, CO, or NO). Therefore, inert gas transfer is predominantly prescribed by perfusion, while the gas transfer of a reactive gas is limited by aqueous-phase diffusion besides perfusion. Inert gases instantaneously reach partial pressure equilibration between capillary blood and alveolar gas so that inert gases are useful for estimating lung V_A/Q distribution. Although O_2 and CO_2 are significantly limited by aqueous-phase diffusion, equilibration of these gases between capillary blood and alveolar gas reaches during passage through capillary networks. CO is limited largely by reaction with Hb/HbO_2 in erythrocytes, thereby, no equilibration between capillary blood and alveolar gas is attained. NO is almost completely absorbed by Hb so that it does not reach equilibration between capillary blood and alveolar gas

2.4.1 Inert Gases

Inert gases in the blood are generally limited neither by aqueous-phase diffusion (plasma layer and inside of the erythrocyte) nor by reaction with Hb, indicating that inert gases are purely perfusion-limited in the capillary blood under most of the pathophysiological conditions. This is true irrespective of gas species such as being soluble or insoluble as well as being low-molecular weight or high-molecular weight [10]. Therefore, inert gases reach, almost instantaneously, the complete equilibration between capillary blood and gas phase in the acinus, which indicates that inert gases are most appropriate for estimating the gas exchange kinetics simply decided by V_A/Q distribution but neither by aqueous-phase diffusion nor by reaction with Hb.

2.4.2 Oxygen (O_2)

The association rate constant of O_2 with reduced Hb is slightly larger than that of NO and tenfold larger than that of CO, indicating that association reaction of O_2 with reduced Hb is fastest among the three reactive gases (O_2, CO, and NO) that bind to Hb [24]. This implies that the contribution of reaction limitation to gas transport is lowest in O_2 among the three reactive gases. On the other hand, the dissociation rate constant of O_2 from oxygenated Hb (HbO_2) is quite large as compared with those of CO and NO, resulting in that the affinity of O_2 to Hb is lowest and many free O_2 molecules are generated in the capillary blood. Therefore, the partial pressure of O_2 (PO_2) in the blood, which is the representative of activity of free O_2 molecules and is approximated by the equilibrium PO_2 estimated from the HbO_2 dissociation curve, increases and reaches the equilibration with alveolar PO_2 during passage through capillary networks. As such, O_2 transport in the capillary blood is predominantly limited by perfusion in combination with aqueous-phase diffusion through alveolocapillary tissue barrier (membrane component of diffusing capacity (D_M)) and inside of the erythrocyte (blood component of diffusing capacity (D_B)), though the effect of reaction evoked by the combination of O_2 with Hb is negligible. Based on these facts, it can be assumed that O_2 is useful for detecting the distribution of V_A/Q and that of D/Q in the lung.

2.4.3 Carbon Dioxide (CO_2)

CO_2 in the blood is present in three different forms; namely, physically dissolved CO_2 in plasma and erythrocyte, bicarbonate ions in plasma and erythrocyte, and carbamino compounds (R-NHCOO) in erythrocyte. Among the three forms, CO_2 in the erythrocyte is transported primarily by the form of bicarbonate ions, which are yielded as a result of hydration of CO_2 catalyzed by the carbonic anhydrase (CA) in the erythrocyte. The hydration reaction is almost instantaneous owing to the presence of CA, which indicates that the time required to attain chemical equilibrium is much smaller than the diffusion time in the erythrocyte. In other words, the uptake of CO_2 by the erythrocyte is mainly prescribed by the aqueous-phase diffusion-limited process. The partial pressure of CO_2 (PCO_2) in the blood, which is yielded by the free CO_2 molecules, may be approximated by the equilibrium PCO_2 estimated from the CO_2 dissociation curve. These free CO_2 molecules are transported by the perfusion-related process, leading to the conclusion that gas transport of CO_2 in the blood is predominantly limited by both perfusion and aqueous-phase diffusion, which indicates that CO_2 is useful for detecting the distribution of V_A/Q as well as that of D/Q in the lung.

2.4.4 Carbon Monoxide (CO)

The association rate constant of CO with reduced Hb is only 10% of that of O_2 so that CO is substantially limited by reaction with reduced Hb in the erythrocyte. On the other hand, the dissociation rate constant of CO from carbon monoxide-bound Hb (HbCO) is very small as compared with that of O_2, i.e., $4 \cdot 10^{-4}$ of O_2 [24], as the result of which the affinity of CO with reduced Hb is 234 times larger than that of O_2. Despite this fact, partial pressure of CO (P_{CO}) in the erythrocyte and thus in the capillary blood is not zero but shows a low but finite value depending on PO_2 surrounding Hb in the erythrocyte [11–13], which indicates that CO has a nature limited by perfusion, as well. P_{CO} in the capillary blood is represented by the equilibrium P_{CO} estimated from the CO dissociation curve constructed in the presence of O_2 [11–13]. Furthermore, it should be noted that under physiological conditions (normoxic PO_2 and low P_{CO}), the uptake of CO by the erythrocyte is predominantly limited by the replacement reaction between CO and HbO_2, which is certified to be much slower than the association reaction of CO with reduced Hb [24]. Hence, CO can be considered as the gas that is limited by the combination of perfusion, aqueous-phase diffusion, and reaction with Hb/HbO_2, among which replacement reaction of CO with HbO_2 may act as a major limiting process for the CO gas transfer in the blood. These facts suggest that CO is promising for estimating the diffusing capacity of the lung (D), particularly the blood component of D (D_B). Furthermore, CO may be useful for predicting the distribution of D/Q in the lung.

2.4.5 Nitric Oxide (NO)

The association rate constant of NO is of the same order of magnitude as that of O_2, but its dissociation rate constant is $1.5 \cdot 10^{-6}$ of O_2, which results in that the equilibrium constant (i.e., affinity) of NO with reduced Hb is $6.25 \cdot 10^5$ times larger than that of O_2 [24]. Therefore, NO is irreversibly absorbed by Hb, resulting in that partial pressure of NO (P_{NO}) in the erythrocyte (thus, in the capillary blood) is maintained at zero independent of erythrocyte PO_2 surrounding Hb; namely, there are no free NO molecules to be transported by perfusion, which indicates that NO is purely limited by aqueous-phase diffusion but neither by reaction with Hb nor by perfusion. These facts suggest that NO is the ideal gas for estimating the aqueous-phase diffusion-limited steps in the lung; that is, the process through the alveolocapillary tissue barrier (including plasma layer), which is represented by the membrane component of diffusing capacity (D_M), and that through the inside of the erythrocyte, which is represented by the diffusion-related process of blood component of diffusing capacity (D_B). However, as NO is not limited by perfusion, it is not useful for predicting the distribution of V_A/Q and/or D/Q in the lung.

3 Quantitation of Ventilation–Perfusion (V_A/Q) Distribution

3.1 Basic Rationale of V_A/Q *Distribution Estimated from Inert Gas Exchange*

Since inert gases are purely prescribed by perfusion but neither by aqueous-phase diffusion nor by reaction with Hb in the capillary blood, they are fundamentally promising for estimating the gas exchange that is simply decided by V_A/Q distribution. Under steady-state conditions in which there is no change in the partial pressure of an inert gas (P) against the elapsed time (t) at any portion of the lung (i.e., $\partial P/\partial t =$ zero), the partial pressure of a foreign inert gas in the gas phase (P_A) and that in blood phase (P_c) in a certain microregion of the lung are described by the simple equation in accordance with the law of mass balance [25];

$$P_A / P_v = P_c' / P_v = \lambda / \left[\lambda + \left(V_A / Q\right)\right] \tag{9.1}$$

where P_c' and P_v denote, respectively, the partial pressure of a given inert gas in end-capillary blood and mixed venous blood, whereas λ is the blood/gas partition coefficient of a gas (mL/mL at a given total pressure and temperature). The eq. (9.1) is valid when an indicator gas is not contained in the inspired air. The eq. (9.1) implies that the partial pressure of an inert gas in a lung microregion is simply decided by the ratio of alveolar ventilation (V_A) against capillary perfusion (Q) distributed there (Fig. 9.2) [7, 8]. Although there are many physiological assumptions required for the establishment of eq. 9.1, they were extensively addressed elsewhere [7, 8, 14]. The important issue that should be emphasized is that eq. (9.1) is implicitly prerequisite for the presence of microregion satisfying the definition of functional gas exchange unit. As extensively argued in the previous section, the microregion given by the HB acinus encounters the definition of functional gas exchange unit in expiration. The multiple inert gas elimination technique (MIGET), which predicts the continuous distribution of V_A/Q in the lung, uses gas-phase concentrations of inert gases collected during expiration (see Sect. 3.2). Therefore, the V_A/Q distribution investigated from this method can be interpreted on the basis of the functional gas exchange unit that corresponds anatomically to the HB acinus over expiration.

What we should be aware of is that the sum of V_A in each functional gas exchange unit amounts to the total effective alveolar ventilation defined by the total expired ventilation (V_E) minus the dead space ventilation. The dead space ventilation is fictitiously assumed for explaining the relationships between inspired, alveolar, and expired gases in the classical gas exchange theory. The dead space ventilation includes the ventilation formed not only by the alveolar regions with infinite V_A/Q

value (i.e., the alveolar dead space) but also by the functional dead space closely related to the anatomical dead space. The detailed arguments on what are the anatomical dead space, functional dead space, and dead space ventilation are deployed in Chap. 10.

3.2 Determination of V_A/Q Distribution by MIGET

Saline containing a small quantity of six foreign inert gases with a wide variety of λ, including sulfur hexafluoride (SF_6), ethane, cyclopropane, halothane, diethyl ether, and acetone (Fig. 9.4), is infused through the peripheral vein at a constant rate of 2 mL/min [7, 8, 26, 27]. After the steady-state is established, expired gas, arterial blood, and mixed venous blood are simultaneously collected and the concentrations of six inert gases in these samples are measured by gas chromatography [28]. Using the data measured for six inert gases, Wagner and colleagues [26, 27] established a novel method allowing for predicting a continuous distribution of V_A/Q in a representative manner. In this method, they assumed 50 functional gas exchange units with different V_A/Q values, including right-to-left shunt with V_A/Q of zero, dead space consisting of alveolar dead space with infinite V_A/Q value and functional dead space, and 48 units with finite values of V_A/Q ranging from 0.005 to 100, all of which are arranged in parallel and are equally spaced on a logarithmic scale. The basic equation used for determining the continuous distribution of fractional Q (q) along the V_A/Q axis is described as [7]:

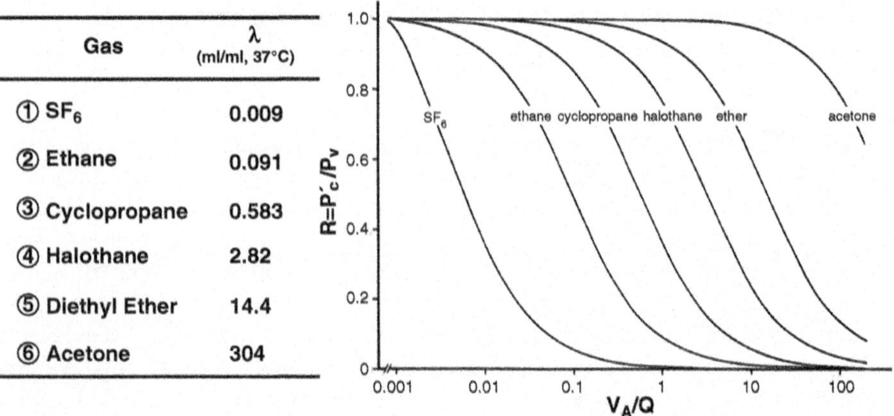

Gas	λ (ml/ml, 37°C)
① SF_6	0.009
② Ethane	0.091
③ Cyclopropane	0.583
④ Halothane	2.82
⑤ Diethyl Ether	14.4
⑥ Acetone	304

Fig. 9.4 Six inert gases used for estimating lung V_A/Q distribution. Retention (R) of an indicator gas is defined as P_c'/P_v, where P_c' and P_v are partial pressures of a gas in end-capillary blood and in mixed venous blood, respectively. R indicates gas exchange efficiency in a functional gas exchange unit. R differs largely depending on blood/gas partition coefficient (λ) and V_A/Q value, which suggests that these six inert gases are promising for separating functional gas exchange units with varied V_A/Q values

$$L = \Sigma W_i \cdot \left[R_i - \Sigma q_j \cdot \left(\lambda_i / \left(\lambda_i + (V_A / Q)_j \right) \right) \right]^2 + \mu \cdot \left(1 - \Sigma q_j \right) + S \cdot \Sigma \varphi_j \cdot q_j^2 \quad (9.2)$$

where L is the objective variable that should be minimized with respect to the fractional perfusion (q_j, $j = 1, 2 \cdots 50$) distributed to each functional gas exchange unit with a certain V_A/Q value. R_i is the measured value of arterial gas concentration for each inert gas divided by its mixed venous concentration (i=1, 2, ⋯ 6). W_i is the coefficient for weight of each inert gas, which is given by the reciprocal of variance of R_i. The second term of the equation denotes the equality constrain for q_j, the sum of which amounts to 1.0. μ is the Lagrange multiplier that is determined and replaced by an adequate value during computation. The third term designates the correction based on the enforced smoothing using the Lagrange multiplier of φ_j, which is the value related to the compartmental weight of each V_A/Q unit, while S is the smoothing term that is empirically assumed to be 40. The enforced smoothing is applied for stabilizing the resultant V_A/Q distribution. Replacing q_j in the eq. (9.2) by fractional ventilation (v_j), the ventilation distribution along the V_A/Q axis is readily decided, as well.

Recently, a new method for measuring the regional V_A/Q distribution has been developed, which is called the proton magnetic resonance imaging (MRI). This MRI technique combines arterial spin labeling measures of regional pulmonary blood flow with oxygen enhanced measures of regional specific ventilation and fast gradient echo measures of regional proton density [29]. Although the spatial resolution of the proton MRI method is high (about 1.0 cm^3), corresponding to the volume composed of 5–6 acini, it is not possible to determine the V_A/Q distribution of the whole lung. Thus, we consider that the newly developed proton MRI method does not surpass the classical MIGET.

3.3 Heterogeneous Distribution of V_A/Q in Diseased Lungs

The V_A/Q distribution estimated from the MIGET reflects the interregional heterogeneity of V_A/Q distribution at the acinar level. The basic consideration on the clinical interpretation of V_A/Q heterogeneity is extensively discussed in our previous paper [15] but briefly summarized in Table 9.1. The lesions forming acini with low V_A/Q values are primarily yielded by obstruction and/or destruction of acinar airways as well as microvascular injury with paralysis of hypoxic pulmonary vasoconstriction (HPV). If the HPV is paralytic in injured microvasculature, the perfusion there is not decreased so that the formation of low V_A/Q acini advances. The HPV paralysis was identified in some specific lung diseases, e.g., hepatopulmonary syndrome (HPS), acute respiratory syndrome (ARDS), and so forth.

The secondary cause for producing low V_A/Q acini is the redistribution of perfusion from affected to non-affected regions. This phenomenon is recognized in patients with microvascular loss or microvascular occlusion without HPV paralysis.

Table 9.1 Pathophysiological causes for generating acini with low or high V_A/Q

Low V_A/Q acini	High V_A/Q acini
Primary lesions	*Primary lesions*
1. Destruction or obstruction of acinar Airways	1. Microvascular loss
2. Microvascular injury with HPV Paralysis	2. Microvascular occlusion without HPV paralysis
3. Aqueous-phase diffusion limitation (diffusional shunt)	3. Aqueous-phase diffusion limitation (diffusional dead space)
Secondary lesions	*Secondary lesions*
1. Shift of perfusion from acini with microvascular loss or microvascular occlusion without HPV paralysis to normal acini	1. Shift of ventilation from affected acini to normal acini
	2. Shift of perfusion from normal acini to those with microvascular injury with HPV paralysis

HPV hypoxic vasoconstriction

However, the primary lesion of microvascular loss or microvascular occlusion without HPV paralysis is the genesis of acini with high V_A/Q values because of reduction in acinar perfusion there.

The secondary cause for forming high V_A/Q acini is the redistribution of ventilation from affected to non-affected regions. This is observed in patients with bronchiolitis obliterans (BO), COPD, cystic disease, pleuroparenchymal fibroelastosis (PPFE), and so forth. In a patient with microvascular injury with HPV paralysis, the perfusion in normal acini may be redistributed to the acini with HPV-paralytic microvascular lesions, thus converting the acini with normal V_A/Q values to those with relatively high V_A/Q values.

Of note, acini simultaneously revealing shunt and dead space effects are generated under conditions where their D/Q values are zero irrespective of V_A/Q values, i.e., complete aqueous-phase diffusion limitation. This issue will be discussed in detail in Sect. 4.

To address the considerations described above in a more clinical fashion, we present the V_A/Q distributions in two cases either with secondary alveolar proteinosis (AP) or pulmonary veno-occlusive disease (PVOD) in combination with pulmonary capillary hemangiomatosis (PCH) (Figs. 9.5 and 9.6).

4 Determination of V_A/Q and D/Q Distribution in the Lung

To deeply understanding the mechanism giving rise to hypoxemia, Piiper [9] elaborated the concept of V_A/Q–D/Q field (Fig. 9.7), in which D indicates aqueous-phase diffusing capacity in different regions of the lung. His theoretical analysis suggests that in addition to V_A/Q heterogeneity, D/Q heterogeneity similarly impairs overall

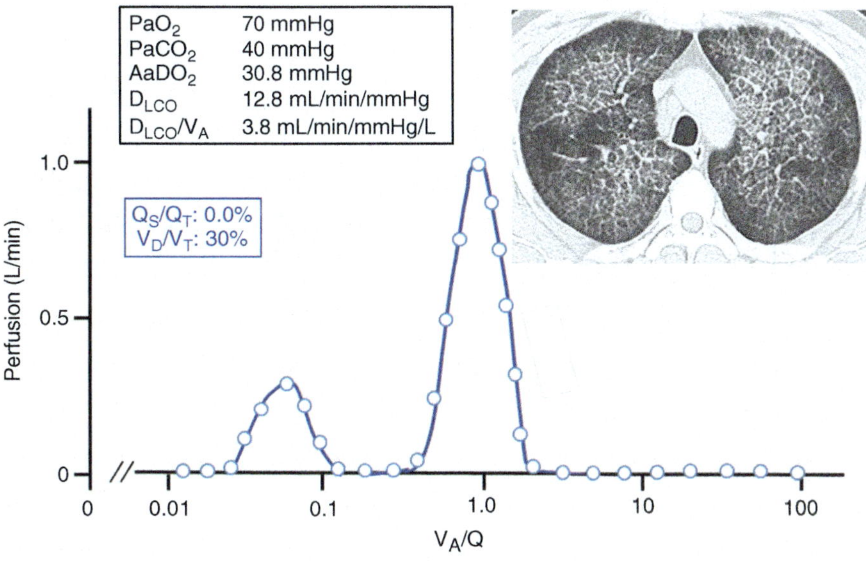

Fig. 9.5 V_A/Q distribution in a patient with secondary alveolar proteinosis (AP). Case is a 56-year-old, nonsmoking woman, who was observed for myelodysplastic syndromes (MDS) for a long time. She exhibited abnormal findings on chest X-P and increased serum CEA (10.2 ng/mL) as well as KL-6 (950 U/mL). Therefore, she was hospitalized for definitive diagnosis. Bronchoalveolar lavage fluid (BALF) contained significant amounts of PAS-stain positive substances, large foamy macrophages, and mononuclear macrophages. However, GM-CSF autoantibody was negative in BALF as well as in blood. Therefore, she was diagnosed as suffering from MDS-induced, secondary AP. Chest CT showed crazy paving appearance in which thickened lobular septa, thickened axial interstitial tissues, and ground-glass opacity (GGO) were identified. Pa_{O2} was decreased in association with widened AaD_{O2}. D_{LCO} was conspicuously decreased but D_{LCO}/V_A was moderately decreased. V_A/Q analysis in terms of MIGET revealed the formation of low V_A/Q acini that may be elicited by fluid accumulation in alveolar spaces and injuries of acinar airways

O_2 gas exchange, thus eliciting arterial hypoxemia. However, many studies done after Piiper's suggestion consistently demonstrated that the aqueous-phase diffusion-associated mechanism contributes minimally to the emergence of hypoxemia in patients with a variety of chronic lung diseases. The exception was the patients with interstitial lung disease during excise [30, 31], in which V_A/Q heterogeneity did not account for all of the hypoxemia and required a small but significant contribution of diffusion impairment. Although the major involvement of diffusion-related mechanism on hypoxemia was denied, it does not imply that the distribution of D/Q is homogeneous over the entire lung in any lung disease, but it simply implies that O_2 gas does not have a high sensitivity to detect the diffusion impairment, including D/Q heterogeneity. As depicted in Fig. 9.3, due to the physical natures of O_2 molecules in the blood, PO_2 easily reaches the equilibrium between alveolar gas and capillary blood even when the local diffusing capacity for O_2 (D_{O2}) is substantially impeded. However, if CO gas is used as the indicator, it does not reach the equilibrium between alveolar gas and capillary blood under most of the pathophysiological

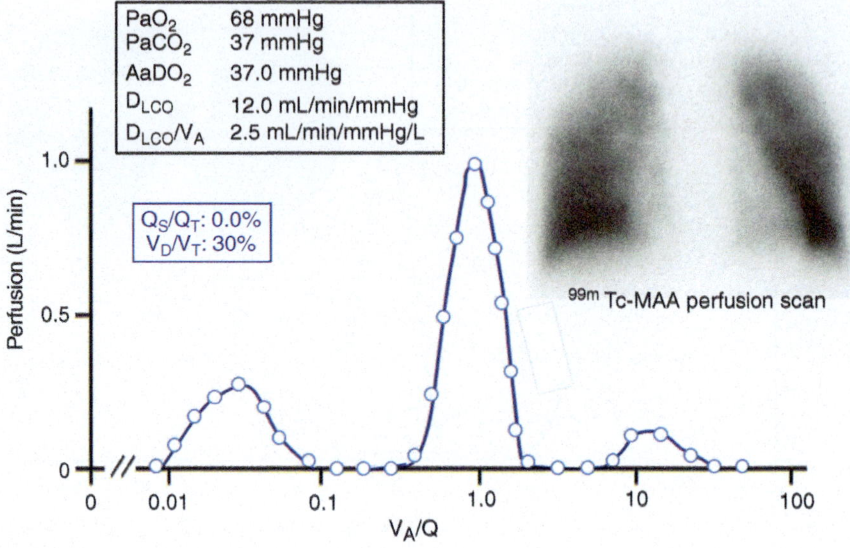

Fig. 9.6 V_A/Q distribution in a patient with pulmonary veno-occlusive disease (PVOD) and pulmonary capillary hemangiomatosis (PCH). Case is a 38-year-old, nonsmoking man, who underwent allogeneic hematopoietic stem cell transplantation (HSCT) against MDS. After HSCT, he was progressively conscious of the difficulty in breathing on exertion in association with arterial hypoxemia. CT examination showed thickened lobular septa with dilated peripheral pulmonary arteries. ^{99m}Tc-MAA perfusion scan showed a slight redistribution of perfusion from upper to lower lung fields. Lung biopsy under video-assisted thoracoscopy (VATS) confirmed the presence of PVOD concomitant with PCH. Pa_{O2} was decreased in association with widened AaD_{O2}. D_{LCO} and D_{LCO}/V_A were significantly decreased. V_A/Q analysis detected both high V_A/Q acini formed by microvascular occlusion (PVOD and PCH) without paralysis of hypoxic pulmonary vasoconstriction (HPV) and low V_A/Q acini generated by redistribution of perfusion from affected regions with microvascular occlusion to non-affected regions

conditions, which indicates that the sensitivity of CO gas for predicting D/Q heterogeneity in the lung is much higher than that of O_2 gas. These facts led the group of Yamaguchi [11–13] to develop the practical method allowing for simultaneously predicting the V_A/Q and D/Q distribution in the lung based on the measurements of gas exchange behaviors of nine indicator gases, including six inert gases (the same as used for MIGET), O_2, CO_2, and CO.

4.1 *Basic Rationale of* V_A/Q *and* D/Q *Distribution*

Yamaguchi et al. [11] took the HB acinus as the functional gas exchange unit, in which in addition to V_A and Q, D is included (Fig. 9.2). D indicates total diffusing capacity for respective indicator gases prescribed by the component of alveolocapillary tissue barrier (D_M) including plasma layer and that of erythrocytes (D_B) residing

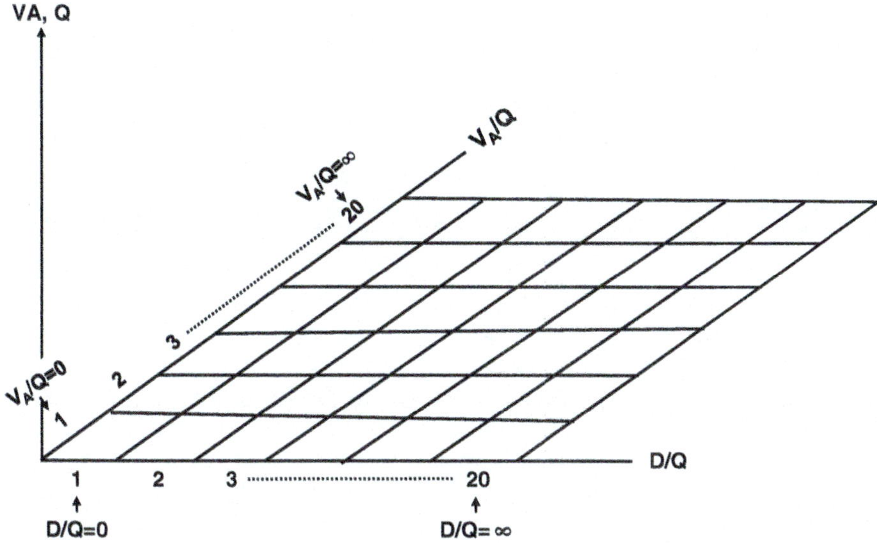

Fig. 9.7 V_A/Q-D/Q field. Functional gas exchange unit is assumed to have two functional parameters including V_A/Q and D/Q for prescribing gas exchange efficiency there. V_A/Q and D/Q values from zero to infinite are assigned to respective gas exchange units. These values are equally spaced on a logarithmic scale. 20 (along V_A/Q axis) × 20 (along D/Q axis) gas exchange units are assumed on V_A/Q-D/Q filed (i.e., 400 gas exchange units in total). Units with V_A/Q = zero and V_A/Q = infinite indicate right-to-left shunt and dead space, respectively. Units with $D/Q = 0$ express both effects of shunt (diffusional shunt) and dead space (diffusional dead space) irrespective of V_A/Q values there. Units with D/Q = infinite indicate those with no diffusion limitation, i.e., they satisfy classical V_A/Q concept

in the alveolar septa of a certain acinus. For the sake of simplicity, we first consider the inert gas exchange in the acinus on the assumption that it is limited by both V_A/Q and D/Q. The restriction of inset gas exchange by aqueous-phase diffusion takes place when D/Q is less than 10^{-4} mL (STPD)/mL/mmHg regardless of V_A/Q values [11]. The mass balance for inert gas exchange limited by both V_A/Q and D/Q in an acinus under steady-state conditions where inspired gas mixture contains no intended inert gas is described as:

$$P_A / P_v = \lambda \cdot (1-K) / [(V_A / Q) + \lambda \cdot (1-K)] \tag{9.3}$$

$$P_c' / P_v = [(V_A / Q) \cdot K + \lambda \cdot (1-K)] / [(V_A / Q) + \lambda \cdot (1-K)] \tag{9.4}$$

$$K = e^{[-D/(\beta \cdot Q)]} \tag{9.5}$$

where λ and β are the blood/gas partition coefficient and capacitance coefficient of a gas, respectively. For inert gases, β equals the Bunsen solubility coefficient (α) corrected for the difference in total pressure ($\beta = \alpha \bullet (760 - 47)/760$), while it equals the slope of each dissociation curve for a Hb-reactive gas such as O_2, CO_2, or CO. P_A and $P_c^{'}$ denote the partial pressure of an inert gas in alveolar space and that in end-capillary blood, respectively, while P_v is that in mixed venous blood. K is the term defining the effect of D/Q on the inert gas exchange. Of note, when D/Q is infinite (i.e., no aqueous-phase diffusion limitation), K equals zero, resulting in $P_A / P_v = P_c^{'} / P_v = \lambda / \left[\lambda + \left(V_A / Q\right)\right]$, which is the same as defined by the eq. (9.2) that describes the inert gas exchange limited by V_A/Q alone. On the other hand, when D/Q is zero (i.e., complete aqueous-phase diffusion limitation), K is 1.0, forming the functional gas exchange units where P_A and $P_c^{'}$ equal zero and P_v, respectively. Hence, the complete aqueous-phase diffusion limitation simultaneously elicits the shunt effect for capillary blood (diffusional shunt) and dead space effect for alveolar gas (diffusional dead space) (Fig. 9.2, Table 9.1).

In the case of reactive gases such as O_2, CO_2, and CO, however, the relationships between P_A and P_c are not straightforward because β values for these gases are not constant and change continuously according to their respective dissociation curves. Furthermore, O_2 and CO are contained in the inspired gas when attempting to decide the V_A/Q and D/Q distributions (see Sect. 4.2). Hence, the mass balance for a reactive gas is assumed as:

$$V_{AI} \bullet P_I - V_A \bullet P_A = Q \bullet \left(C_c - C_v\right) / \beta_g \tag{9.6}$$

$$\Delta M = \left(P_A - P_c\right) \bullet \Delta D = Q \bullet \Delta C_c \tag{9.7}$$

where V_{AI} and V_A are inspired and expired alveolar ventilation, respectively, while P_I, P_A, and P_c are partial pressure of a reactive gas in an inspired gas mixture, that in expired alveolar gas, and that at a certain point of capillary blood, respectively. C_c and C_v are contents of a reactive gas in capillary blood and mixed venous blood, respectively. βg indicates the capacitance coefficient of a gas in the alveolar gas phase, which generally equals 1/(760-47) at the sea level. ΔM is the transfer rate of a gas through ΔD. Eq. (9.7) is the differential equation that expresses the relationship between P_A and P_c at a certain microregion of Δ in the alveolocapillary tissue barrier. The Bohr integration from mixed venous point (P_v) to end-capillary point ($P_c^{'}$) using eq. (9.7) describes the V_A/Q and D/Q-prescribed equilibration kinetics between P_A and $P_c^{'}$ (Fig. 9.8). The O_2 gas exchange is significantly restricted by aqueous-phase diffusion and arterial hypoxemia emerges when D/Q is less than 10^{-3}. Therefore, the units with the D/Q of less than 10^{-3} is defined as the low D/Q units.

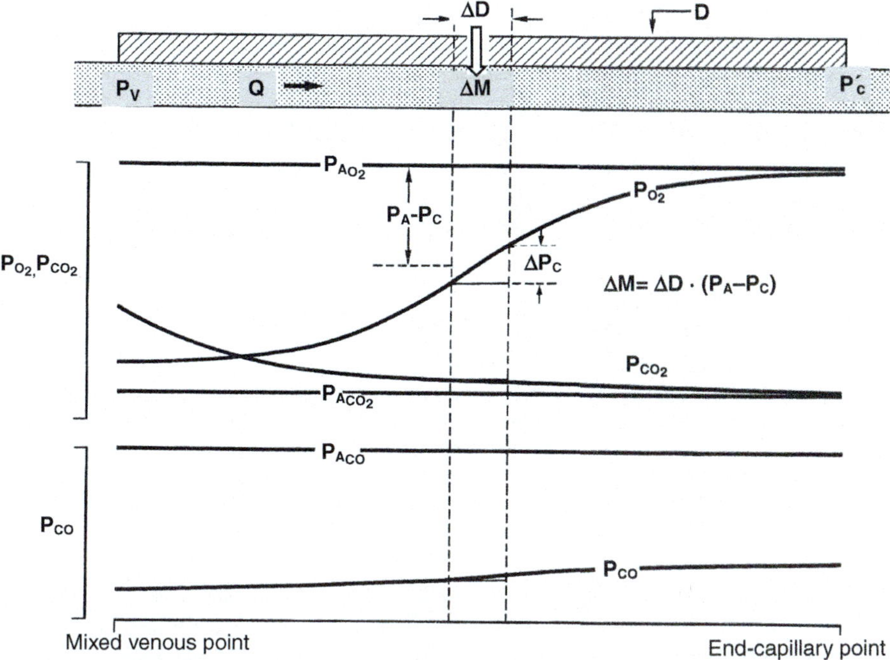

Fig. 9.8 Bohr integrations for O_2, CO_2, and CO. O_2, CO_2, and CO interact in capillary blood in a complicated manner and the capacitance coefficient of each gas is not constant, changing depending on their partial pressures. Therefore, the decision of partial pressure profiles of these gases in capillary blood requires Bohr integration between mixed venous blood and end-capillary blood. For this integration, respective dissociation curves for O_2, CO_2, and CO, in which interactions between three gases are taken into consideration, should be used

4.2 *Determination of* V_A/Q *and* D/Q *Distributions*

To estimate the V_A/Q and D/Q distribution in the lung, Yamaguchi et al. [11–13] infused six inert gases that are used for the MIGET (SF_6, ethane, cyclopropane, halothane, diethyl ether, and acetone) while inspiring the gas mixture containing 21% O_2 and 0.1% CO. Although CO_2 is not contained in the inspired gas, the CO_2 produced in the body was used as one of the indicator gases. The physical properties of nine indicator gases are depicted in Table 9.2. After the steady-state was established, expired gas, arterial blood, and mixed venous blood were simultaneously collected and the concentrations and/or contents of nine indicator gases in respective samples were measured with the appropriate methods.

The first crucial issue in the analysis of V_A/Q and D/Q distributions is how to define the diffusing capacity (D) of each indicator gas in the resistive alveolar septa

Table 9.2 Physical properties of nine gases used for determining V_A/Q and D/Q distributions

Gas species	λ (mL/mL)	MW (g/mol)	Tissue diffusivity (d, cm^2/s)	Tissue solubility (α, mL/mL/mmHg)
SF_6	0.009	146	$1.1 \cdot 10^{-5}$	$1.03 \cdot 10^{-5}$
Ethane	0.091	30	$2.4 \cdot 10^{-5}$	$1.06 \cdot 10^{-4}$
Cyclopropane	0.583	42	$2.0 \cdot 10^{-5}$	$6.76 \cdot 10^{-4}$
Halothane	2.82	197.5	$0.9 \cdot 10^{-5}$	$3.27 \cdot 10^{-3}$
Diethyl ether	14.4	74	$1.5 \cdot 10^{-5}$	$1.67 \cdot 10^{-2}$
Acetone	304	58	$1.7 \cdot 10^{-5}$	$3.53 \cdot 10^{-1}$
O_2	Variable	32	$2.3 \cdot 10^{-5}$	$2.93 \cdot 10^{-5}$
CO_2	Variable	44	$2.0 \cdot 10^{-5}$	$6.34 \cdot 10^{-4}$
CO	Variable	28	$2.5 \cdot 10^{-5}$	$2.49 \cdot 10^{-5}$

λ: blood/gas partition coefficient. λ values are not constant but variable for O_2, CO_2, and CO. This is because O_2, CO_2, and CO are combined with Hb in erythrocytes, thus producing dissociation curves, the slopes of which (effective solubilities) change dependent on partial pressures of O_2, CO_2, and CO surrounding Hb. λ values of O_2, CO_2, and CO are much larger than those of inert gases. MW: molecular weight. d: diffusion coefficient in resistive tissue barrier. α: Bunsen solubility coefficient in resistive tissue barrier. Values are expressed at 37 °C

composed of the alveolocapillary membrane, plasm layer, and erythrocytes in a functional gas exchange unit. As discussed in Sect. 2.4, inert gases and O_2 are predominantly limited by aqueous-phase diffusion in functional gas exchange units where D/Q is less than 10^{-4} for inert gases and less than 10^{-3} for O_2. Therefore, under the condition with D/Q of less than 10^{-4}, D for an inert gas (D_{inert}) is calculated from D_{O2} as follows:

$$D_{inert} = D_{O2} \cdot \left(KR_{inert} / KR_{O2} \right) \tag{9.8}$$

where KR_{inert} and KR_{O2} are, respectively, the Krogh diffusion constants for inert gases and O_2, which are expressed by the products of Bunsen solubility coefficients (α) and diffusion coefficients (d) in the resistive tissue barrier (Table 9.2). However, interrelationships between D_{O2}, D_{CO2}, and D_{CO} are complicated because these gases react with Hb in qualitatively different manners. Therefore, D_{CO2}/D_{O2} and D_{CO}/D_{O2} were adopted from the values measured experimentally under the rebreathing condition where functional heterogeneities influencing the diffusing capacities for O_2, CO_2, and CO were greatly diminished [32, 33]. Thus, D_{CO2}/D_{O2} and D_{CO}/D_{O2} were, respectively, postulated to be 3.3 and 0.85 in the analysis, implying that D_{O2} was used as the representative of diffusing capacities for all nine indicator gases.

The second crucial issue is how to treat nonlinear dissociation curves for O_2, CO_2, and CO. These dissociation curves are needed when Bohr integrations for O_2, CO_2, and CO in capillary blood are conducted. In respect of O_2 and CO dissociation curves for human blood, Yamaguchi et al. [11, 13] applied those constructed in the presence of both gases, in which CO_2-related Bohr effect and intracellular 2,3-diphosphoglycerate (DPG) in addition to competitive reaction between CO and O_2 were correctly considered. On the other hand, they used the CO_2 dissociation curve reported by Kelman [34] on the assumption that the ratio determining the distribution of CO_2 between the erythrocyte and the plasma was a function of pH and total saturation of Hb with O_2 and CO.

As for the determination of V_A/Q and D/Q distributions, 20 units were, respectively, assigned along the V_A/Q and D/Q axis (Fig. 9.7); that is, the lung is assumed to be consisting of 400 functional gas exchange units in total. To minimize the sum of squares of deviation (L) between the measured arterial concentrations of nine indicator gases and the values predicted from the V_A/Q–D/Q lung model, the enforced smoothing technique qualitatively the same as that used by Wagner and colleagues [26, 27] for deciding the V_A/Q distribution (eq. 9.2) was applied:

$$L = \Sigma W_{\mathrm{i}} \bullet \left[C_{\mathrm{ai}} - \Sigma q_{\mathrm{j}} \bullet C'_{\mathrm{ci,j}} \right]^2 + \mu \bullet \left(1 - \Sigma q_{\mathrm{j}} \right) + \Sigma \varphi_{\mathrm{j}} \bullet q_{\mathrm{j}}^2 \qquad (9.9)$$

where C_{ai} is the measured value of arterial content concerning each indicator gas ($i = 1, 2, \cdots 9$), while $C'_{\mathrm{ci,j}}$ is the predicted content of each indicator gas in the end-capillary blood of each functional gas exchange unit ($j = 1, 2, \cdots 400$). W_{i} is the coefficient for weight of each indicator gas, which is given by the reciprocal of variance of C_{ai}. However, the Lagrange multiplier of φ_{j}, which is the value related to the weight of each functional gas exchange unit, was not identical to that incorporated in the MIGET. Instead, it was empirically decided in the V_A/Q–D/Q analysis [11]:

$$\varphi_{\mathrm{j}} = \left[20 \bullet \left\{ 1 + \left(Q_{\mathrm{T}} / V_{\mathrm{E}} \right)^2 \bullet \left(V_{\mathrm{A}} / Q \right)_{\mathrm{j}}^2 \right\} \bullet \left\{ 1 + 10^{n} \bullet \left(Q_{\mathrm{T}} / V_{\mathrm{E}} \right)^2 \bullet \left(D / Q \right)_{\mathrm{j}}^2 \right\} \bullet \left\{ 1 + 10^{-4} / \left(D / Q \right)_{\mathrm{j}} \right\}^2 \right]^{1/2} \qquad (9.10)$$

where $n = 3$ when D/Q is less than 0.1 and $n = 5$ when D/Q is over 0.1. Q_T and V_E denote total perfusion and total ventilation of the lung, respectively.

4.3 Heterogeneous Distribution of V_A/Q *and* D/Q *in Diseased Lungs*

The V_A/Q–D/Q analysis [13] demonstrated that the low D/Q units (D/Q is less than 10^{-3}) that impair O_2 gas exchange and give rise to arterial hypoxemia are found in idiopathic pulmonary fibrosis (IPF) with active alveolitis or extensive fibrosis (Fig. 9.9). In the IPF, about 20% of impaired O_2 gas exchange is attributed to the low D/Q units. These findings are explicable from the heterogeneous infiltration of inflammatory substances into alveolar interstitium and alveolar space (i.e., alveolitis) or from the heterogeneous fibrosis of alveolar interstitium leading to increased thickness and decreased surface area of alveolar septa (i.e., extensive fibrosis). The significant D/Q heterogeneity in patients with IPF is anticipated to be enhanced during exercise that further increases the D/Q heterogeneity [30]. Canine model with ARDS also revealed the prominent D/Q heterogeneity [12]. This may be explained

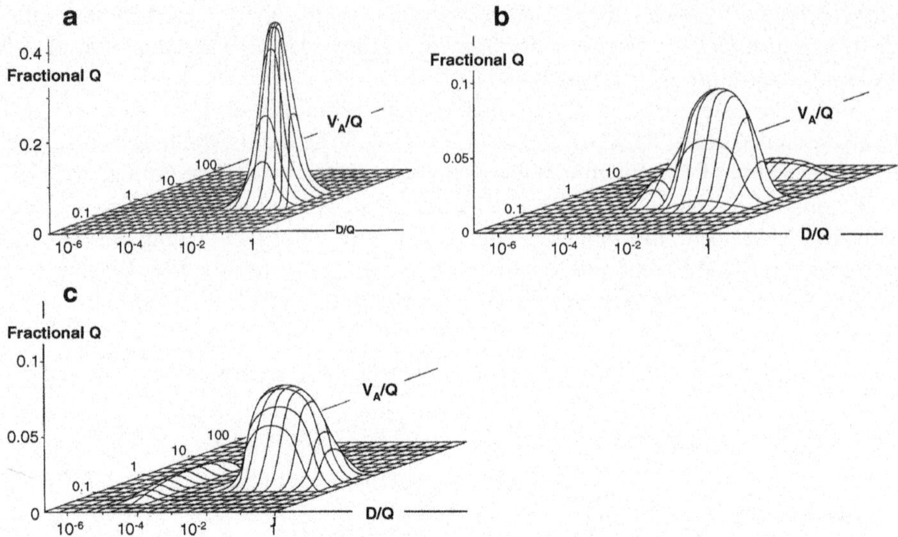

Fig. 9.9 V_A/Q and D/Q distributions in various lung diseases. V_A/Q and D/Q distributions in normal lung (**a**), in lung with COPD (predominantly emphysema) (**b**), and in lung with idiopathic pulmonary fibrosis (IPF) having active alveolitis (**c**). COPD patient with predominant emphysema revealed V_A/Q heterogeneity with high V_A/Q regions but minimal regions with very low D/Q values. Patient with IPF in association with active alveolitis exhibited minimal V_A/Q heterogeneity but very low D/Q regions that yield diffusional shunt effect that gives rise to hypoxemia. Very low D/Q regions may be caused by decreased surface area and/or increased thickness of the alveolocapillary membrane. Adopted from Yamaguchi et al. [13] with permission of the publisher

by the accumulation of edematous fluid into alveolar interstitium and airspace, which augments the aqueous-phase diffusion limitation in a heterogeneous manner. On the other hand, there are no low D/Q units in COPD, suggesting that the major factor for eliciting hypoxemia in COPD is V_A/Q heterogeneities but not D/Q heterogeneities.

5 Conclusion Remarks

The functional gas exchange unit conforms to the anatomical gas exchange defined as the acinus during expiration but strictly not during inspiration. However, as most of the parameters, including V_A/Q and D/Q distributions allowing for assessing the gas exchange dynamics in the lung, are based on the measurements of gas samples collected over expiration, the gas exchange abnormalities can be interpreted on the basis of structural alterations of acinar airways, alveoli, microcirculation, and alveolar interstitial organizations, the conclusive judgments of them being not possible even when the high-precision diagnostic imaging technologies are applied. The V_A/Q distribution decided with the multiple inert gas elimination technique is clinically useful for detecting the abnormalities of airways and microcirculation in the acinus of a variety of lung diseases. On the other hand, the distribution of V_A/Q and D/Q predicts the abnormalities residing in alveolar interstitial tissues in addition to those in acinar airways and microcirculation. The V_A/Q-D/Q analysis demonstrated that the aqueous-phase diffusion exerts a small but significant impact on O_2 gas exchange and induces arterial hypoxemia in patients with IPF but not with COPD.

References

1. Keith A. Further advances in physiology. In: Hill L, editor. The mechanism of respiration in man. London: Arnold; 1909.
2. Rohrer F. Der Strömungswiderstand in den menschlichen Atemwegen und der Einfluss der unregelmässigen Verzweigung des Bronachialsystems auf den Atmungsverlauf in verschiedenen Lungernbezirke. Pfluegers Arch Ges Physiol. 1915;162:225–9.
3. Krogh A, Lindhard J. The volume of the dead space in breathing and the mixing of gases in the lungs of man. J Physiol (Lond). 1917;51:59–90. https://doi.org/10.1113/jphysiol.1917.sp001785
4. Haldane JS. Respiration. New Haven, CT: Yale University Press; 1922.
5. Fenn WO, Rahn H, Otis AB. A theoretical study of the composition of alveolar air at altitude. Am J Physiol. 1946;146:637–53. https://doi.org/10.1152/ajplegacy.1946.146.5.637.
6. Riley RL, Cournand A. "Ideal" alveolar air and the analysis of ventilation/perfusion relationships in the lungs. J Appl Physiol. 1949;1:825–47. https://doi.org/10.1152/jappl.1949.1.12.825.
7. Wagner PD, West JB. Ventilation-perfusion relationships. In: West JB, editor. Pulmonary gas exchange, Ventilation, blood flow, and diffusion, vol. I. New York: Academic; 1980. p. 219–62.
8. West J, Wagner P. Chap. 126. Ventilation-perfusion relationships. In: Crystal RG, West JB, Barnes PJ, Weibel ER, editors. The LUNG Scientific Foundations, vol. II. 2nd ed. Philadelphia, New York: Lippincott-Raven Publishers; 1997. p. 1693–709.
9. Piiper J. Variations of ventilation and diffusing capacity to perfusion determining the alveolar-arterial O_2 difference: theory. J Appl Physiol. 1961;16:507–10. https://doi.org/10.1152/jappl.1961.16.3.507.
10. Scheid P, Piiper J. Chap. 125. Diffusion. In: Crystal RG, West JB, Barnes PJ, Weibel ER, editors. The LUNG Scientific Foundations, vol. II. 2nd ed. Philadelphia,·New York: Lippincott-Raven Publishers; 1997. p. 1681–91.
11. Yamaguchi K, Kawai A, Mori M, Asano A, Takasugi T, Umeda A, et al. Distribution of ventilation and of diffusing capacity to perfusion in the lung. Respir Physiol. 1991;86:171–87. https://doi.org/10.1016/0034-5687(91)90079-X.
12. Yamaguchi K, Mori M, Kawai A, Takasugi T, Asano K, Oyamada Y, et al. Ventilation-perfusion inequality and diffusion impairment in acutely injured lungs. Respir Physiol. 1994;98:165–77. https://doi.org/10.1007/BF01277020.
13. Yamaguchi K, Mori M, Kawai A, Takasugi T, Oyamada Y, Koda E. Inhomogeneities of ventilation and the diffusing capacity to perfusion in various chronic lung diseases. Am J Respir Crit Care Med. 1997;156:86–93. https://doi.org/10.1164/ajrccm.156.1.9607090.
14. Yamaguchi K, Tsuji T, Aoshiba K, Nakamura H, Abe S. Anatomical backgrounds on gas exchange parameters in the lung. World J Respirol. 2019;9(2):8–28. https://doi.org/10.5320/wjr.v9.i2.8.
15. Yamaguchi K, Tsuji T, Aoshiba K, Nakamura H, Abe S. Insights into anatomical basis prescribing ventilation-perfusion distribution in lung periphery. Austin J Pulm Respir Med. 2019;6(1):1060.
16. Haefeli-Bleuer B, Weibel ER. Morphometry of the human pulmonary acinus. Anat Rec. 1988;220:401–14. https://doi.org/10.1002/ar.1092200410.

17. Sobin SS, Fung YC, Lindal RG, Tremer HM, Clark L. Topology of pulmonary arterioles, capillaries, and venules in the cat. Microvasc Res. 1980;19:217–33. https://doi.org/10.1016/0026-2862(80)90042-4.
18. König MF, Lucocq JM, Weibel ER. Determination of pulmonary vascular perfusion by electron and light microscopy. J Appl Physiol. 1993;75:1877–83. https://doi.org/10.1152/jappl.1993.75.4.1877.
19. Tanabe N, Todoran TM, Zenk GM, Bunton BR, Wagner WW Jr, Presson RG Jr. Perfusion heterogeneity in the pulmonary acinus. J Appl Physiol. 1998;84:933–8. https://doi.org/10.1152/jappl.1998.84.3.933.
20. Paiva M. Gas transport in the human lung. J Appl Physiol. 1973;43:401–10. https://doi.org/10.1111/j.1939-0025.1973.tb00811.x.
21. Swan AJ, Tawhai MH. Evidence for minimal oxygen heterogeneity in the healthy human pulmonary acinus. J Appl Physiol. 2011;110:528–37. https://doi.org/10.1152/japplphysiol.00888.2010.
22. Sapoval B, Filoche M, Weibel ER. Smaller is better - but not too small: a physical scale for the design of the mammalian pulmonary acinus. Proc Natl Acad Sci USA. 2002;99:10411–6. https://doi.org/10.1073/pnas.122352499.
23. Felici M, Filoche M, Straus C, Similowski T, Sapoval B. Diffusional screening in real 3D human acini - a theoretical study. Respir Physiol Neurobiol. 2005;145:279–93. https://doi.org/10.1016/j.resp.2004.10.012.
24. Chakraborty S, Balakotaiah V, Bidani A. Diffusing capacity reexamined: relative roles of diffusion and chemical reaction in red cell uptake of O_2, CO, CO_2, and NO. J Appl Physiol. 2004;97:2284–302. https://doi.org/10.1152/japplphysiol.00469.2004.
25. Kety S. The theory and applications of the exchange of inert gas at the lungs and tissues. Pharmacol Rev. 1951;3:1–41. https://doi.org/10.1136/bmj.1.4696.41-b.
26. Wagner PD, Saltzman H, West JB. Measurement of continuous distributions of ventilation-perfusion ratios: Theory. J Appl Physiol. 1974;36:588–99. https://doi.org/10.1152/jappl.1974.36.5.588.
27. Wagner PD, Laravuso RB, Uhl RR, West JB. Continuous distributions of ventilation-perfusion ratios in normal subjects breathing air and 100% O_2. J Clin Invest. 1974;54:54–68. https://doi.org/10.1172/JCI107750
28. Wagner PD, Naumann PF, Laravuso RB. Simultaneous measurement of eight foreign gases in blood by gas chromatography. J Appl Physiol. 1974;36:600–5. https://doi.org/10.1152/jappl.1974.36.5.600.
29. Sá RC, Henderson AC, Simonson T, Arai TJ, Wagner H, Theilmann RJ, et al. Measurement of the distribution of ventilation-perfusion ratios in the human lung with proton MRI: comparison with the multiple inert-gas elimination technique. J Appl Physiol. 2017;123:136–46. https://doi.org/10.1016/B978-3-437-21455-4.00009-X.
30. Wagner P. Ventilation-perfusion inequality & gas exchange during exercise in lung disease. In: Dempsey JA, Reed CE, editors. Muscular exercise and the lung. Madison, WI: The University of Wisconsin Press; 1977. p. 345–56.
31. Agustí AG, Roca J, Gea J, Wagner PD, Xaubet A, Rodriguez-Roisin R. Mechanisms of gas-exchange impairment in idiopathic pulmonary fibrosis. Am Rev Respir Dis. 1991;143:219–25. https://doi.org/10.1164/ajrccm/143.2.219.
32. Piiper J, Meyer M, Marconi C, Scheid P. Alveolar-capillary equilibration kinetics of $^{13}CO_2$ in human lungs studied by rebreathing. Respir Physiol. 1980;42:29–41. https://doi.org/10.1080/00150198008009004.
33. Meyer M, Scheid P, Riepl G, Wagner HJ, Piiper J. Pulmonary diffusing capacities for O_2 and CO measured by a rebreathing technique. J Appl Physiol. 1981;51:1643–50. https://doi.org/10.1152/jappl.1981.51.6.1643.
34. Kelman GR. Digital computer procedure for the conversion of PCO_2 into blood CO_2 content. Respir Physiol. 1967;22:659–74. https://doi.org/10.1152/jappl.1967.22.4.659.

Chapter 10
Physiological Basis of Effective Alveolar–Arterial O_2 Difference (AaD_{O2}) and Effective Physiological Dead Space (V_D/V_T): Are AaD_{O2} and V_D/V_T Supported by Structurally and Physiologically Correct Knowledges?

Kazuhiro Yamaguchi

Abstract Although determination of ventilation–perfusion (V_A/Q) distribution and/or simultaneous determination of V_A/Q and diffusing capacity–perfusion (D/Q) distribution are needed to precisely diagnose gas exchange abnormalities occurring in the lung periphery, these analyses cannot be straightforwardly performed in clinical situation. On the other hand, effective alveolar–arterial P_{O2} difference (effective AaD_{O2}) and physiological dead space (V_D/V_T), both of which are appraised as a base of the classical gas exchange theory, can be measured with ease. In classical gas exchange theory, effective alveolar gas compositions (effective P_{AO2} and P_{ACO2}), effective AaD_{O2}, and physiological V_D/V_T are clinically important while diagnosing gas exchange abnormalities concerning O_2 and CO_2 in a first approximation. Key assumption allowing classical gas exchange theory to be applied for clinical practice is that effective P_{ACO2} can be replaced with arterial P_{CO2} (P_{aCO2}) that is measured routinely at bedside. If this assumption is not adopted, measurements of P_{O2} and P_{CO2} in mixed venous blood, which are only possible through right heart catheterization, are indispensable for determining representatives of P_{AO2} and P_{ACO2}. When effective P_{ACO2} (= P_{aCO2}) is known, other indicators, including effective P_{AO2}, effective AaD_{O2}, and physiological V_D/V_T, can be easily decided. Effective AaD_{O2} functions as the indicator for identifying regions with low V_A/Q values, including right-to-left shunt (V_A/Q = zero), while physiological V_D/V_T is useful for detecting regions with high V_A/Q values, including alveolar dead space (V_A/Q = infinite).

K. Yamaguchi (✉)
Department of Respiratory Medicine, Tokyo Medical University, Tokyo, Japan

Division of Respiratory Medicine, Tohto Clinic, Kenko-Igaku Association, Tokyo, Japan
e-mail: yamaguc@sirius.ocn.ne.jp

K. Yamaguchi (ed.), *Structure-Function Relationships in Various Respiratory Systems*, Respiratory Disease Series: Diagnostic Tools and Disease Managements, https://doi.org/10.1007/978-981-15-5596-1_10

Keywords Classical gas exchange theory · Mean or ideal alveolar gas · Effective alveolar gas · Alveolar–arterial P_{O2} difference (effective AaD_{O2}) · Physiological dead space (V_D/V_T)

1 Introduction

About 100 years ago, Krogh et al. [1] first described the important principle in the field of respiratory physiology; namely, the gas exchange taking place in any lung region is determined by neither the ventilation (V_A) nor the blood flow (Q) but by the ratio of one another. Shortly after this, Haldane [2] recognized that ventilation–perfusion (V_A/Q) heterogeneity could cause hypoxemia. These pioneering works led Fenn and Rahn [3], Riley and Cournand [4], and others into establishment of classical gas exchange theory in a more precise fashion. Rahn [5] proposed the concept of "mean alveolar gas," while Riley et al. [4, 6] proposed the concept of "ideal alveolar gas." These two terms are physiologically identical. However, to calculate the mean (ideal) alveolar gas, it is necessary to measure P_{O2} and P_{CO2} in the mixed venous blood. As the mixed venous blood sampling is extremely cumbersome in a clinical setting, Riley et al. [6] proposed the new concept of "effective alveolar gas," in which the mean (ideal) alveolar P_{CO2} (P_{ACO2}) is taken to equal the arterial P_{CO2} (P_{aCO2}), thus allowing for estimating alveolar P_{O2} (P_{AO2}) and P_{ACO2} with no information of P_{O2} and P_{CO2} in the mixed venous blood. Ever since, the concept of effective alveolar gas has got a great deal of support from physicians at bedside. This concept has led the alveolar–arterial P_{O2} difference (AaD_{O2}) and the physiological dead space for CO_2 (V_D/V_T) to be useful in a clinical situation for rating the impaired gas exchange of O_2 and CO_2, respectively. More recently, Wagner and colleagues [7, 8] established the intriguing method for measuring the continuous distribution of V_A/Q in the lung, i.e., the multiple inert gas elimination technique (MIGET). Furthermore, Yamaguchi et al. [9–11] developed the method by which the distribution of V_A/Q and diffusing capacity–perfusion (D/Q) in the lung is simultaneously decided. Unfortunately, however, these elaborate methods also require the mixed venous blood sampling through the right heart catheter inserted into the main pulmonary artery, which indicates that the applications of these methods are largely limited in a clinical situation while diagnosing the pathophysiological conditions of a wide variety of lung diseases. Based on these historical backgrounds, the present chapter highlights the following issues: (1) basic rationale of classical gas exchange theory, including mean and ideal alveolar gas; (2) effective alveolar air and effective alveolar–arterial P_{O2} difference (AaD_{O2}); and (3) physiological dead space for CO_2 (V_D/V_T for CO_2).

2 Basic Rationale of Classical Gas Exchange Theory: Mean and Ideal Alveolar Gas

The important assumption underlying the classical gas exchange theory is that the compositions of alveolar gas (P_{AO2} and P_{ACO2}) are primarily prescribed by the metabolic state of the whole body irrespective of the pathophysiological conditions in the lung. The circumstance of body metabolism is represented by the total O_2 consumption (M_{O2}) and the total CO_2 production (M_{CO2}), which define the respiratory quotient that is alternatively called the CO_2/O_2 gas exchange ratio ($R = M_{CO2}/M_{O2}$). Under the steady-state condition with room air-breathing, M_{O2}, M_{CO2}, and, thus, R are identical at any portion of the body, including peripheral organs and tissues, systemic circulation, pulmonary circulation, alveolar region, and mouth level (Fig. 10.1). Therefore, R in the pulmonary capillary blood (R_b) is equal to that in the alveolar gas phase (R_g) and that at the mouth level. Furthermore, as N_2 is inactive

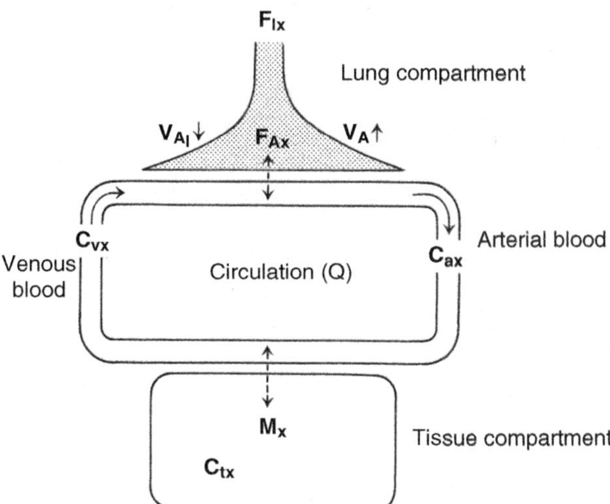

Fig. 10.1 Metabolic demand in the whole body under steady-state condition M_x: O_2 consumption or CO_2 production. M_{N2} is zero. C_{tx}, C_{vx}, and C_{ax}: contents of O_2, CO_2, or N_2 in tissues/organs, mixed venous blood, and arterial blood, respectively. Q: cardiac output. V_{AI} and V_A: inspiratory and expiratory alveolar ventilation, respectively. F_{Ix} and F_{Ax}: fractional concentrations of O_2, CO_2, or N_2 in inspiratory and alveolar gas, respectively. M_{O2} and M_{CO2} are constant anywhere in the body under steady-state conditions, resulting in that R defined as M_{CO2}/M_{O2} is also constant at any site. The concept of metabolic demand of the whole body under steady-state conditions decides representatives of alveolar gas compositions for O_2, CO_2, and N_2 irrespective of morbid states of the lung, which make V_A/Q distribution in the lung heterogenous

and not metabolized in the body, consumption or production of N_2 (M_{N2}) is zero under the steady-state condition [12, 13]. These considerations establish the following equations under the steady-state condition during room air-breathing in which inspired CO_2 fraction (F_{ICO2}) is zero:

$$M_{O2} = V_{AI} \cdot F_{IO2} - V_A \cdot F_{AO2} = Q \cdot (C_{cO2} - C_{vO2}) \quad (10.1)$$

$$M_{CO2} = V_A \cdot F_{ACO2} - V_{AI} \cdot F_{ICO2} = V_A \cdot F_{ACO2} = Q \cdot (C_{vCO2} - C_{cCO2}) \quad (10.2)$$

$$M_{N2} = V_{AI} \cdot F_{IN2} - V_A \cdot F_{AN2} = Q \cdot (C_{cN2} - C_{vN2}) = \text{zero} \quad (10.3)$$

$$F_{IN2} = 1 - F_{IO2} \quad (10.4)$$

$$F_{AN2} = 1 - F_{AO2} - F_{ACO2} \quad (10.5)$$

$$R_g = M_{CO2} / M_{O2} = V_A \cdot F_{ACO2} / (V_{AI} \cdot F_{IO2} - V_A \cdot F_{AO2}) \quad (10.6)$$

$$R_b = M_{CO2} / M_{O2} = (C_{vCO2} - C_{cCO2}) / (C_{cO2} - C_{vO2}) \quad (10.7)$$

$$R_g = R_b \quad (10.8)$$

where V_{AI} and V_A are inspired and expired alveolar ventilation, respectively, while Q is pulmonary blood flow. F_I and F_A denote inspired and expired alveolar gas fractions of O_2, CO_2, and N_2, respectively. C_c and C_v are gas contents of O_2, CO_2, and N_2 in capillary blood and mixed venous blood, respectively. Solving these equations, P_{AO2} (= F_{AO2}·(P_B-47), in which P_B indicates barometric pressure) and P_{ACO2} (= F_{ACO2}·(P_B-47)) can be expressed by the single value of V_A/Q of the whole lung under the condition where inspired fractions and mixed venous contents of O_2 and CO_2 are fixed. In other words, the intersection of R_b and R_g lines, constructed by changing the relationship between M_{O2} and M_{CO2}, defines the single V_A/Q value and the corresponding unique pair of P_{O2} and P_{CO2} in alveolar gas and capillary blood of a certain lung. Such graphical analysis has been called the O_2-CO_2 diagram (alternatively, the Fenn diagram) or the V_A/Q line (Fig. 10.2) [14, 15]. The P_{AO2} and P_{ACO2} values thus estimated were defined as the mean alveolar gas by Rahn [5] or the ideal alveolar gas by Riley et al. [4, 6], both of which are the same in a physiological sense.

Of note, although the mean (ideal) P_{AO2} and P_{ACO2} satisfactorily meet the metabolic state of the whole body, it does not secure that these P_{AO2} and P_{ACO2} can

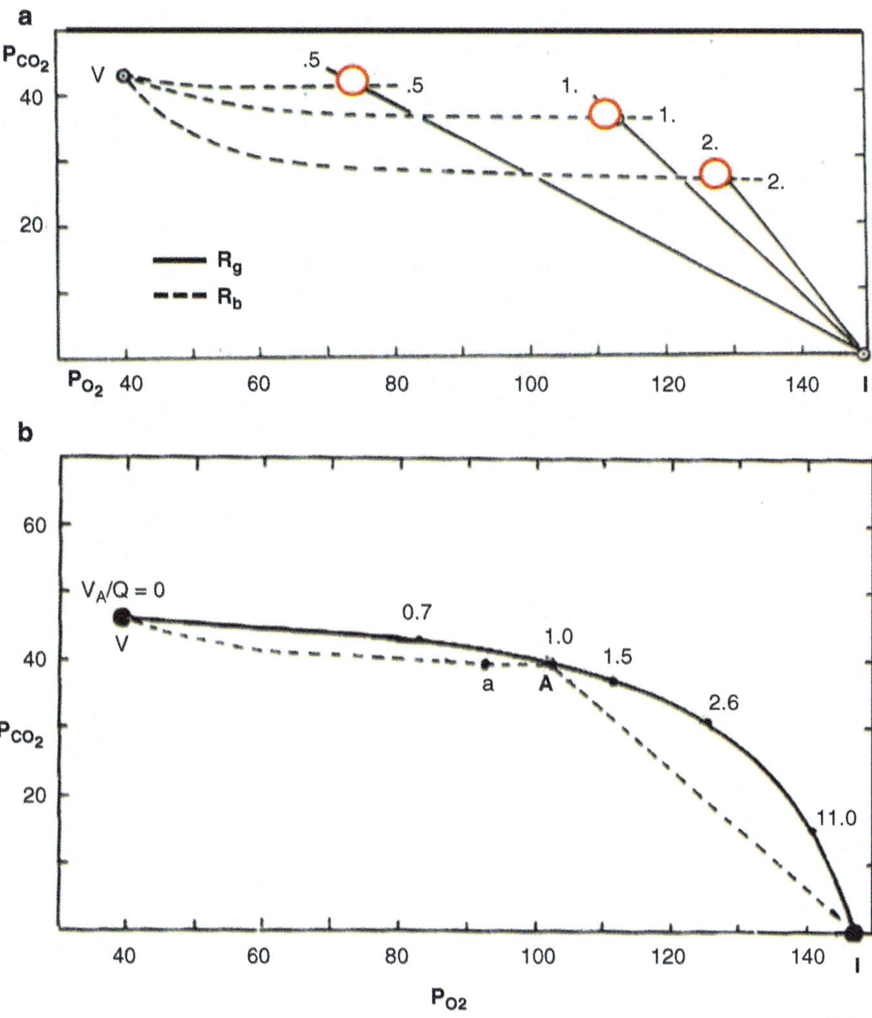

Fig. 10.2 O_2-CO_2 diagram and V_A/Q line (**a**): O_2-CO_2 diagram, in which solid and dotted lines indicate gas R lines (R_g, black solid line) and blood R lines (R_b, black dotted line), respectively. Intersect of these two lines denoted by red circle provide single V_A/Q value and corresponding unique pair of P_{O2} and P_{CO2}. (**b**): V_A/Q line constructed by connecting each intersection of R_g and R_b. I: inspired point, A: alveolar point, a: arterial point, v: mixed venous point. Adopted from Yamaguchi et al. [17] with permission of the publisher

function as the representatives of alveolar gas compositions even in diseased lungs with considerable V_A/Q heterogeneities. This is because the concept of mean (ideal) alveolar gas is largely lacking in the considerations on complicatedly intertwined lung anatomy and intricate physiological gas transfer mechanisms along the airways and in the lung periphery, which are gravely influenced by the lung diseases (see Sect. 3).

3 Effective Alveolar Air and Effective Alveolar–Arterial P_{O2} Difference (AaD_{O2})

3.1 Basic Rationale of Effective Alveolar Air and Effective AaD_{O2}

To calculate the mean (ideal) alveolar gas, it is necessary to measure P_{O2} and P_{CO2} in the mixed venous blood (P_{vO2} and P_{vCO2}). As the mixed venous blood sampling is not easy in a clinical setting, Riley et al. [6] proposed a clinically important concept of "effective alveolar gas," in which the mean (ideal) alveolar P_{CO2} (P_{ACO2}) is taken to equal the arterial P_{CO2} (P_{aCO2}). Because of the relative flatness of the blood CO_2 dissociation curve under a physiological condition, the difference between P_{ACO2} and P_{aCO2} is remarkably small (Fig. 10.2), which indicates that it is not unreasonable to replace P_{ACO2} with P_{aCO2}. Hence, P_{aCO2} is defined as "effective alveolar P_{CO2}" and alveolar P_{O2} calculated therefrom is called "effective alveolar P_{O2}." As it is not necessary to measure the gas pressures in mixed venous blood, the concept of effective alveolar gas by Riley et al. [6] has been proactively applied for clinical practice in the field of respirology. Based on the alveolar gas equation, the effective alveolar P_{O2} (effective P_{AO2}) is defined as:

$$\text{Effective } P_{AO2} = P_{IO2} + F_{IO2} \cdot P_{aCO2} \cdot (1-R)/R - P_{aCO2}/R \quad (10.9)$$

Simplifying the right side of the equation:

$$\text{Effective } P_{AO2} = P_{IO2} - P_{aCO2}/R \quad (10.10)$$

where P_{IO2} and F_{IO2} denote inspiratory partial pressure and inspiratory fraction of O_2, respectively. The respiratory quotient, *R*, is taken to be 0.83 in general. This value is valid even in patients suffering from a variety of lung diseases as far as their metabolic states are not impaired. Using Eq. 10.10, one can easily predict the effective alveolar–arterial P_{O2} difference (effective AaD_{O2}) at bedside as:

$$\text{Effective } AaD_{O2} = \text{effective } P_{AO2} - \text{measured } P_{aO2} \quad (10.11)$$

The first, major study concerning the physiological significance of AaD_{O2} was undertaken by Lilienthal, Riley, and coworkers in 1946 [16], who measured effective AaD_{O2} under resting and exercise conditions in young, normal subjects while breathing the gas containing air or low concentration of O_2. They found that the effective AaD_{O2} averages 9 mm Hg at rest and 17 mm Hg at exercise in the young subjects. Expanding these findings, they considered that the effective AaD_{O2} may function as an important parameter for assessing pulmonary gas exchange kinetics of O_2 in normal subjects with no lung diseases. However, the study by Lilienthal et al. [16] does not assure the matter that effective AaD_{O2} is also useful for identifying O_2 gas exchange abnormalities in patients with a variety of lung diseases.

3.2 Validation of Effective P_{ACO2} and P_{AO2} in Diseased Lungs

Differing from the V_A/Q theory that promoted the development of multiple inert gas elimination technique (MIGET) in the 1970s [7, 8], the classical gas exchange theory is not sufficiently warranted by the anatomical and physiological facts in the lung. Hence, it is necessary for certifying the issue of whether the effective P_{ACO2} (= P_{aCO2}) and P_{AO2} indeed act as the representatives of P_{CO2} and P_{O2} in the mixed gas exhaled from all functional gas exchange units in patients with various lung diseases. For answering this question, it is indispensable to compare the results obtained from the classical gas exchange theory with those from the MIGET. We first examined what kinds of V_A/Q units are sensitively detected by O_2 gas exchange. We calculated P_{AO2} in each V_A/Q unit on the assumption that P_{O2} equilibration between alveolar gas and capillary blood is complete (Fig. 10.3). We found that P_{AO2} changes sharply in the units having V_A/Q ranging from 0.1 to 1.0 but not in other V_A/Q units, resulting in that effective AaD_{O2} is sensitive to gas exchange abnormalities emerging in regions with moderately low V_A/Q values but insensitive to those in regions with high V_A/Q values, including true alveolar dead space (V_A/Q = infinite). Furthermore, the effective AaD_{O2} may not discriminate right-to-left shunt (V_A/Q = zero) from regions with low V_A/Q values (V_A/Q = low but finite). These findings indicate that the ability of effective AaD_{O2} for detecting V_A/Q heterogeneity is substantially limited and does not cover the whole range of V_A/Q heterogeneity as the MIGET does. Next, we investigated the coincidence between P_{ACO2} and P_{AO2} in mixed expired gas estimated from V_A/Q distribution recovered from the MIGET and effective P_{ACO2} (= P_{aCO2}) and P_{AO2} calculated from Eq. 10.10 in patients with interstitial pneumonia or COPD (Fig. 10.4). The coincidence between effective P_{ACO2} and P_{ACO2} in mixed expired gas is satisfactory. Qualitatively the same tendency was observed for the relationship between effective P_{AO2} and P_{AO2} in mixed expired gas obtained by means of the MIGET, as well. These results suggest that effective P_{AO2} and, thus, effective AaD_{O2} can accordingly change reflecting the formation of moderately low V_A/Q regions that are related to the emergence of hypoxemia. Hence, we concluded that effective P_{AO2} and effective AaD_{O2} are practically useful in a clinical setting for predicting gas exchange abnormalities elicited by low V_A/Q regions.

3.3 Significance of aAD_{CO2} and aAD_{N2}

In addition to AaD_{O2}, arterial–alveolar difference for CO_2 (aAD_{CO2}) and that for N_2 (aAD_{N2}) have also been proposed. Since retention of CO_2 in regions with low V_A/Q values is balanced by increased elimination of CO_2 from regions with high V_A/Q

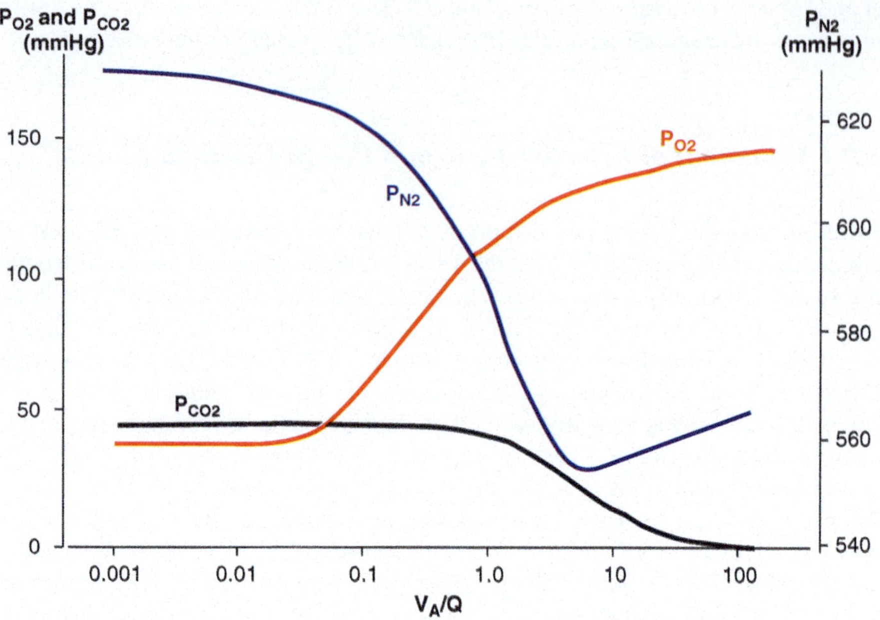

Fig. 10.3 V_A/Q regions detected by gas exchange kinetics of O_2, CO_2, and N_2
Calculations were made under steady-state conditions with air-breathing. P_{O2} in alveolar gas (= end-capillary blood) changes sharply at V_A/Q ranging between 0.1 and 1.0, suggesting that O_2 is sensitive to detection of gas exchange in regions with moderately low V_A/Q values. Meanwhile, CO_2 is susceptible to gas exchange in regions with moderately high V_A/Q values ranging from 1.0 to 10. P_{N2} changes sharply at V_A/Q raging between 0.1 and 1.0, the tendency of which is qualitatively the same as that of O_2. However, the ability of N_2 to separate right-to-left shunt (V_A/Q = zero) from regions with low V_A/Q values (low but not zero) is considered substantially higher than that of O_2. This is because P_{N2} changes appreciably, but P_{O2} changes insignificantly, in low V_A/Q regions ranging from zero to 0.1

values, P_{aCO2} is close to the ideal (mean) value of P_{ACO2}. However, in a strict sense, there exists a small but significant aAD_{CO2}, which is due mainly to the fact that regions with high V_A/Q values contribute proportionally more than others to the expirate and thereby lower P_{ACO2}. Thus, aAD_{CO2} is manifested by a drop of P_{ACO2} and reflects regions with high V_A/Q values (Fig. 10.3). However, in a practical sense, the concept of effective, but not ideal (mean), alveolar gas has been applied for estimating the effective AaD_{O2} that is the most important parameter for detecting gas exchange abnormality in the lung. Therefore, aAD_{CO2} must be taken to be zero in a clinical setting. When needed, aAD_{CO2} should be evaluated using the other value that represents alveolar P_{CO2} such as end-tidal P_{CO2}. Instead of aAD_{CO2}, it is better to apply the physiological dead space for CO_2 when attempting to assess the abnormal gas exchange of CO_2 in the lung. This is because the physiological dead space for CO_2 is also calculated on the basis of the concept of effective alveolar gas (see Sect. 4).

The P_{N2} value in regions with high V_A/Q values remains close to the ideal (mean) value of P_{AN2} because in high V_A/Q regions, the increase in P_{O2} is accompanied by a

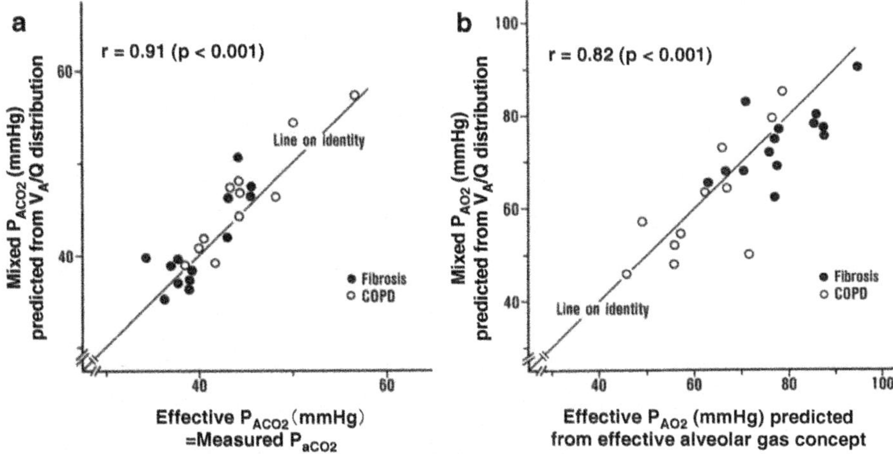

Fig. 10.4 Validation of effective alveolar gas compositions of CO_2 and O_2. (**a**) Comparison of effective P_{ACO2} (= P_{aCO2}) and P_{ACO2} in mixed alveolar gas decided by MIGET. (**b**) Comparison of effective P_{AO2} and P_{AO2} in mixed alveolar gas decided by MIGET. Comparisons were made for patients with interstitial pneumonia ($n = 14$) and COPD ($n = 11$)

decrease in P_{CO2}, leaving P_{N2} close to P_{IN2} (Fig. 10.3). Blood returning from low V_A/Q regions has an elevated P_{N2} and predominates in mixed venous blood. Therefore, gas exchange of N_2 is opposite of that of CO_2, indicating that aAD_{N2} is mainly due to a shift in blood away from regions with ideal (mean) V_A/Q values, leading to reflection of gas exchange kinetics in low V_A/Q regions like AaD_{O2}. P_{AN2} can be calculated based on the equation described below:

$$\text{Effective}\,P_{AN2} = (P_B - 47) - \text{effective}\,P_{AO2} - \text{effective}\,P_{ACO2} \quad (10.12)$$

Although the ability of effective aAD_{N2} (P_{aN2} − effective P_{AN2}) for predicting regions with right-to-left shunt (V_A/Q = zero) is higher than that of effective AaD_{O2}, effective aAD_{N2} has not been routinely utilized in a clinical setting because of the difficulty in measuring P_{N2} in arterial blood.

4 Physiological Dead Space for CO_2

4.1 Anatomical and Functional Boundaries Between Dead Space and Alveolar Region

The anatomical boundary between dead space and alveolar region is simple, lying at the level of terminal bronchiole or transitional bronchiole having no alveoli, corresponding to the entrance of acinus of Loeschcke or that of Haefeli-Bleuer and

Weibel, respectively [17, 18]. Namely, the dead space is anatomically conceptualized as the space configured by upper and lower conductive airways up to terminal or transitional bronchiole [19]. The volume of anatomically determined dead space varies from 140 to 260 mL depending on the extent of lung inflation and age [20–22]. The morphological studies consistently showed that ~50% of the dead space exists in extrathoracic upper airways, indicating that the volume of lower conductive airways, including central and peripheral airways up to terminal or transitional bronchiole, ranges between 70 and 130 mL.

The functional boundary between dead space and alveolar gas exchange region has been examined by many investigators based on single-breath measurements using CO_2, N_2, and foreign inert gases [23–26]. The functional boundary may be better understood to use the term "imaginary boundary." Under resting conditions with air-breathing, the convective flow velocity falls about 100-fold from the trachea to the acinar airways and the contribution of diffusion to gas transport significantly surpasses that of convection within the acinus [27, 28]. Paiva [29] as well as Swan and Tawhai [30] identified that during inspiration, the front of a gas, which is the concentration profile of a gas having the sigmoid shape against the axial distance along the acinar airways, lags far behind the front of tidal volume. The front of tidal volume reaches the 22th generation of airways (i.e., the alveolar duct) at end-inspiration, whereas the gas front remains almost stationary in the second-order respiratory bronchiole during inspiration. The findings demonstrated by these authors suggest that the dead space can be functionally described as the inspiratory volume decided by the gas front, but not by the front of tidal volume. The classical gas exchange theory assumes that the dead space has a sharp, all-or-nothing transition boundary as if it is formed only by the ventilation-elicited convection (i.e., the effective alveolar ventilation). The effective alveolar ventilation may be defined by the total ventilation minus the ventilation estimated from converting the effect of diffusion to the equivalent effect of convection (Fig. 10.5). The functional (imaginary) boundary between dead space and alveolar gas exchange region has been shown to reside in the second-order respiratory bronchiole, which is slightly distal to the terminal or transitional bronchiole forming the entrance of anatomical acinus. Based on these facts, we consider that the term "functional dead space" is physiologically more suitable than the term "anatomical dead space."

Although this is out of reality, the classical gas exchange theory tactically hypothesizes that the reduced contribution of ventilation-induced convection to gas transport in the acinus, which is attributed to the predominant contribution of diffusive mixing process, can be decided if the concept of dead space ventilation (dead space V) is introduced as (Fig. 10.6):

$$V_{\mathrm{AI}} = V_{\mathrm{I}} - \text{dead space}\, V \ \left(\text{during inspiration}\right) \tag{10.13}$$

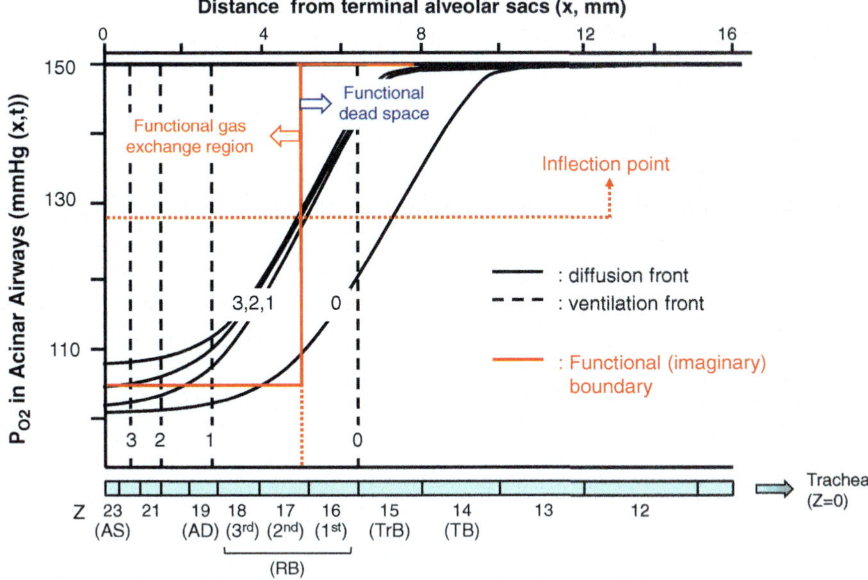

Fig. 10.5 Boundary between functional dead space and gas exchange region in the acinus during inspiration. Time-dependent, spatial distributions of P_{O2} were calculated based on geometrical data of human acinus reported by Haefeli-Bleuer and Weibel [34]. O_2 is assumed to be not absorbed through the alveolocapillary tissue barrier. Calculations were done under conditions in which convection and axial diffusion were taken as main mechanisms for gas mixing. Time zero: just after beginning of inspiration, 1 and 2: midways of inspiration, 3: end-inspiration. To know the representative point of diffusion front at a certain time, inflection point (intersection of red dotted line and black solid lines) was obtained by solving equation of $\partial^2 C/\partial x^2 = 0$ under steady-state conditions ($\partial C/\partial t = 0$). The inflection point gives functional boundary (red solid line) between dead space and gas exchange region, which resided at second-order respiratory bronchiole within the acinus. *TB* terminal bronchiole, *TrB* transitional bronchiole, *RB* respiratory bronchiole, *AD* alveolar duct, *AS* alveolar sac. Modified from Yamaguchi et al. [18] with permission of the publisher

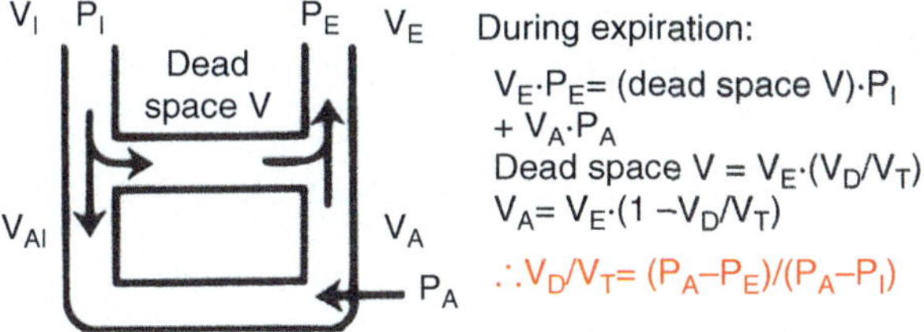

Fig. 10.6 Relationships of partial pressure of a gas between that measured at mouth level and that at alveolar gas exchange region based on classical gas exchange theory. V_{AI} and V_A: inspiratory and expiratory effective alveolar ventilation, respectively. V_I and V_E: inspiratory and expiratory total ventilation, respectively. Dead space V: fictitious ventilation that is defined as V_I or V_E multiplied by relative volume of dead space (V_D/V_T). V_A: effective alveolar ventilation expressed by V_I or V_E multiplied by ($1 - V_D/V_T$)

$$V_A = V_E - \text{dead space } V \text{ (during expiration)} \tag{10.14}$$

where V_{AI} and V_A indicate effective alveolar ventilation during inspiration and that during expiration, respectively. V_I and V_E denote total ventilation in inspiration and that in expiration, respectively. Dead space V indicates dead space ventilation that is defined as V_I or V_E multiplied by relative volume of dead space (V_D/V_T, in which V_D and V_T are dead space volume and tidal volume of ventilation). One should pay attention to the fact that the "dead space" actually exists but the "dead space ventilation" does not in a real lung, i.e., the dead space ventilation is fictitious. Using Eq. (10.14) in combination with the law of mass balance, the relationship between partial pressure of a gas measured at mouth level during expiration and that at alveolar gas exchange region can be predicted with no physiological inconsistency.

4.2 Physiological Dead Space for CO_2

Considering the mass balance of an indicator gas during expiration on the basis of the classical gas exchange theory, Bohr [31] proposed, in 1891, the equation from which the physiological dead space volume (V_D) can be calculated if the expired tidal volume (V_T) and the partial pressures of a gas in inspired (P_I), expired (P_E), and alveolar gases (P_A) are known (Fig. 10.6):

$$V_D / V_T = (P_A - P_E)/(P_A - P_I) \tag{10.15}$$

Although Eq. 10.15 (i.e., the "Bohr equation") can be applied for any indicator gas, the measurement of physiological dead space was greatly facilitated when Enghoff [32] used CO_2 as the indicator and ideal (mean) P_{ACO2} was replaced with P_{aCO2} (effective P_{ACO2}) in Eq. 10.15. The idea of Enghoff is basically the same as the concept of effective alveolar gas introduced by Riley and colleagues [6]. The modified equation by Enghoff was termed the "Bohr–Enghoff equation" and has been in constant use so far in a clinical practice. The physiological V_D/V_T for CO_2 estimated from the Bohr–Enghoff equation is expressed as (P_{ICO2} = zero):

$$V_D / V_T = (P_{aCO2} - P_{ECO2})/P_{aCO2} \tag{10.16}$$

Differing from effective AaD_{O2} that is sensitive to the detection of low V_A/Q regions, physiological V_D/V_T is useful for identifying the abnormal gas exchange elicited by high V_A/Q regions, which prescribe the level of alveolar P_{CO2}. The physiological V_D/V_T for CO_2 may not surely discriminate between the high V_A/Q regions (V_A/Q is finite) and the true dead space whose V_A/Q is infinite. Thus, the physiological V_D/V_T for CO_2 estimated from the Bohr–Enghoff equation includes the anatomical dead space, which is the functional dead space in a strict sense (see Sect. 4.1), and the alveolar dead space formed by the regions with infinite V_A/Q value (i.e., true alveolar dead space) and high but finite V_A/Q values (i.e., alveolar dead space-like effect) in patients with a variety of lung diseases.

4.3 *Primary and Secondary Alveolar Hypoventilation*

The increased P_{CO2} in the blood (i.e., hypercapnia) is caused by the primary alveolar hypoventilation or by the secondary alveolar hypoventilation. The primary alveolar hypoventilation occurs in the impediment of any part of the signal transmission from the integrated brainstem respiratory motoneurons (including the dorsal respiratory group (DRG) and the ventral respiratory group (VRG)) to the respiratory muscles [33]. The signals from the central and peripheral chemoreceptors modify the discharges of the integrated brainstem respiratory motoneurons, as well. The primary alveolar hypoventilation is generally not related to the lung disease. Thereby, the primary alveolar hypoventilation is not caused by the increased physiological V_D/V_T. On the other hand, the secondary alveolar hypoventilation is evoked by the lung disease-associated increase in physiological V_D/V_T. However, the hypercapnia generated by the augmented physiological V_D/V_T is mostly compensated by the increased output of the integrated brainstem respiratory motoneurons, which is caused by the hypercapnic stimulation upon the central chemoreceptor. Thereby, simple inspection of P_{aCO2} level does not allow one to judge whether the gas exchange of CO_2 in the lung is indeed impeded. In this case, the measurement of physiological V_D/V_T serves for a definitive diagnosis of abnormal CO_2 gas exchange. Furthermore, the measurement of effective AaD_{O2} helps the differential diagnosis of the hypoventilation, i.e., the primary alveolar hypoventilation with no conspicuous lung disease reveals the hypercapnia with no increase in physiological V_D/V_T in association with normal effective AaD_{O2}. On the other hand, the secondary alveolar hypoventilation exhibits an increase in physiological V_D/V_T in association with widened effective AaD_{O2} if the lung disease forms both high and low V_A/Q regions. However, if the lung disease does not form low V_A/Q regions, increased physiological V_D/V_T but normal effective AaD_{O2} is found.

The examples with primary alveolar hypoventilation and secondary alveolar hypoventilation are depicted in Table 10.1 and Fig. 10.7, respectively.

Table 10.1 Results of cardiopulmonary function tests in a patient with spinal progressive muscular atrophy (SPMA)

Pulmonary function tests	Blood gases, D_{LCO}, and cardiac parameters during awakening and rest	Blood gases at spontaneous hyperventilation	Blood gases during sleep
VC: 1.4 L	P_{aO2}: 41 mm Hg	P_{aO2}: 109 mm Hg	P_{aO2}: 33 mm Hg
FVC: 1.2 L	P_{aCO2}: 83 mm Hg	P_{aCO2}: 47 mm Hg	P_{aCO2}: 100 mm Hg
FEV_1: 0.99 L	AaD_{O2}: 8.7 mm Hg	AaD_{O2}: −16 mm Hg*	AaD_{O2}: −3.8 mm Hg*
FEV_1/FVC: 80%			
	P_{ECO2}: 60 mm Hg		
	V_D/V_T: 28%		
TLC: 3.9 L			
FRC: 2.8 L	D_{LCO}: 25		
RV: 2.2 L	D_{LCO}/V_A: 4.5		
RV/TLC: 61%			
	PA: 40/22 mm Hg		
	Mean PA: 28 mm Hg		
	P_{cw}: 4 mm Hg		
	Q: 8.3 L/min		

Blood gases were measured at room air-breathing. D_{LCO}: mL/min/mmHg. D_{LCO}/V_A: mL/min/mmHg/L. PA: systolic and diastolic pressures of pulmonary artery. Mean PA: mean pulmonary artery pressure. P_{cw}: pulmonary capillary wedge pressure. Q: cardiac output. Case is a 48-year-old, nonsmoking women. Based on various neurological examinations (muscle biopsy of biceps, electromyogram, etc.), her disease was confirmed to be SPMA accompanied by predominant insufficiency of expiratory muscles (i.e., TLC and FRC are maintained but RV and RV/TLC are increased remarkably). Although CT images showed no abnormality in lung fields, cardiac size was enlarged due to chronic right heart failure evoked by alveolar hypoventilation-induced pulmonary arterial hypertension. Effective AaD_{O2} and physiological V_D/V_T for CO_2 were normal under awaked, resting conditions, suggesting little impairment of pulmonary gas exchange. Hence, the cause of alveolar hypoventilation is primary, which is attributed to respiratory muscle insufficiency. AaD_{O2} at spontaneous hyperventilation and that during sleep showed negative values (*). This may be explained from the fact that steady-state conditions were not definitely attained under these peculiar conditions

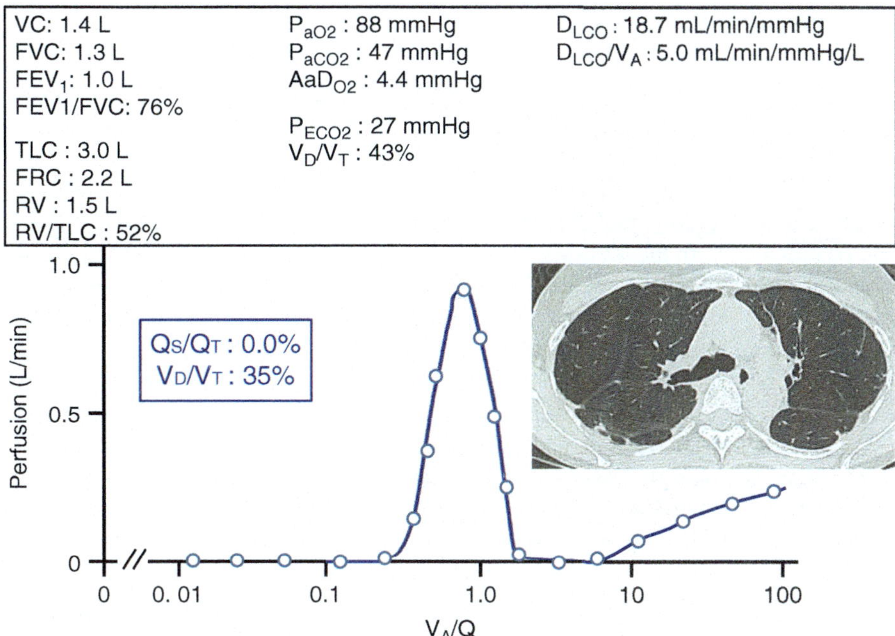

Fig. 10.7 Secondary alveolar hypoventilation in a patient with pleuroparenchymal fibroelastosis. Case is a 42-year-old, nonsmoking woman. Twenty years ago, she underwent allogeneic hematopoietic stem cell transplantation (HSCT) against chronic myelogenous leukemia (CML). A couple of years ago, she was conscious of the difficulty in breathing on exertion in association with sustained dry cough. Chest X-P indicated volume loss in the upper lung field but overinflation in the lower lung field. CT examination showed fibrotic changes and/or scars just beneath pleura in the upper lung field. Lung biopsy targeting right upper lobe under video-assisted thoracoscopy (VATS) confirmed lesion with pleuroparenchymal fibroelastosis (PPFE). Thus, the patient was diagnosed as suffering from HSCT-induced PPFE. Pulmonary function test demonstrated a decrease in TLC and VC but an increase in RV and RV/TLC. FVC and FEV_1 were reduced due to a decrease in lung volumes but FEV_1/FVC was normal. D_{LCO} was slightly decreased but D_{LCO}/V_A (Krogh factor) was maintained, suggesting that PPFE-elicited impediment of alveolocapillary tissue barrier (decrease in D_{LCO}) was compensated by decreased lung volume that minimized change in Krogh factor. Although effective AaD_{O2} was kept within normal range, physiological V_D/V_T for CO_2 was augmented, which led to the emergence of slight hypercapnia. Analysis of V_A/Q distribution by MIGET identified that augmented physiological V_D/V_T was attributed to the formation of high V_A/Q regions, which may be caused by redistribution of alveolar ventilation from the PPFE region (upper lung field) to relatively normal region (lower lung field)

5 Conclusion Remarks

The classical gas exchange theory is still important even today, not only from the physiological aspect but also from the clinical point of view. Particularly, the concept of effective alveolar gas has facilitated the diagnosis of abnormal gas exchange in the lung at bedside. The distinguished assumption underlying the concept of effective alveolar gas is that the representative of alveolar P_{CO2} (i.e., effective P_{ACO2}) can be replaced with the measured arterial P_{CO2} (P_{aCO2}). In connection with this consideration, the representative of alveolar P_{O2} (effective P_{AO2}) and alveolar–arterial P_{O2} difference (effective AaD_{O2}) can be simultaneously decided. Effective AaD_{O2} is essentially important in a clinical practice for identifying the abnormal gas exchange of O_2, which is specifically caused by the low V_A/Q regions, including the right-to-left shunt (V_A/Q = zero). However, the concept of effective alveolar gas refuses the estimation of arterial–alveolar P_{CO2} difference (aAD_{CO2}). This is because the concept of effective alveolar gas assumes that aAD_{CO2} is invariably zero under any condition. Fortunately, however, the concept of effective alveolar gas enables one to estimate the physiological V_D/V_T. Based on the measurements of physiological V_D/V_T, one can predict the abnormal gas exchange of CO_2 elicited by the anatomical dead space (strictly, the functional dead space) and the high V_A/Q regions (i.e., the alveolar dead space), including the true dead space where V_A/Q value is infinite.

References

1. Krogh A, Lindhard J. The volume of the dead space in breathing and the mixing of gases in the lungs of man. J Physiol Lond. 1917;51:59–90. https://doi.org/10.1113/jphysiol.1917.sp001785.
2. Haldane JS. Respiration. New Haven, CT: Yale University Press; 1922.
3. Fenn WO, Rahn H, Otis AB. A theoretical study of the composition of alveolar air at altitude. Am J Phys. 1946;146:637–53. https://doi.org/10.1152/ajplegacy.1946.146.5.637.
4. Riley RL, Cournand A. "Ideal" alveolar air and the analysis of ventilation/perfusion relationships in the lungs. J Appl Physiol. 1949;1:825–47. https://doi.org/10.1152/jappl.1949.1.12.825.
5. Rahn H. A concept of mean alveolar air and the ventilation-blood flow relationships during pulmonary gas exchange. Am J Phys. 1949;158:21–30. https://doi.org/10.1152/ajplegacy.1949.158.1.21.
6. Riley RL, Lilienthal JL, Proemmel DD, Franke RE. On the determination of the physiologically effective pressures of oxygen and carbon dioxide in alveolar air. Am J Phys. 1946;147:191–8. https://doi.org/10.1152/ajplegacy.1946.147.1.191.
7. Wagner PD, Saltzman H, West JB. Measurement of continuous distributions of ventilation-perfusion ratios: theory. J Appl Physiol. 1974;36:588–99. https://doi.org/10.1152/jappl.1974.36.5.588.
8. Wagner PD, Laravuso RB, Uhl RR, West JB. Continuous distributions of ventilation-perfusion ratios in normal subjects breathing air and 100% O_2. J Clin Invest. 1974;54:54–68. https://doi.org/10.1172/JCI107750.
9. Yamaguchi K, Kawai A, Mori M, Asano A, Takasugi T, Umeda A, et al. Distribution of ventilation and of diffusing capacity to perfusion in the lung. Respir Physiol. 1991;86:171–87. https://doi.org/10.1016/0034-5687(91)90079-X.

10. Yamaguchi K, Mori M, Kawai A, Takasugi T, Asano K, Oyamada Y, et al. Ventilation-perfusion inequality and diffusion impairment in acutely injured lungs. Respir Physiol. 1994;98:165–77. https://doi.org/10.1007/BF01277020.
11. Yamaguchi K, Mori M, Kawai A, Takasugi T, Oyamada Y, Koda E. Inhomogeneities of ventilation and the diffusing capacity to perfusion in various chronic lung diseases. Am J Respir Crit Care Med. 1997;156:86–93. https://doi.org/10.1164/ajrccm.156.1.9607090.
12. Klocke RA, Gurtner GH, Farhi LE. Gas transfer across the skin in man. J Appl Physiol. 1963;18:311–6. https://doi.org/10.1152/jappl.1963.18.2.311.
13. Herron JM, Saltzman HA, Hills BA, Kylstra JA. Differences between inspired and expired minute volumes of nitrogen in man. J Appl Physiol. 1973;35:546–51. https://doi.org/10.1152/jappl.1973.35.4.546.
14. Yamaguchi K. O_2-CO_2 diagram and its physiological significance. Kokyu. 1985;4:405–14. (in Japanese)
15. Rahn H, Fenn WO. A graphical analysis of the respiratory gas exchange. Washington, DC: The American Physiological Society; 1955. p. 1–41.
16. Lilienthal JL Jr, Riley RL, Proemmel DD, Franke RE. An experimental analysis in man of the oxygen pressure gradient from alveolar air to arterial blood during rest and exercise at sea level and at altitude. Am J Phys. 1946;147:199–216. https://doi.org/10.1152/ajplegacy.1946.147.1.199.
17. Yamaguchi K, Tsuji T, Aoshiba K, Nakamura H, Abe S. Anatomical backgrounds on gas exchange parameters in the lung. World J Respirol. 2019;9(2):8–29. https://doi.org/10.5320/wjr.v9.i2.8.
18. Yamaguchi K, Tsuji T, Aoshiba K, Nakamura H, Abe S. Insights into anatomical basis prescribing ventilation-perfusion distribution in lung periphery. Austin J Pulm Respir Med. 2019;6(1):1–12.
19. Anthonisen NR, Fleetham JA, Fishman AP, Farhi LE, Tenney SM, Geiger SR. Ventilation: total, alveolar, and dead space. In: Handbook of physiology, section 3: The respiratory system, vol IV: Gas exchange. Bethesda, MD: American Physiological Society; 1987. p. 113–29.
20. Loewy A. Uber die Bestimmung der Grosse des "Sehadlichen Luftraumes" im thorax und der alveolaren Sauersttoffs-panung. Arch Gesamte Physiol Menschen Tiere. 1894;58:416–27. https://doi.org/10.1007/BF01672254.
21. Rohrer F. Der Strömungswiderstand in den menschlichen Atemwegen und der Einfluss der unregelmässigen Verzweigung des Bronachialsystems auf den Atmungsverlauf in verschiedenen Lungernbezirke. Pfluegers Arch Ges Physiol. 1915;162:225–9.
22. Nunn JF, Campbell EJM, Peckett BW. Anatomical subdivisions of the volume of respiratory dead space and effect of position of the jaw. J Appl Physiol. 1959;14:174–6. https://doi.org/10.1152/jappl.1959.14.2.174.
23. Fowler WS. Lung function studies. V. The respiratory dead space in old age and in in pulmonary emphysema. J Clin Invest. 1950;29:1439–44. https://doi.org/10.1172/JCI102383.
24. Bartels J, Severinghaus JW, Forster RE, Briscoe WA, Bates DV. The respiratory dead space measured by single breath analysis of oxygen, carbon dioxide, nitrogen or helium. J Clin Invest. 1954;33:41–8. https://doi.org/10.1172/JCI102868.
25. Young AC. Dead space at rest and during exercise. J Appl Physiol. 1955;8:91–4. https://doi.org/10.1152/jappl.1955.8.1.91.
26. Lacquet LM, Van Der Linden, Paiva M. Transport of H_2 and SF_6 in the lung. Respir Physiol. 1975;25:157–73. https://doi.org/10.1016/0042-207X(75)91452-9.
27. Hsia CCW, Hyde DM, Weibel ER. Lung structure and the intrinsic challenges of gas exchange. Compr Physiol. 2016;6:827–95. https://doi.org/10.1002/cphy.c150028.
28. Verbanck S, Paiva M. Gas mixing in the airways and airspaces. Compr Physiol. 2011;1:809–34. https://doi.org/10.1002/cphy.c100018.
29. Paiva M. Gas transport in the human lung. J Appl Physiol. 1973;35:401–10. https://doi.org/10.1152/jappl.1973.35.3.401.

30. Swan AJ, Tawhai MH. Evidence for minimal oxygen heterogeneity in the healthy human pulmonary acinus. J Appl Physiol. 2011;110:528–37. https://doi.org/10.1152/japplphysiol.00888.2010.
31. Bohr C. Über die Lungenatmung. Skand Arch Physiol. 1891;2:236–68. https://doi.org/10.1111/j.1748-1716.1891.tb00581.x.
32. Enghoff H. Volumen inefficax. Bemerkungen zur Frage des schädlichen Raumes. Upsala Laekarefoeren Foerh. 1938;44:191–218. doi.org/10.1002/j.2050-0416.1938.tb05855.x
33. von Euler C. Neural organization and rhythm generation. In: Crystal RG, West JB, Weibel ER, Barnes PJ, editors. The LUNG scientific foundations. Philadelphia: Lippincott-Raven; 1997. p. 1711–24.
34. Haefeli-Bleuer WER. Morphometry of the human pulmonary acinus. Anat Rec. 1988;220:401–14. https://doi.org/10.1002/ar.1092200410.

Chapter 11
Oxygen Transport from Air to Tissues as an Integrated System: What Limits Maximal O_2 Consumption in Health and Disease?

Peter D. Wagner

Abstract Oxygen moves from the air to the mitochondria following a well-known sequence of steps comprising (a) ventilation, (b) diffusion across the pulmonary blood:gas barrier, (c) vascular transport to each tissue, and (d) diffusion from the tissue microcirculation to its mitochondria, where (e) oxidative phosphorylation uses that O_2 to sustain cell function. O_2 transport is therefore a system of interconnected processes, where each of the steps functions in a manner that is affected by each of the other four. Therefore, to understand O_2 transport, and especially its limits, every step of the pathway must be considered simultaneously and not separately. Based on the overarching principle governing O_2 transport—that conservation of mass of O_2 is maintained at every step—simple mass conservation equations can be written describing O_2 transport for each step. These equations can then be solved simultaneously to provide a quantitative understanding of how the transport processes function together, how each affects overall transport, and in particular, how O_2 transport reaches a limit. This quantitative systems analysis can be applied not only in health—to understand limits O_2 transport places on exercise—but also in chronic diseases to identify which steps may be abnormal.

Keywords Ventilation · Perfusion · Diffusion · Oxygen delivery · Oxygen extraction

P. D. Wagner (✉)
University of California San Diego, La Jolla, CA, USA
e-mail: pdwagner@ucsd.edu

K. Yamaguchi (ed.), *Structure-Function Relationships in Various Respiratory Systems*, Respiratory Disease Series: Diagnostic Tools and Disease Managements, https://doi.org/10.1007/978-981-15-5596-1_11

1 Introduction

From at least the early twentieth century, and likely long before that, there has been sustained interest in knowing what limits exercise capacity, and what physiological properties enable one individual to outpace another and "win the gold medal." Interest has not been limited to exercise physiologists seeking scientific understanding. The athletic population in general is very much taken with this concept, and one of the most common questions on the table remains "*What determines* $V_{O2}max$?" (VO_2max being the maximal rate of oxygen utilization by a subject performing at his or her limits).

Exercise capacity and VO_2max however must not be equated. VO_2max is only one determinant of exercise capacity. Moreover, the importance of VO_2max as a determinant of exercise capacity varies greatly with the nature of the exercise undertaken. Thus, the capacity for brief sprint-, field-, or strength-based sports does not depend on VO_2max. On the other hand, the capacity to undertake endurance events is very much related to VO_2max. But even there, other factors determining exercise capacity are crucial: including technical expertise, courage (pain tolerance), will, competition tactics, and how close to VO_2max an athlete can come before blood lactate levels begin to rise.

This chapter will deal just with the question: what determines VO_2max? The central theme is that one must consider the entire O_2 transport/utilization system [1] to address this question. Efforts to answer this question by comparing the individual performance of the different tissues and organs involved in moving oxygen from the air to the mitochondria constitute the wrong approach. Because O_2 movement from air to mitochondria is accomplished by a series of interdependent processes acting as an integrated system [2–4], the question can only be answered by a systems analysis of O_2 transport. It is such an analysis that will be presented here.

To an engineer, the preceding may be obvious, raising the question of why most physiologists have not understood the need for a systems approach even after decades of discussion. It likely stems from the well-known Fick Principle, defined by the following equation as follows:

$$VO_2 = Q \times \left[CaO_2 - CvO_2\right] \tag{11.1}$$

This equation conveys only conservation of mass: The amount of O_2 consumed per unit time, VO_2, can be expressed as the product of blood flow (Q) and the oxygen concentration difference between arterial and venous blood (Ca O_2 and CvO_2 respectively) of the tissue in question. It can be applied at local levels or globally to the entire body.

When one compares the whole-body values found in elite endurance athletes to those found in healthy nonathletes, each group measured at their respective maximal exercise capacities, one finds that VO_2 and Q are significantly higher in the elite athletes but that $[CaO_2 - CvO_2]$ is very similar in the two groups. Thus, VO_2max

correlates with cardiac output across subjects. This has been interpreted to support the idea that cardiac output (i.e., blood flow) is *the* determinant of VO_2max [5, 6]. Furthermore, the similarity in $[CaO_2 - CvO_2]$ between groups is taken to imply that tissue O_2 extraction capacity is similar between elite athletes and healthy nonathletes.

Although the correlation between VO_2max and peak cardiac output does of course exist, the interpretation that this proves that cardiac output is the sole determinant of VO_2max is deeply flawed. This becomes very apparent when the systems analysis described in this chapter is applied to the problem, as will be shown.

Equation (11.1) can be modified slightly by multiplying both numerator and denominator by CaO_2, as follows:

$$VO_2 = [Q \times CaO_2] \times [(CaO_2 - CvO_2) / CaO_2] \tag{11.2}$$

The term $[Q \times CaO_2]$ is called O_2 delivery. It is the amount of O_2 delivered per unit time from the left ventricle to the tissue or organ(s) under study. For that tissue or organ, Q remains the blood flow to that tissue or organ. The term $[(CaO_2 - CvO_2)/CaO_2]$ is called O_2 extraction and is a dimensionless fraction. It represents the fraction of O_2 delivered to the tissues or organ that has been extracted from the blood and moved to, and used by, the tissue mitochondria.

It may seem like a small step to then claim that O_2 delivery in eq. (11.2) represents the "central" cardiothoracic contribution to overall mitochondrial O_2 availability and utilization while O_2 extraction represents the tissue's ability to transfer O_2 from the blood to the mitochondria and thus represents the "peripheral" contribution to O_2 availability in eq. (11.2). This simple interpretation would be fair only if the several O_2 transport steps and processes that constitute the system were independent of each other. As soon as one understands that all of the transport processes influence each other's function, it becomes clear that one may no longer consider eq. (11.2) alone as the basis of separating "central" from "peripheral" factors in their quantitative contributions to mitochondrial O_2 availability.

Perhaps the most obvious example of this concept is in the effect of blood flow (Q) on O_2 transport: The higher the blood flow, the higher will be O_2 delivery—in direct proportion to blood flow, just as eq. (11.2) suggests—*unless the higher blood flow impairs arterial oxygenation (transport step 2 below) and/or tissue O_2 extraction (step 4 below).* The higher blood flow does in fact reduce extraction because the red cells spend less time unloading O_2 from the tissue microvessels and therefore are unable to release all of the O_2 which they would have had the microvascular transit time been longer. In other words, higher blood flow directly reduces O_2 extraction, offsetting to some extent the benefit of the higher flow. This concept can be generalized: Interactions occur between each pair of steps in O_2 transport—with all having effects that offset gains in function at any one step by losses in function at other steps. Note that higher blood flow not only reduces microvascular transit time in the tissues as mentioned above but also in the pulmonary capillaries, which may reduce arterial PO_2 and thus also CaO_2 [7]. Thus, an increase in Q commonly results in a reduction in CaO_2 plus an increase in CvO_2, reducing extraction and thus mitochondrial O_2 availability. In this way, blood flow affects O_2 transport not only via its conventional "central" mass transport function but also in its effects on arterial and venous O_2 concentration. This example should serve to point out that the traditional mapping of Q to "central" function and CaO_2—CvO_2 to "peripheral" function is deeply flawed.

Finally, this point of view is not just of academic interest: It has recently been shown that considering O_2 transport as an integrated system allows identification and separation of truly central from truly peripheral contributions to O_2 transport in both health and disease [8–11]. While the traditional view of O_2 transport would have suggested that Q is its dominant determinant, a systems analysis allowing for interactions such as described above identifies peripheral factors as often equally important.

These concepts are now developed quantitatively.

2 O_2 Transport as a System of Five Linked Processes

There are four fundamental physiological steps in O_2 transport to tissues and a fifth step describing mitochondrial O_2 utilization [1]. The four transport steps are:

1. Ventilation: inspiration of O_2 in air by breathing, which transports O_2 from the external environment to the alveolar gas within the lungs. This is mostly a convective process.
2. Diffusion: transport by diffusion across the alveolar blood:gas barrier from alveolar gas into pulmonary capillary blood.
3. Perfusion: another convective process whereby the heart pumps blood with its O_2-containing red cells from the lungs to the tissues.
4. Diffusion: transport by diffusion from red cells within tissue microvessels to tissue mitochondria.

These four steps form the integrated O_2 transport system. Each step is governed by the requirement to conserve O_2 mass at every step. In fact, this simple overarching principle is sufficient for the understanding of not just each step of the transport system but also the interactions among all four steps, as will become evident in the following. Each step is now considered one at a time.

These four steps deliver O_2 from the environment to the mitochondria but stop short of describing the mitochondrial use of O_2. Moreover, the above four steps are usually considered "physiological" processes, while mitochondrial use of O_2 is considered a "biochemical" process (i.e., oxidative phosphorylation). This distinction does not prevent mitochondrial use of O_2 from being considered in the O_2 system framework as the fifth step embodying the biochemistry of mitochondrial respiration, and this will be addressed as step 5 as the analysis continues by describing each step.

2.1 *Step 1: Ventilation*

It is sufficient for O_2 transport analysis to consider the mass conservation aspect of ventilation. What sets the values for tidal volume and respiratory frequency is an entirely different issue, well beyond the scope of this chapter. Thus, for a given level of ventilation we can write the following mass conservation equation:

$$VO_2 = VI x FIO_2 - VA x FAO_2 \quad (11.3)$$

Here, VO_2 is the same variable as in eqs. (11.1) and (11.2), i.e., the net amount of O_2 taken up from the air, transported to, and used by, the mitochondria per unit time. Inspired and expired ventilation (VI and VA, respectively) are the volumes of gas inhaled and exhaled per minute, and FIO_2 and FAO_2 are the fractional concentrations of O_2 in the inspired air and expired alveolar gas, respectively. The term VI x FIO_2 is therefore the amount of O_2 inhaled per unit time while the term VA x FAO_2 is the amount of O_2 exhaled per unit time. In other words, eq. (11.3) says simply that, over any time period, subtracting the amount of O_2 exhaled from the amount inhaled over that period must signify the amount of O_2 taken out of the air and transferred into the pulmonary capillary blood over the same period: Said differently, all the O_2 which leaves the environment ends up in the pulmonary capillary blood, none is "lost" along the way, and O_2 mass is therefore conserved in eq. (11.3).

One may ask why VI and VA are not identical such that the equation can be simplified to eq. (11.4). In fact, it can, with negligible error. Thus, VI can be equated to VA so that:

$$VO_2 = VA x [FIO_2 - FAO_2] \quad (11.4)$$

Technically, VI and VA are not generally identical because more O_2 is taken up per minute than is CO_2 eliminated. Typically, resting VO_2 is about 300 ml/min while VCO_2 is about 240 ml/min. The only other gas, N2, is not metabolized so that there is no net uptake or elimination of N2. Hence, VI is usually about (300–240), or 60 ml/min, greater than VA. With VA usually being about 6 L/min, this makes for a 1% error when one uses eq. (11.4) in place of eq. (11.3). This is an acceptable tradeoff in the description of O_2 transport as a system such as herein, but it should be noted that this minor approximation can be accounted for with additional considerations.

2.2 *Step 2: Diffusion in the Lung*

It is well accepted that O_2 moves from alveolar gas to pulmonary capillary blood by simple, passive, diffusion from a region of high to a region of low O_2 partial pressure (i.e., from alveolar gas to capillary blood). This process can be described by another mass conservation equation commonly known as the Fick law of diffusion (originally described by Adolph Fick in 1855, [12]):

$$VO_2 = DL \times [PAO_2 - PcapO_2] \tag{11.5}$$

Here, VO_2 is again the same variable as in eqs. (11.1) and (11.2), while PAO_2 and $PcapO_2$ are the alveolar and pulmonary capillary O_2 partial pressures, respectively. DL is the total lung O_2 diffusing capacity, so that eq. (11.5) states simply that the amount of O_2 that is transferred by diffusion across the blood:gas barrier can be considered as the product of the diffusional transport characteristics of the blood:gas barrier (expressed collectively as DL, the amount of O_2 that can diffuse per unit time per mm Hg driving O_2 partial pressure), and the P O_2 difference "driving" diffusion from alveolar gas to capillary blood, $[PAO_2 - PcapO_2]$, which is expressed in mm Hg.

There is however a complexity innate to this simple diffusion equation that must be addressed: Capillary PO_2 rises from venous to arterial values from beginning to end of the pulmonary capillary, begging the question of which value of $PcapO_2$ belongs in eq. (11.5). Specifically, as red cells return from the tissues after O_2 has been extracted from them and they enter a pulmonary capillary, their PO_2 ($PcapO_2$) is low, usually about 40 mm Hg, while alveolar PO_2 (PAO_2) is high, usually about 100 mm Hg. Thus, thinking of an individual red cell at the beginning of the capillary, O_2 flux from gas into that red cell is rapid, as the PO_2 difference is 60 mm Hg. However, this means that PO_2 in that red cell ($PcapO_2$) must rise quickly, so that PAO_2–$PcapO_2$ is reduced. This in turn then reduces, via eq. (11.5), instantaneous O_2 flux from alveolar gas into that red cell as it continues to move along the capillary. If DL is high enough, and if there is time enough spent by the red cell in the pulmonary capillary, $PcapO_2$ can rise to a value essentially equal to PAO_2. This complexity can be resolved formally by recognizing that eq. (11.5) is actually a simple differential equation that must be integrated from start to end of the capillary to reflect the overall diffusive process. This results in eq. (11.6):

$$VO_2 = DL \times [PAO_2 - PmpcO_2] \tag{11.6}$$

in which $PmpcO_2$ is the *m*ean *p*ulmonary *c*apillary PO_2 averaged over the course of the red cell's transit through the lung microvasculature. It must have a value greater than that at red cell entry into the lung, but less than that at the end of the capillary where the red cell leaves the gas exchange region and starts to move to the left atrium.

In addition, in this analysis, the PO_2 at the end of the capillary is taken to equal the PO_2 that will be measured in the systemic arterial blood, PaO_2. This approximation assumes that the lung is functionally homogeneous and that there is no physiologically significant post-pulmonary venous admixture from bronchial or Thebesian venous blood reaching the left circulation. If functional inhomogeneity does occur, as is common in most lung diseases, complexity is further increased, but this can be resolved by modern computational approaches. Physiological studies have not revealed significant bronchial or Thebesian venous admixture (reviewed in [13]).

In eq. (11.6), PAO_2 is usually considered to be constant, but an astute reader may say that this would violate the underlying principle of mass conservation as: As O_2 leaves the alveolar gas and enters the capillary blood, PAO_2 should fall because O_2 molecules are leaving the alveolar gas. This would be true in a closed system, but ventilation continues and this maintains an essentially constant PAO_2 by replenishing alveolar gas with O_2 every breath. Importantly, this underscores the integrative and interdependent nature of O_2 transport as a system. Thus, it should be evident that the first two steps, ventilation and diffusion, are linked (since alveolar PO_2 (PAO_2) is common to both), and these steps cannot be properly considered separately: Ventilation affects PAO_2 and PAO_2 affects diffusion.

There is another complexity to consider: All of the preceding can be included in the treatment of diffusion as a step in O_2 transport, but in practice there is another complexity: The shape of the O_2Hb dissociation curve, which is sufficiently complicated that the solution of the above diffusion equation cannot be accomplished algebraically by equations as simple as (11.5) or (11.6): It requires numerical analytical approaches, thankfully not problematic to create and use in this computer age. But if one does simplify the O_2Hb dissociation curve by making a linear approximation (as done by Piiper and Scheid [14] who assigned a slope to the dissociation curve which they called "β"), an algebraic solution to the diffusion equation can be obtained, as follows:

$$PAO_2 - PaO_2 = (PAO_2 - PvO_2) \times \exp(-DL/(\beta \times Q)) \quad (11.7)$$

This equation is introduced at this time, despite it being an approximation, because it beautifully lays out the way in which the diffusive process determines the arterial PO_2 (PaO_2). It states that for given values of both alveolar PO_2 (PAO_2) and venous PO_2 (PvO_2), arterial PO_2 (PaO_2) is determined by the compound term DL/(β x Q), which is the ratio of the diffusing capacity of the lung (DL) to the capacity (β x Q) of the vascular system to carry O_2 on Hb (β) and then transport it by blood flow (Q).

2.3 *Step 3: Perfusion*

This third step in O_2 transport describes the quantitative transport of O_2, essentially all attached to Hb, from the lungs to the tissues by the circulation. Anatomically this transport process involves all blood vessels between the lungs and the tissues, both arterial and venous, as well as the heart itself as the pump. Despite anatomical complexity, its function can be described simply by the Fick principle, as set out in Eq. 11.1 and reproduced here:

$$VO_2 = Q x [CaO_2 - CvO_2] \quad (11.1)$$

This equation is also nothing more than an expression of O_2 mass conservation as O_2 flows from lungs to tissues. Analogous to eq. (11.3) for ventilation, it states that, over any given time period, the amount of O_2 used by the tissues (VO_2) is the difference between the amount delivered from the lungs (Q x CaO_2) in the arterial blood and the amount not extracted by the tissues and therefore leaving the in the venous blood (Q x CvO_2).

2.4 *Step 4: Diffusion in the Tissues*

In the tissue or organ microvessels, O_2 dissociates off Hb in the red cells and then moves by diffusion to the mitochondria in the various cells of the tissue. This process can be described by another mass conservation equation also based on the Fick law of diffusion, just as for diffusion in the lung, as in eqs. (11.5), (11.6), and (11.7). In fact the diffusion processes in the lungs and tissues are conceptually and procedurally considered identical, and only the anatomy and conditions are different. Thus:

$$VO_2 = DT x [PcapO_2 - PmO_2] \quad (11.8)$$

Here, VO_2 has the same meaning and definition as previously, DT is the diffusing capacity of the tissue under consideration, and $PcapO_2$ and PmO_2 are tissue microvascular and mitochondrial O_2 partial pressures, respectively. Microvascular PO_2 begins at the (high) arterial value and falls along the microvessels to end up at the (low) venous PO_2 as the red cells move into the venules and veins for transport back to the lungs for reoxygenation. As with the lung, a *m*ean *t*issue *c*apillary PO_2 can be defined as $PmtcO_2$, and the integrated form of Eq. 11.8 is then given by:

$$VO_2 = DT x [PmtcO_2 - PmO_2\} \quad (11.9)$$

The same concepts, rules, and complexities apply in using eqs. (11.8) and (11.9) as in the lung (eqs. (11.5) and (11.6)) and will not be restated here, except to note that the shape of the O_2Hb dissociation curve again precludes a simple algebraic treatment of tissue diffusion unless one approximates the O_2Hb dissociation curve by a straight line with slope β as discussed above [15]. When one makes this approximation and sets up and solves the corresponding differential equation to that for diffusion in the lung in eq. (11.5), the result is as follows:

$$PvO_2 - PmO_2 = (PaO_2 - PmO_2) x \exp(-DT / (\beta x Q)) \quad (11.10)$$

This solution, analogous to eq. (11.7) for diffusion in the lung, does assume a temporally and spatially constant value for PmO_2, implying homogeneity of the ratio of O_2 supply to O_2 utilization. Equation (11.10) states that, for a given PaO_2 and PmO_2, the venous PO_2 is determined by the ratio of DT to (β x Q), analogous to the conclusion for the lung where DL replaces DT.

2.5 Step 5: Mitochondrial Utilization of O_2

The final step of the O_2 pathway is mitochondrial use of O_2 by means of oxidative phosphorylation. The underlying biochemistry is complex, but fortunately it is not necessary to incorporate basic individual biochemical equations encompassing each of the many steps describing both glycolysis and the citric acid cycle to account for the effect of this last step on O_2 transport itself. The key is that, due to the law of mass action, mitochondrial VO_2 depends on mitochondrial PO_2 [16] in a hyperbolic manner (other factors constant), as follows:

$$VO_2 = Vmax \, x \, PmO_2 / (PmO_2 + P50m) \tag{11.11}$$

Here, VO_2 again represents O_2 transported to and used by the mitochondria while Vmax is the maximal rate of mitochondrial O_2 use when PO_2 is in excess, meaning that it greatly exceeds P50m. P50m is the PO_2 at which VO_2 is half of Vmax.

3 Integrating the Entire O_2 Transport Pathway

3.1 Conceptual Basis

Given the five steps in O_2 transport described above, in which each step is described by a single equation, the overall system behavior can be analyzed. The five individual equations are restated here:

$$STEP\,1, VENTILATION: \; VO_2 = VA \, x \left[FIO_2 - FAO_2\right] \tag{A}$$

$$STEP\,2, LUNG\ DIFFUSION: \; VO_2 = DL \, x \left[PAO_2 - PmpcO_2\right] \tag{B}$$

$$STEP\,3, PERFUSION: \; VO_2 = Q \, x \left[CaO_2 - CvO_2\right] \tag{C}$$

$$STEP\ 4, TISSUE\ DIFFUSION: \quad VO_2 = DT \times [PmtcO_2 - PmO_2\} \qquad (D)$$

$$STEP\ 5, MITOCHONDRIAL\ RESPIRATION: \quad VO_2 = Vmax \times PmO_2 / (PmO_2 + P50m) \qquad (E)$$

To mitigate confusion, these five equations have been retagged A through E, but are identical to eqs. (11.4), (11.6), (11.1), (11.9), and (11.11), respectively. How are these five equations used to understand not just O_2 transport at each step, but how the five steps come together as a system to govern O_2 transport?

The example of large muscle mass (i.e., whole body) exercise will be used from now on as the platform on which O_2 transport will be analyzed. O_2 transport to, and use by, all of the other tissues and organs of the body can be incorporated into such an analysis, but for simplicity, this will be neglected in the following, which means that Q represents cardiac output and PvO_2 represents not just muscle venous PO_2 but also mixed venous PO_2 in the pulmonary artery. This is reasonable because muscle O_2 use during intense exercise accounts for the great majority of total body O_2 use.

To begin the analysis, one must first envision how the body undertakes exercise.

First, the intensity of exercise is governed by conscious decisions. We run at the speed we wish to—we do not think to target a particular muscle contractile effort (or cardiac output or level of ventilation). The chosen speed results in the necessary degree of muscle contractile response. Inspired O_2 fraction is also a given as a choice we can make.

Second, there are strong, well-defined empirical relationships between exercise intensity and the values of ventilation (VA) and cardiac output (Q) seen at any exercise intensity. While it still remains uncertain just how both VA and Q are regulated to support a given exercise intensity, there are likely many complementary stimuli involved, something that is well beyond the scope of this chapter to review. Suffice it to say that both VA and Q are dictated by the chosen exercise intensity, are measurable, and thus considered as known variables in the system.

Third, the muscles possess innate structural and biochemical content determining both Vmax and P50m (which may of course vary with age, sex, training status, disease, and so on), so that these are also taken as given (but alterable) parameters of the system.

Fourth, the O_2Hb dissociation curve and [Hb] itself are also considered as known parameters that change in known ways upon exercise as pH, PCO_2, and temperature change. The slope of the O_2Hb curve (β), if the linear approximation is used, is also considered to be known.

Fifth, the structural properties of the lungs and muscles that determine their respective diffusing capacities, DL and DM (note the change from DT to DM in moving from considering *t*issues in general to *m*uscle in particular) are fixed (over the short term) and are also considered as given parameters.

What is left unknown in the above five equations? There are five unknown (i.e., dependent) variables in total: VO_2, PAO_2, PaO_2, PvO_2, and PmO_2. This is because both $PmpcO_2$ and $PmmcO_2$ (note the change from $PmtcO_2$ to $PmmcO_2$ also reflecting the focus on *m*uscle in particular rather than *t*issues in general) are determined by PaO_2 and PvO_2, and are not themselves additional dependent variables. The five equations thus contain five unknowns, which implies there will a unique solution in which only one value of each of the five can simultaneously satisfy the five equations. The five variable values will of course differ from each other, but there can be only one solution value for each variable. The solution is not difficult to reach with modern computer methods, which have been published, and this aspect is therefore not further presented herein.

From the multiple appearances of both given (independent) and unknown (dependent) variables across the five equations, it can be explicitly seen how each step interacts with the others to affect the performance of the entire system. Thus, unexpected performance at any one step may not just be due to a functional difference within that step, but to interactions among the steps.

In summary, the O_2 transport system is governed by five O_2 transport equations in which the relevant functions of each system component are known, resulting in a unique solution for both VO_2 enabled by the system and the unique PO_2 values that must exist in alveolar gas, arterial and venous blood, and the mitochondria. Thus, the values of PO_2 at each step can be calculated, and overall system performance (VO_2) estimated knowing the values for FIO_2, VA, QT, DL, DM, [Hb], the parameters describing the O_2Hb dissociation curve, mitochondrial Vmax, and mitochondrial P50.

3.2 A Graphical Approach to O_2 Transport System Function

For most of us, the foregoing textual description of a complex system underwritten by a series of mathematical equations is not wholly satisfying as a way to comprehend the system's behavior. To mitigate this shortcoming, a graphical analysis of the O_2 transport system has been developed [17, 18] in which the key equations are shown pictorially and in a way that allows their interactions as a system to be easily appreciated.

To begin with, it will be assumed that in normoxia mitochondrial oxidative capacity (Vmax in Eq. (E) above) exceeds the capacity to transport O_2 to the mitochondria. This is most commonly the case, since VO_2max can be increased by breathing 100% O_2 [19] but when it is not the case, such that O_2 transport capacity exceeds mitochondrial oxidative capacity, the analysis can still be used, as will be described later herein.

However, before proceeding, it is important to understand that while the system described and its underlying equations apply equally at rest, during submaximal, and during maximal exercise, it is only during intense, (near) maximal exercise that the following graphical approach is applicable with little need for approximation. Specifically this means that mitochondrial PO_2 (PmO_2 in the equations) is low enough to neglect in the analysis, for which there is good, albeit indirect, evidence [20–22]. With PmO_2 in the low single digits at (near) maximal exercise, and muscle microvascular PO_2 ranging between arterial (~90–100 mm Hg) and venous (~20–40 mm Hg), the mean value of PO_2 along the capillaries is commonly between 40 and 50 mm Hg. Thus, whether PmO_2 is taken to be zero or in the low single digits will not impact the graphical analysis below. In what follows, PmO_2 is therefore approximated to zero, and it thus applies only to intense exercise.

In contrast, at rest, PmO_2 may be relatively high, in the order of 20–30 mm Hg [23], and approximating PmO_2 to zero would lead to significant errors. Note also that the experimental finding that PmO_2 during intense exercise is very close to zero, while PO_2 in the microvasculature is one to two orders of magnitude greater, by itself leads to an important conclusion—that the diffusive transport capacity for O_2 from the microvessels to the mitochondria is limited and may impede mitochondrial O_2 availability. This is because, were there no diffusion limitation, PmO_2 would be very close to venous PO_2 just as arterial PO_2 is very close to alveolar PO_2 in the absence of pulmonary diffusion limitation.

Figure 11.1 exemplifies the basic concept of this graphical approach (adapted from [18]) and it depicts the loci of two of the equations in the transport system. It has been called the "Fick" diagram because both equations (as written in the legend box and shown on the figure as lines) reflect the Fick Principle as stated in eq. (11.1) (curved line) and the Fick law of diffusion (straight line). The latter is described by

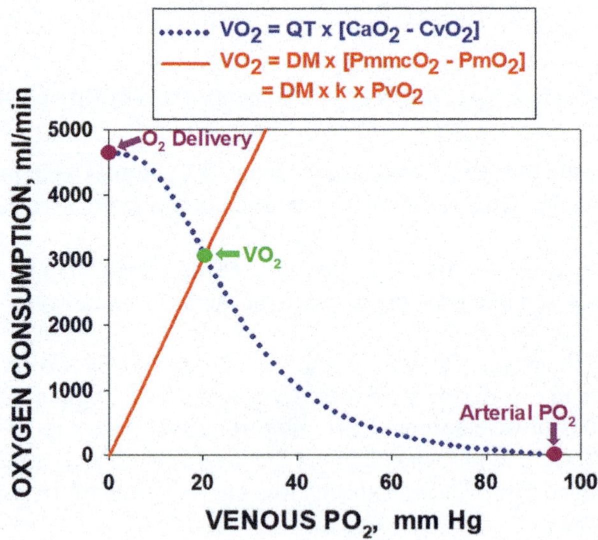

Fig. 11.1 The Fick diagram. As laid out in the legend box of the figure, the curved line shows the relationship between VO_2 and venous PO_2 required to satisfy the Fick Principle of mass conservation (i.e., $VO_2 = QT \times [CaO_2 - CvO_2]$) while the straight line shows the relationship between VO_2 and venous PO_2 required to satisfy the Fick Law of diffusion. The only place on the entire *X–Y* plane where both relationships allow for the same VO_2 and venous PO_2 is at their point of intersection (solid green circle), which marks the unique values for VO_2 and PvO_2 for the given conditions. See text for the definition of symbols and further explanation

eq. (11.9) restated to focus on the muscles (i.e., replacing T (and t) by M (and m) in the terms of the equation):

$$VO_2 = DM x [PmmcO_2 - PmO_2] \quad (11.12)$$

where DM is muscle diffusing capacity, $PmmcO_2$ is *m*ean *m*uscle *c*apillary PO_2 and PmO_2 is mitochondrial PO_2 (as previously defined). In accordance with the text above, PmO_2 is taken to be zero, so that:

$$VO_2 = DM x PmmcO_2 \quad (11.13)$$

Additionally, it can be shown that mean muscle capillary PO_2 and muscle venous (i.e., end capillary) PO_2 (PvO_2) vary in direct proportion to each other over the physiological range encountered, so that mean PO_2 ($PmmcO_2$ in eq. (11.13)) can be replaced by a constant of proportionality (call it "k") multiplying PvO_2. Thus, $PmmcO_2 = k x PvO_2$. Making this substitution then harmonizes both lines shown so that both the straight and curved lines can be plotted as a function of the same variable—venous PO_2. This action, i.e., replacing $PmmcO_2$ by k x PvO_2, is required only for the graphical analysis, and not for the formal mathematical analysis of the five eqs. (A) through (E) above. Thus, for visual presentation purposes, eq. (11.13) is rewritten as:

$$VO_2 = k x DM x PvO_2 \quad (11.14)$$

In the end, it is eqs. (11.1) and (11.14) that are plotted in Fig. 11.1 (and in several of the subsequent figures. It is important to understand the information conveyed by both lines in Fig. 11.1.

The *curved* line shows, from Eq. (11.1), what VO_2 (*Y*-axis) would have to be in order that mass be conserved if the venous PO_2 (*X*-axis) took any value between its obvious lower limit of zero and upper limit of arterial PO_2. It is a curved line because it plots O_2 concentration as a function of PO_2, and thus takes the exact shape of the O_2HB dissociation curve, albeit inverted because of the negative sign in eq. (11.1) before the venous [O_2] term. Mathematically it defines the complete range of possible VO_2 values across the range of possible PvO_2 values: VO_2 must lie somewhere on the line, or mass conservation is violated. The approximation that PmO_2 be zero is immaterial to this equation and line because PmO_2 is not a variable within the Fick Principle Eq. (C).

The *straight* line represents, via eq. (11.14), the diffusive process. It shows what VO_2 (*Y*-axis) would have to be in order that O_2 mass be conserved if the venous PO_2 (*X*-axis) took any value between its lower limit of zero and upper limit of arterial PO_2. Mathematically it also defines the complete range of possible VO_2 values across the range of possible PvO_2 values: VO_2 must lie somewhere on the straight line, or mass conservation is violated.

The inescapable conclusion is that the only place on the entire X-Y plane of Fig. 11.1 *where mass conservation is guaranteed by both equations simultaneously is at their point of intersection.* That point defines not only the unique VO_2 that must exist but the unique coexisting venous PO_2 value that must be present to be compatible with that VO_2.

A discerning reader will see that Fig. 11.1 displays only two of the five equations. That is true, however, the equations representing ventilation and diffusion in the lungs (Eqs. (A) and (B) above) are implicit to the figure because of their role in determining arterial PO_2. In addition, for the purposes of presentation, the final equation has been ignored as explained previously because the value of PmO_2 is so low that it can be taken to be zero in the context of the scale of Fig. 11.1.

Therefore, while a very useful way to visualize the O_2 transport pathway during intense exercise as an interactive system, Fig. 11.1 remains an approximation to the more formal mathematical approach defined above using Eqs. (A)–(E), in which there is no requirement for the approximation that mitochondrial PO_2 is zero, nor that there is a proportional relationship between mean muscle capillary PO_2 and muscle venous PO_2.

It is very important to note the determinants of the pathways taken by the two lines in Fig. 11.1. The curved line depends on:

- Blood flow (QT).
- [Hb].
- The shape and position of the O_2Hb dissociation curve.
- Arterial PO_2. Arterial PO_2 is, as stated, in part determined by ventilation (VA) and diffusion (DL), via Eqs. (A) and (B).

The straight line depends only on muscle diffusion via DM as in Eq. (D).

The inescapable conclusion is that VO_2 is affected by the function of every part of the O_2 pathway—lungs, heart, blood, and muscle. There can be NO SINGLE VARIABLE that uniquely determines (or limits) VO_2 (such as cardiac output). Since the figure represents O_2 transport at maximal exercise and uses the values of independent variables determined at maximal exercise, it shows what the value of maximal VO_2 must be, given the values of those independent variables. It also shows what the associated venous PO_2 must be in order that mass is conserved. From a practical standpoint, the figure can be applied "in reverse" to use measured values of PvO_2 and VO_2 from an individual subject to estimate the subject's muscle diffusing capacity (DM) from the slope of the line joining the origin to the measured data point (VO_2, PvO_2) on the diagram.

It is very helpful to further understand the behavior of O_2 transport as a system to ask how would things change when only *one* individual contributing factor was altered? First, the effects of acute changes in inspired PO_2 are examined, assuming no other changes in O_2 transport variables have occurred. Fig. 11.2 (adapted from [18]) shows how moderate and extreme hypoxia (i.e., reduced PIO_2) would change the analysis, all other factors unchanged. The straight line would be unaffected because DM is dependent on muscle structural factors such as capillarity and muscle fiber size and not on PIO_2. However, the curved line will change

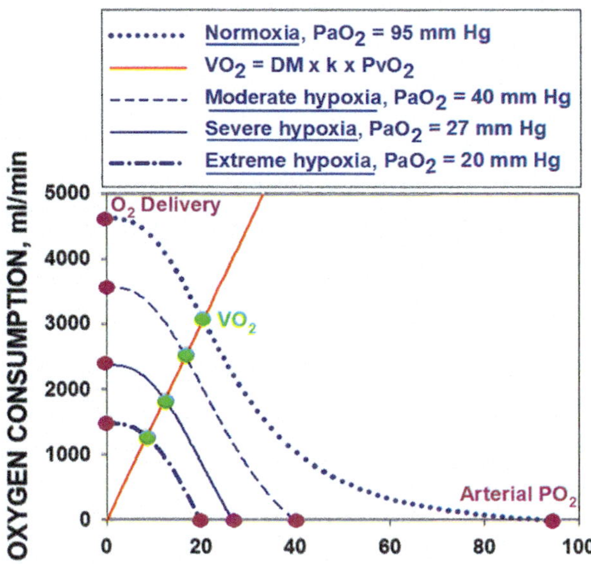

Fig. 11.2 The Fick diagram showing how VO_2 and PvO_2 must change when the subject breathes gas of low O_2 content (normoxia and three levels of hypoxia are displayed). The key conclusion is that VO_2 and PvO_2 fall with increasing hypoxia in direct proportion to one another (solid green circles)

because arterial O_2 concentration (CaO_2, Eq. (C)) must be lower as PIO_2, and thus PO_2 at every step of the O_2 pathway falls with hypoxia. In the examples shown, CaO_2 is reduced by about 25% with moderate hypoxia, by about 50% with severe hypoxia and even more with extreme hypoxia. There is no change to [Hb] or QT in Fig. 11.2 because the objective is to ask how the drop in PIO_2 alone will change the behavior of the system. As the figure shows, the points of intersection between the straight and curved lines progressively fall with greater hypoxia. The key observation is that the four intersection points (solid green circles) lie on the single (red) diffusion line. The result is that maximal VO_2 and the associated PvO_2 fall together *and do so in proportion*, with increasing hypoxia. This is an easily tested prediction, requiring only measurement of both maximal VO_2 and PvO_2 simultaneously as PIO_2 is acutely reduced. This experiment has been performed and follows the prediction very closely in both isolated canine muscle [24] and in intact humans [25], as described later herein. Note that dependence of VO_2max on both ventilation and pulmonary diffusion (via Eqs. (A) and (B) above) will have qualitatively similar effects as changes in PIO_2 because all three factors act through altering arterial PO_2.

Figure 11.3 shows how changing another of the variables in the system—cardiac output (or muscle blood flow)—will affect V O_2max and PvO_2. It can be seen that again, that the straight line representing diffusion in unaffected because blood flow is not a part of the diffusion equation per se, which depends only on DM. However, the curved line representing the Fick Principle is directly affected because it contains blood flow, QT, as the figure legend box indicates. Arterial PO_2 is unaffected in this example because the question is about the effect of a change in QT only on the system. Should changes in QT affect the completeness of diffusional transport

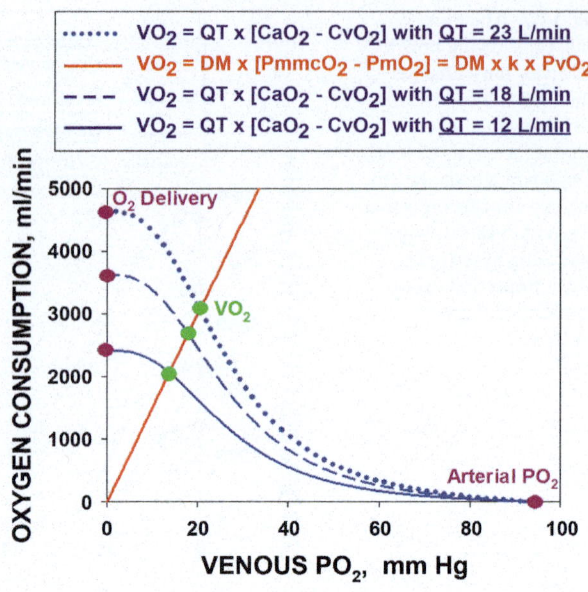

Fig. 11.3 The Fick diagram showing how VO_2 and PvO_2 must change should cardiac output (QT) be progressively reduced (normal and two lower levels of blood flow are shown). As with hypoxia (Fig. 11.2), VO_2 and PvO_2 again fall in direct proportion to one another (solid green circles)

in either the lungs or muscles, these secondary effects can be included in the figure, but are omitted in Fig. 11.3 in order that the pure effect of blood flow change can be appreciated. Although the curved lines underlying each value of QT clearly differ in configuration from those seen when only PIO_2 is reduced (see Fig. 11.2), the final result is quite similar to what is seen when PIO_2 is reduced: there is a linear and proportional relationship between VO_2max and QT as QT is altered, as found in isolated canine muscle [26]. The same outcome was seen if [Hb] was altered [27] (again without changes in any other transport component) because [Hb] can be factored out of the terms CaO_2 and CvO_2 and thus has the same numerical effect on transport as QT:

$$VO_2 = QT \times [Hb] \times [SaO_2 - SvO_2] \tag{11.15}$$

Equation (11.15) does assume that the only way in which O_2 is transported in the blood is bound to Hb, but it is common knowledge that in room air, ignoring the second mode of O_2 carriage in blood (i.e., as physically dissolved O_2) imparts negligible error due to very low solubility of O_2 in plasma and water.

Finally, a change in DM can be examined within the structure of the Fick diagram, and this is shown in Fig. 11.4. As DM falls, so does VO_2max, but since the parameters of the curved line do not depend on DM, that line remains unchanged. Hence, as the slope of the diffusion line falls (because it is proportional to DM), PvO_2 increases as DM falls, the opposite of how the system responds when PIO_2 or QT are changed.

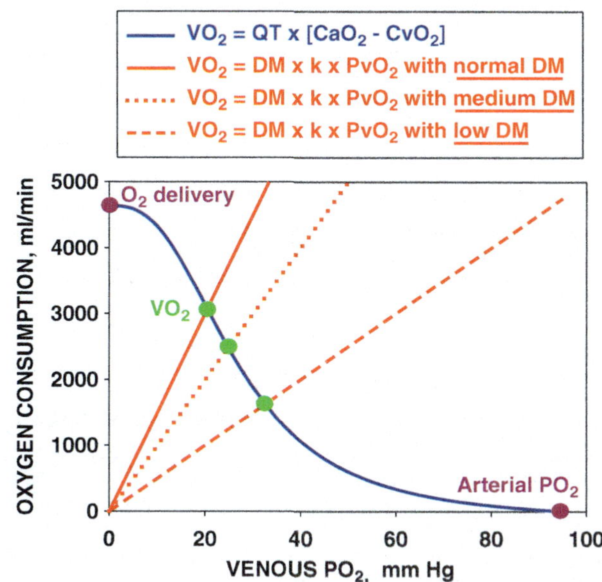

Fig. 11.4 The Fick diagram showing how VO_2 and PvO_2 must change when the capacity to transport O_2 by diffusion from red cells to mitochondria (DM) is reduced. Three conditions are displayed, using normal and two lower levels of DM. In this case, VO_2 falls but PvO_2 increases as DM is reduced (solid green circles)

The outcome of this analysis is that by examining how VO_2max and PvO_2 change as conditions change, or when comparing different subjects, great insight is gained into the individual steps of the O_2 transport system, not just qualitatively but also quantitatively. For example, it would be easy to examine how exercise training affects the O_2 transport system in terms of relative contributions from every one of the transport variables involved by comparing changes in VO_2 and PvO_2 on the Fick diagram and plotting the underlying equations (as in Fig. 11.1) for both pre-training and post-training measurements. This will be addressed in the final section of this chapter.

The preceding discussion raises the question of how quantitatively important each step of the O_2 transport chain is in affecting VO_2max, were that step to change, while all other steps remained unchanged. This question cannot be realistically answered in vivo because it is not possible to change just one step independently of all others. The question can however be addressed theoretically, as Figs. 11.2–11.4 suggest. The outcome is shown in Fig. 11.5. This figure (adapted from [28]) shows calculated VO_2max changes when the function of each step of the O_2 pathway is increased or decreased from typical normal values, one step at a time. The abscissa of this figure is normalized to % values of each variable in order to plot the results on a single graph. Perhaps surprisingly, the effects of each step are very similar to one another. Importantly, raising the function of any step conveys very little benefit. On the other hand, reducing the function of any one step has a substantial negative impact on VO_2max. This is exactly the kind of behavior expected of an in-series "bucket brigade" type of system: Suppose several people are passing buckets of water from one to the next. Having one person who is stronger than all the rest does

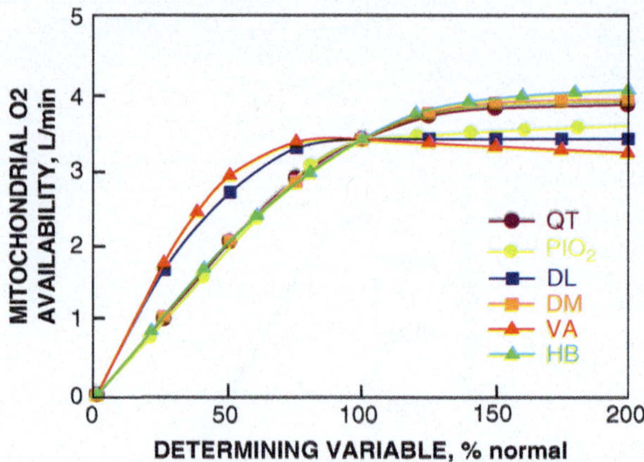

Fig. 11.5 Calculated VO_2 (equivalent to mitochondrial O_2 availability) as each major determinant of O_2 transport is increased or lowered from normal (i.e., 100%) values, without changing the values of any other variables in the system. The main findings are that (1) every step of the O_2 pathway has a similar effect on VO_2 when its ability to transport O_2 is changed and (2) enhancing any one variable does little to raise VO_2 while reducing any one variable significantly compromises VO_2

not raise bucket throughput over that seen when all persons are of similar strength; however, having one person weaker than all the others materially reduces throughput.

The inclusion of how mitochondrial ability to use delivered O_2 affects the system is now addressed. Recall that it was mentioned above that Fig. 11.1 assumes that mitochondrial Vmax exceeds the capacity to deliver O_2 to the mitochondria. The figure and analysis can still be applied should the converse be the case—that O_2 transport exceeds Vmax (the capacity of the mitochondria to use O_2). Note that both conditions (i.e., Vmax > O_2 transport as in Fig. 11.1, and Vmax < O_2 transport) result in measurable O_2 remaining in the muscle venous blood, and may or may not show the same VO_2. This implies that looking at one pair of VO_2 and PvO_2 values alone is insufficient to distinguish between these two possibilities. The two possibilities and the associated uncertainty can be resolved by the construct of Fig. 11.6. This figure simply adds two horizontal straight lines to Fig. 11.1. Each line depicts the values of VO_2 at one of two hypothetical values of mitochondrial Vmax: a) where Vmax < O_2 transport (lower green horizontal line) and b) the converse, where Vmax > O_2 transport (as assumed for Figs. 11.1–11.4, upper purple horizontal line). For the case where Vmax > O_2 transport capacity to deliver O_2, the outcomes depicted in Figs. 11.1–11.4 apply, as reflected by the solid purple circle. However, in the event that Vmax < O_2 transport capacity, VO_2max can only equal and not exceed Vmax, by definition. At the same time, O_2 mass conservation must still apply, and therefore, the operating point of the subject must also lie on the curved line for the Fick Principle. The result is that VO_2 and PvO_2 lie at the point indicated by the solid green circle.

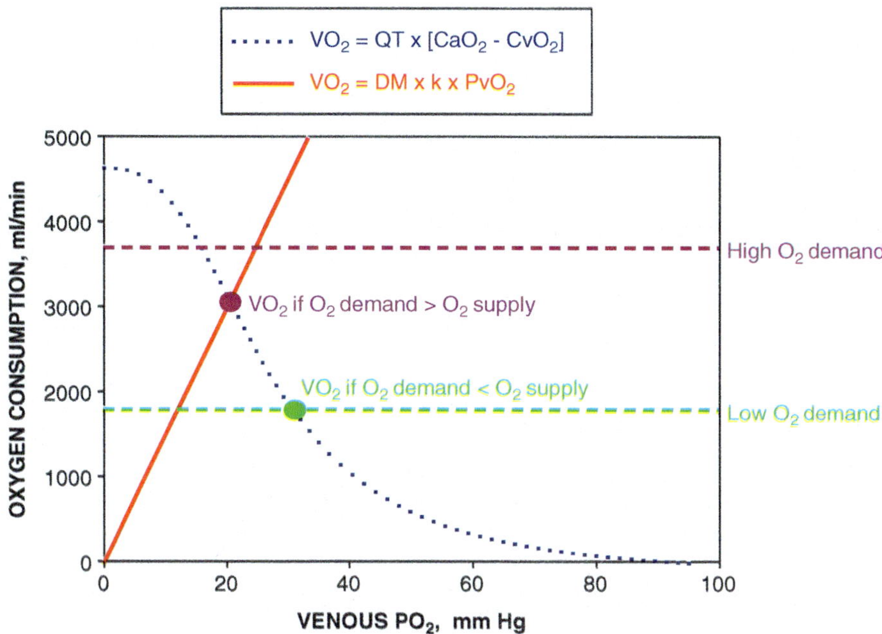

Fig. 11.6 Adding the concept of metabolic capacity to consume O_2 to the Fick diagram. Maximal metabolic capacity may exceed the capacity to deliver O_2 (purple line, text, and circle) or it may be less than the capacity to deliver O_2 (green line, text, and circle). The solid circles show in each case the requisite values of VO_2 and PvO_2

The Fick diagram can be used to determine experimentally whether Vmax > O_2 transport or whether Vmax < O_2 transport, as follows. The experiment would be to measure VO_2max and PvO_2 over as wide a range of arterial PO_2 as possible and plot the results on the Fick diagram. The simplest way to do this is to have the subject breathe O_2/N_2 mixtures having a range of PIO_2, within the limits of safety. Figure 11.7 shows the expected result of such a study, were the severe levels of hypoxemia depicted in the figure able to be tolerated. The key result is that once O_2 supply exceeds Vmax, VO_2 will not increase as PIO_2 is increased (three right-most solid green circles in Fig. 11.7) but when Vmax exceeds O_2 supply, VO_2 will fall along the red line for diffusive transport (two left-most solid green circles in Fig. 11.7). Thus, Figs. 11.2 and 11.7 contrast the behavior of VO_2 as PIO_2 is altered according to the relationship between Vmax and O_2 supply capacity. Studies based on these expectations are described in the next section.

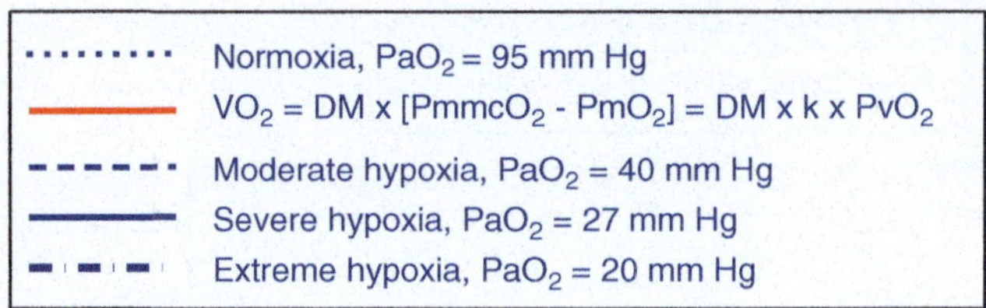

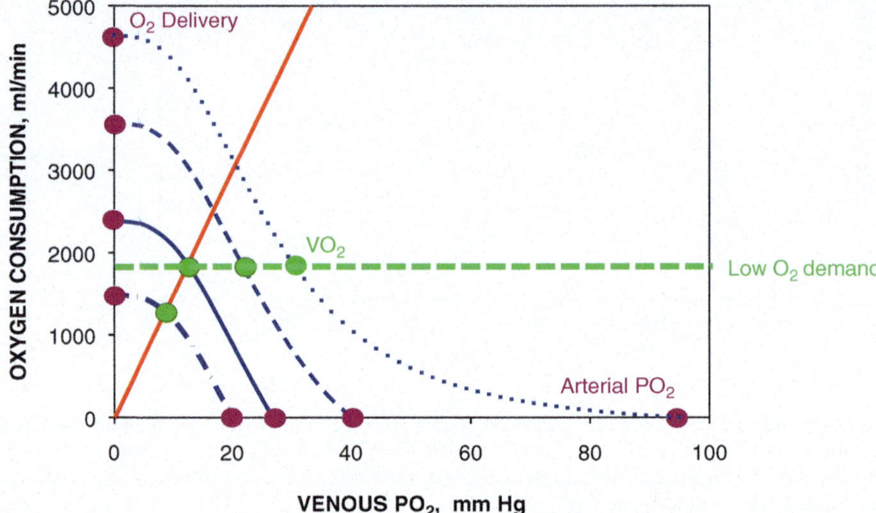

Fig. 11.7 The Fick diagram showing the effects of altering inspired O_2 levels on VO_2 and PvO_2 (solid green circles) when the mitochondrial capacity to use O_2 is less than O_2 deliverable. The nonlinear result should be contrasted with the outcome in Fig. 11.2 which reflects the same measurements when the mitochondrial capacity to use O_2 exceeds the capacity to deliver it

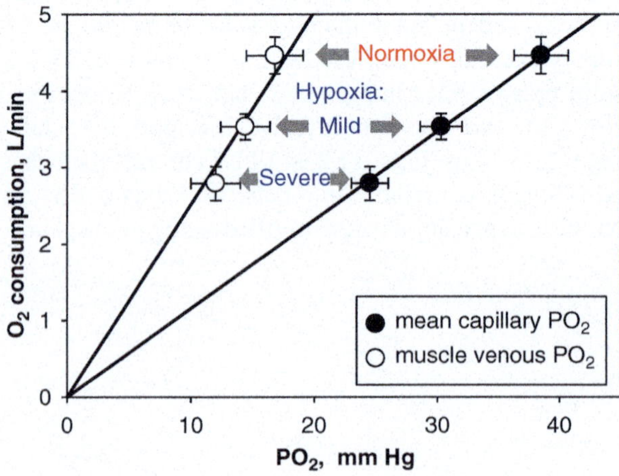

Fig. 11.8 Experimental confirmation in healthy humans that VO_2 and mean capillary PO_2 do in fact fall together as lower inspired O_2 levels are breathed. Also shown is the similar relationship between VO_2 and PvO_2. See text for more detail

4 Experimental Observations Supporting the Systems Approach to O_2 Transport

Measuring VO_2max in normal subjects over a range of acutely altered values of PIO_2 to test the prediction shown in Fig. 11.2 that VO_2max and muscle venous PO_2 fall in proportion is feasible. This experiment was done in isolated canine muscle by Hogan et al. [24] and in exercising humans by Roca et al. [25]. The results of the human study are shown in Fig. 11.8. Here, the values of VO_2max are plotted against both mean muscle capillary PO_2 (closed circles) and muscle venous PO_2 (open circles). The former embodies eq. (11.13), while the latter reflects eq. (11.14). The relationship between mean capillary PO_2 and VO_2max is closely proportional, as expected from eq. (11.13) and is also approximated well by using muscle venous PO_2. Had the data of Fig. 11.8 not fallen on a line of proportionality, this experiment would have cast doubt on the way the O_2 transport system is described above. However, compatibility with the predicted behavior is not per se proof of correctness.

Two additional interventional studies lend support to the way in which O_2 transport is described in the preceding text. The concept in these studies was to change the affinity of O_2 for Hb while maintaining both muscle blood flow and arterial O_2 concentration constant. Changing affinity should change mean capillary PO_2 (high affinity leading to lower mean capillary PO_2 and vice versa). This required an in situ, pump-perfused, maximally electrically stimulated contracting muscle preparation using normal blood and also blood for which HbO_2 affinity (P50) had been changed pharmacologically. Maintaining 100% O_2 saturation by inhalation of 100% O_2 assured constant O_2 concentration independent of P50.

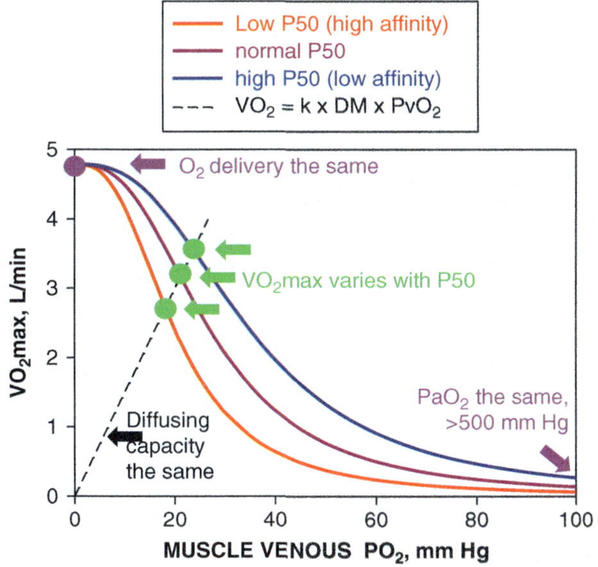

Fig. 11.9 Fick diagram showing expected changes in VO_2 and PvO_2 (green circles) as hemoglobin P50 (affinity for O_2) is changed while keeping O_2 delivery constant. Hb with high O_2 affinity should reduce VO_2 and Hb with low affinity should increase VO_2—both in proportion to the changes in venous PO_2

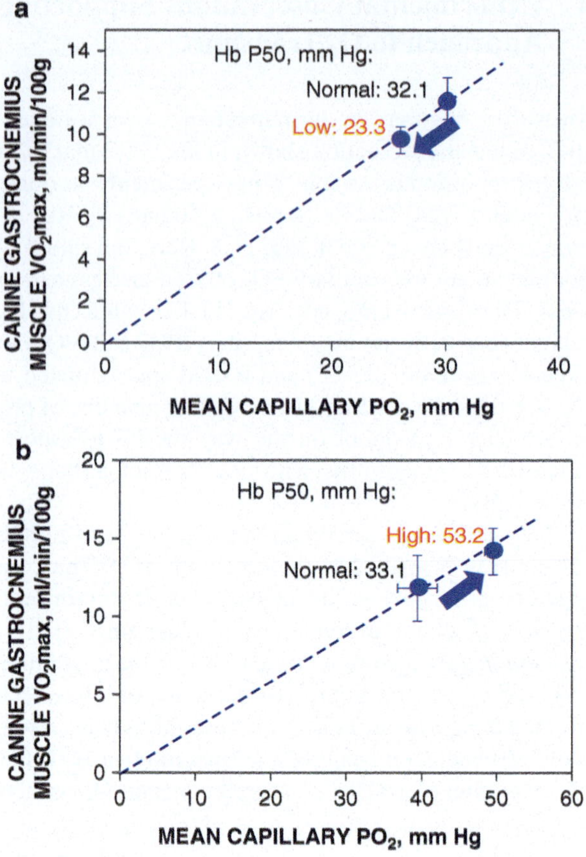

Fig. 11.10 Experimental confirmation that altering hemoglobin P50 alters VO_2 in proportion to mean capillary (or venous) PO_2. Top panel shows effect of reducing P50 (i.e, increasing HbO2 affinity); bottom panel shows effect of increasing p50. See text for more detail

In this way, O_2 delivery into the muscle vascular bed could be held constant as P50 was altered. Figure 11.9 shows the concept underlying this experiment and how it predicts systematic changes in VO_2max as a function of P50. First, Hogan et al. [29] lowered P50 and separately, Richardson et al. raised P50 [30]. The results are shown in Fig. 11.10a and b: At constant O_2 delivery into the muscle bed, VO_2max indeed fell with increased Hb O_2 affinity and increased with reduced affinity.

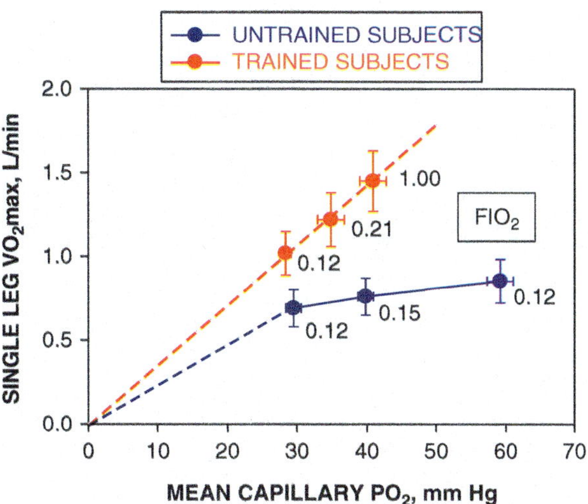

Fig. 11.11 Effects of altering inspired O_2 levels on maximal VO_2 in normal subjects. Trained subjects are shown red, untrained in blue. Referring to Figs. 11.2 and 11.7, the results in Fig. 11.11 suggest that maximal VO_2 in trained athletes is O_2 supply-limited, while untrained subjects are close to their mitochondrial limit to use O_2 rather than being limited by O_2 supply

5 Applications in Health and Disease

In this final section are shown some applications of the systems analysis of O_2 transport in health and cardiopulmonary disease. Returning to the issue of whether VO_2max is determined by O_2 supply or by mitochondrial oxidative capacity, Fig. 11.11 shows experimental data [31, 32] from two groups of subjects—trained athletes (red symbols) and untrained sedentary subjects (blue symbols). The athletes clearly show how VO_2max increases with PIO_2, at least up to that seen during pure O_2 breathing, proving that O_2 availability, and not mitochondrial oxidative capacity, determines VO_2max. In untrained subjects, the picture resembles that of Fig. 11.7, suggesting that the subjects were approaching maximal mitochondrial metabolic rate as PIO_2 was increased. The implication of these two data sets is that exercise training results in a relatively greater increase in mitochondrial oxidative capacity than in O_2 supply capability.

A question of longstanding interest has been to understand how some healthy humans (i.e., elite mountaineers) can ascend the summit of Mt. Everest without breathing supplemental O_2. While there is no question that courage, determination, and skill are all important, the feat cannot be accomplished by most climbers without supplemental O_2. However, when Oelz et al. [33] examined several elite climbers in the laboratory, they could find no differences, compared to control subjects, in common physiological parameters important to O_2 transport such as ventilation,

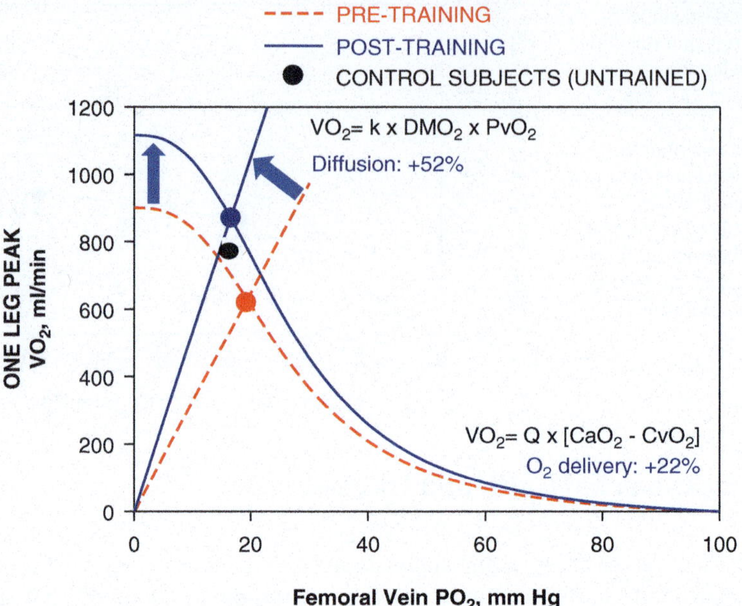

Fig. 11.12 Fick diagram in patients with heart failure (low ejection fraction) before (red) and after (blue) muscle training. Normal untrained subjects are shown by the black circle. Pre-training, patients show subnormal blood flow as expected, but also subnormal diffusing capacity. Both perfusion and diffusion are enhanced after training, bringing the patients into the normal range

diffusion, and arterial O_2 saturation. That said, they noted that muscle capillary density was 40% greater than in controls. This finding may well answer the question, as higher capillary density increases muscle O_2 diffusive conductance (DM in Fig. 11.4).

Systems analysis of O_2 transport has been applied using the Fick figure approach in patients with heart failure—both those with reduced [34] and with maintained [35] ejection fraction. In both groups a substantial factor contributing to the low VO_2peak was reduction in muscle diffusing capacity for O_2. Thus, it was not only impaired cardiac function that lowered O_2 delivery to muscle but also impaired intramuscular O_2 diffusion. When muscle exercise training was then employed (in patients with reduced ejection fraction) and the measurements repeated, exercise capacity improved, along with muscle O_2 diffusing capacity, also with some improvement in central cardiac function. Figure 11.12 shows pre- and post-training outcomes in those patients [34, 36]. It is important to note that muscle venous PO_2 was essentially unchanged by training, which might lead some to suggest a lack of improvement in muscle O_2 extractability. The figure shows however that this component of O_2 transport was increased by 52%, while O_2 delivered to the muscle was

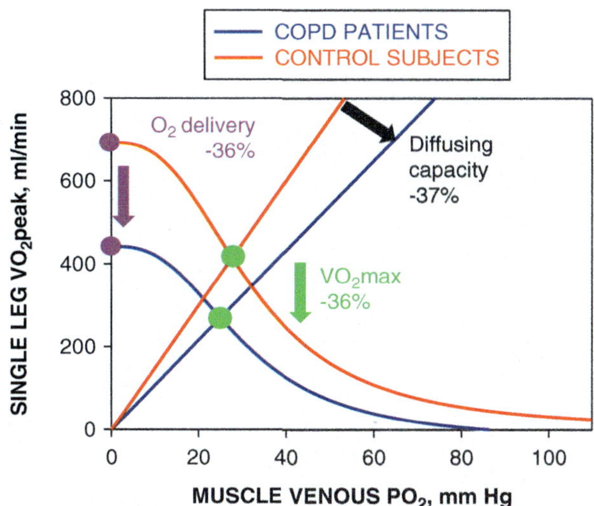

Fig. 11.13 Fick diagram in patients with COPD (blue) and healthy controls (red). As with heart failure, both perfusion and diffusion are impaired. Unpublished data show that after muscle training, the diffusive component normalizes

increased much less (by 22%). It is evident in the figure that the gain in peak VO_2 shown in the figure was due more to the diffusive improvement than to O_2 delivery changes.

A similar outcome has been found in patients with COPD as shown in Fig. 11.13—a substantial contribution of impaired muscle O_2 diffusive transport to low exercise capacity that is reversible with exercise training despite no improvement in central cardiorespiratory function [37].

6 Conclusions

The traditional framework for considering the determinants of maximal VO_2 has been to examine each step of the well-known O_2 transport pathway and ask "which step of the pathway is the limiting factor to VO_2max?." The O_2 pathway is however an integrated system of transport steps in which every step has the power to affect VO_2max. This chapter has laid out the framework for understanding the O_2 transport pathway as an integrated system—both qualitatively and quantitatively. Critical to understanding is that the various steps in the pathway interact with, and affect, one another. While this makes analysis more complex that it would be by studying each step separately, it is amenable to a pictorial analysis that enables understanding of the role of each pathway step in determining VO_2max. Importantly, the pictorial approach provides a practical basis for quantifying the transport function of any step and determining how each step may be impacted as a result of exercise training, or by common disease states such as heart failure and COPD. Properly identifying the site(s) of impaired O_2 transport is the first step toward generating targeted therapies in these complex diseases.

References

1. Weibel ER. The pathway for oxygen. Structure and function in the mammalian respiratory system. Cambridge, MA: Harvard University Press; 1984.
2. Shephard RJ. A non-linear solution of the oxygen conductance equation. Applications to performance at sea-level and at an altitude of 7350 ft. Int Z Angew Physiol. 1969;27:212–25.
3. di Prampero PE. Factors limiting maximal performance in humans. Eur J Appl Physiol. 2003;90:420–9.
4. Wagner PD. Determinants of maximal oxygen transport and utilization. In: Massaro D, editor. Annual reviews of physiology. Palo Alto, CA: Annual reviews; 1996. p. 21–50.
5. Barclay JK, Stainsby WN. The role of blood flow in limiting maximal metabolic rate in muscle. Med Sci Sports Exerc. 1975;7:116–9.
6. Horstman DH, Gleser M, Delehunt J. Effects of altering O2 delivery on VO2 of isolated, in situ working muscle. J Appl Physiol. 1976;230:327–34.
7. Dempsey JA, Wagner PD. Exercise-induced arterial hypoxemia. J Appl Physiol (1985). 1999;87:1997–2006.
8. Wagner PD. A re-analysis of the 1968 Saltin et al. "Bedrest" paper. Scand J Med Sci Sports. 2015;25:83–7.
9. Houstis NE, et al. Exercise intolerance in heart failure with preserved ejection fraction: Diagnosing and ranking its causes using personalized O2 pathway analysis. Circulation. 2018;137:148–61. Among authors: Wagner PD
10. Broxterman RM, et al. Determinants of the diminished exercise capacity in patients with chronic obstructive pulmonary disease: looking beyond the lungs. J Physiol. 2020;598:599–610. Among authors: Wagner PD
11. Wagner PD. Operation Everest II and the 1978 Habeler/Messner ascent of Everest without bottled O2: what might they have in common? J Appl Physiol (1985). 2017;123:1682–8.
12. Fick A. Über diffusion. Pogg Ann. 1855;94:59–86.
13. Hopkins SR, Wagner PD. The multiple inert gas elimination technique (MIGET)The American Physiological Society; 2017.
14. Piiper J, Scheid P. Model for capillary-alveolar equilibration with special reference to O2 uptake in hypoxia. Respir Physiol. 1981;46:193–208.
15. Piiper J, Scheid P. Modeling oxygen availability to exercising muscle. Respir Physiol. 1999;118:95–101.
16. Wilson DF, Erecinska M, Drown C, Silver IA. Effect of oxygen tension on cellular energetics. Am J Phys. 1977;233:C135–40.
17. Wagner PD. An integrated view of the determinants of maximum oxygen uptake. In: Gonzalez NC, Fedde MR, editors. Oxygen transfer from atmosphere to tissues, vol. 227. New York: Plenum; 1988. p. 245–56.
18. Wagner PD. Determinants of maximal oxygen transport and utilization. Annu Rev Physiol. 1996;58:21–50.
19. Welch HG. Hyperoxia and human performance: a brief review. Med Sci Sports Exerc. 1982;14:253–62.
20. Gayeski TE, Honig CR. O2 gradients from sarcolemma to cell interior in red muscle at maximal VO2. Am J Phys. 1986;251(4 Pt 2):H789–99.
21. Groebe K, Thews G. Calculated intra- and extracellular PO2 gradients in heavily working red muscle. Am J Phys. 1990;259(1 Pt 2):H84–92.

22. Richardson RS, Noyszewski EA, Kendrick KF, Leigh JS, Wagner PD. Myoglobin O_2 desaturation during exercise: evidence of limited O_2 transport. J Clin Invest. 1995;96:1916–26.
23. Richardson RS, Duteil S, Wary C, Wray DW, Hoff J, Carlier PG. Human skeletal muscle intracellular oxygenation: the impact of ambient oxygen availability. J Physiol. 2006;571:415–24.
24. Hogan MC, Roca J, Wagner PD, West JB. Limitation of maximal O2 uptake and performance by acute hypoxia in dog muscle in situ. J Appl Physiol (1985). 1988;65:815–21.
25. Roca J, Hogan MC, Story D, Bebout DE, Haab P, Gonzalez R, Ueno O, Wagner PD. Evidence for tissue diffusion limitation of V· O_2MAX in normal humans. J Appl Physiol (1985). 1989;67:291–9.
26. Hogan MC, Bebout DE, Wagner PD. Effect of blood flow reduction on maximal O2 uptake in canine gastrocnemius muscle in situ. J Appl Physiol (1985). 1993;74:1742–7.
27. Hogan MC, Bebout DE, Wagner PD. Effect of hemoglobin concentration on maximal O2 uptake in canine gastrocnemius muscle in situ. J Appl Physiol (1985). 1991;70:1105–12.
28. Wagner PD. A theoretical analysis of factors determining VO2 MAX at sea level and altitude. Resp Physiol. 1996;106:329–43.
29. Hogan MC, Bebout DE, Wagner PD. Effect of increased Hb-O2 affinity on VO2max at constant O2 delivery in dog muscle in situ. J Appl Physiol (1985). 1991;70:2656–62.
30. Richardson RS, Tagore K, Haseler L, Jordan M, Wagner PD. Increased Vo2max with right shifted Hb O2 dissociation curve at a constant O2 delivery in dog muscle in situ. J Appl Physiol (1985). 1998;84:995–1002.
31. Cardus J, Marrades RM, Roca J, Barbera JA, Diaz O, Masclans JR, Rodriguez Roisin R, Wagner PD. Effects of Fio2 on leg V. o2 during cycle ergometry in sedentary subjects. Med Sci Sports Exerc. 1998;30:697–703.
32. Richardson RS, Grassi B, Gavin TP, Haseler LJ, Tagore K, Roca J, Wagner PD. Evidence of O2 supply-dependent VO2 max in the exercise-trained human quadriceps. J Appl Physiol (1985). 1999;86:1048–53.
33. Oelz O, Howald H, Di Prampero PE, Hoppeler H, Claassen H, Jenni R, Bühlmann A, Ferretti G, Brückner JC, Veicsteinas A, et al. Physiological profile of world-class high-altitude climbers. J Appl Physiol (1985). 1986;60:1734–42.
34. Esposito F, Mathieu-Costello O, Shabetai R, Wagner PD, Richardson RS. Limited maximal exercise capacity in patients with chronic heart failure: partitioning the contributors. J Am Coll Cardiol. 2010;55:1945–54.
35. Houstis NE, Eisman AS, Pappagianopoulos PP, Wooster L, Bailey CS, Wagner PD, Lewis GD. Exercise intolerance in heart failure with preserved ejection fraction: diagnosing and ranking its causes using personalized O2 pathway analysis. Circulation. 2018;137:148–61.
36. Esposito F, Mathieu-Costello O, Reese V, Shabetai R, Wagner PD, Richardson RS. Isolated quadriceps training reduces exercise intolerance in chronic heart failure: the role of skeletal muscle convective and diffusive oxygen transport. J Am Coll Cardiol. 2011;58:1353–62.
37. Broxterman RM, Hoff J, Wagner PD, Richardson RS. Determinants of the diminished exercise capacity in patients with chronic obstructive pulmonary disease: looking beyond the lungs. J Physiol. 2020;598:599–610. https://doi.org/10.1113/JP279135. Epub 2020 Jan 19

Part V
Alveolar Wall

Chapter 12
Pathophysiological and Clinical Implication of Diffusing Capacity for CO (DLco) and Krogh Factor (Kco): How Do D_{LCO} and K_{CO} Differentiate Various Lung Diseases?

Kaoruko Shimizu

Abstract The single-breath diffusing capacity of carbon monoxide (DL_{CO}) provides independent measurements of kCO, the rate constant of removable CO from the alveolar gas during the breath-holding time, and accessible lung volume (V_A) calculated by the dilution of helium (He), a nonabsorbable gas, in the lung at the time kCO is measured. Those two components of kCO and V_A determines the value of DLco. DLco/V_A, i.e., Kco is the rate constant for carbon monoxide uptake per unit barometric pressure, which is different from kCO in three constant factors: barometric pressure, 1000, and 1.2. Therefore, kCO and Kco are physiologically equivalent, and Kco does not correct DLco by lung volume. Changes in Kco is explained by changes in V_A/the pulmonary capillary volume.(Vc), i.e., Vc/V_A because V_A/the membrane diffusing capacity (DMco) remains almost constant. Due to the pathogenesis, high, normal, or low Kco may be found together with low V_A. Therefore, DLco should be in practical use, combined with the proper interpretation of Kco and V_A.

Keywords Diffusing capacity of carbon monoxide (DLco) · Kco (DLco/V_A) · kCO

K. Shimizu (✉)
Department of Respiratory Medicine, Faculty of Medicine and Graduate School of Medicine, Hokkaido University, Sapporo, Hokkaido, Japan
e-mail: okaoru@med.hokudai.ac.jp

K. Yamaguchi (ed.), *Structure-Function Relationships in Various Respiratory Systems*, Respiratory Disease Series: Diagnostic Tools and Disease Managements, https://doi.org/10.1007/978-981-15-5596-1_12

1 Introduction

The single-breath diffusing capacity of carbon monoxide (DL_{CO}) (known in Europe as the transfer factor, TL_{CO}), which was introduced by Krogh in 1915 [1], is one of the primary tested parameters of pulmonary function tests. kCO is the rate constant of removable CO from the alveolar gas during the breath-holding time. DLco is determined by the two independent components: kCO (the rate of uptake of CO from alveolar gas) and accessible lung volume (V_A) [2, 3].

2 Components of the Carbon Monoxide Diffusing Capacity

2.1 *kCO*

kCO is the measurement of the exponential decay in fractional concentration of CO during breath-holding time.

$$\mathrm{kCO} = \left(\log e\left(\mathrm{COo}/\mathrm{COt}\right)\right)/\ \text{breath} - \text{holding time} \tag{12.1}$$

where COo and COt are the alveolar CO concentrations at the beginning and completion of the breath-holding time, respectively (Fig. 12.1).

2.2 *Alveolar Volume*

The lung volume during breath-holding is calculated by the dilution of helium (He), a nonabsorbable gas, in the lung at the time kCO is measured [4]. The measurement starts from residual volume, completes at maximal inspiration near total lung capacity (TLC), which is defined as the inspired volume (VI). V_A is derived from the helium dilution volume of a single breath, calculated by subtracting the estimated anatomic dead space from VI.

Mean of V_A measured in a single breath divided by TLC measured in multiple breaths is 93.5 ± 6.6% (mean ± SD) in normal subjects. Heterogeneity in ventilation augments incomplete gas mixing during the ten seconds of breath-holding compared with multiple breath methods and causes low V_A/TLC ratio.

2.3 *kCO and DLco/V_A (Kco)*

The total CO transfer of the entire lungs is then calculated as:

$$\mathrm{DLco} = \left(\mathrm{kCO} * \mathrm{V_A\ STPD}\right)/\left(\mathrm{P_B} - \mathrm{P_{H2O}}\right) \tag{12.2}$$

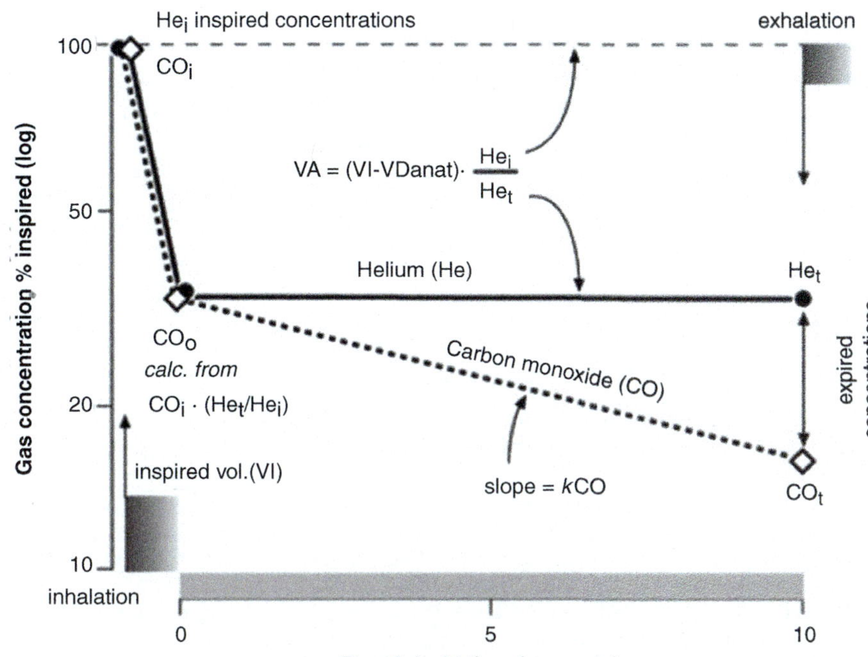

Fig. 12.1 Carbon monoxide (CO) and helium (He) kinetics in a single breath. Concentrations of the marker gases CO and He after rapid inspiration from residual volume to total lung capacity (TLC) were plotted against breath-holding time. Definitions of abbreviations: BHT: breath-holding time; CO0, COt: alveolar concentration of CO at the beginning and completion of the breath-holding time, respectively; COi: inspired concentration of carbon monoxide; DLco: carbon monoxide diffusing capacity; Hei, Het: inspired and expired concentrations of inert marker gas helium, respectively; kCO: rate constant for carbon monoxide uptake; K_{CO}; rate constant for carbon monoxide uptake per unit barometric pressure (kCO/Pb* ~ DLco/V_A); V_A: alveolar volume; VI: inspired volume; VDanat: anatomical dead space. Calculations: The rate constant kCO = log e (CO0/COt)/BHT; for points $t = 0$ and $t = 10$ seconds

where P_B and P_{H2O} are barometric pressure and water vapor pressure at 37°C, respectively.

STPD: standard temperature and pressure, dry

$$\mathrm{DLco} / \mathrm{V_A}\ \mathrm{L\,BTPS} = \left(\mathrm{kCO} / \mathrm{Pb}\right) * 1000 / 1.2 = \mathrm{Kco} \tag{12.3}$$

BTPS: body temperature and pressure, saturated.

Kco is the rate constant for carbon monoxide uptake per unit barometric pressure, which is different from kCO in three constant factors: Pb*, 1000, and 1.2. Therefore, kCO and Kco are physiologically equivalent.

DLco/V_A should be recognized as the pressure-adjusted rate constant for alveolar CO uptake, other than the CO diffusing capacity per unit alveolar volume, which could be misinterpreted as DLco/V_A is DLco corrected for lung volume. While DLco linearly changes to different levels of V_A/V_{ATLC}, DLco/V_A; however, it does not

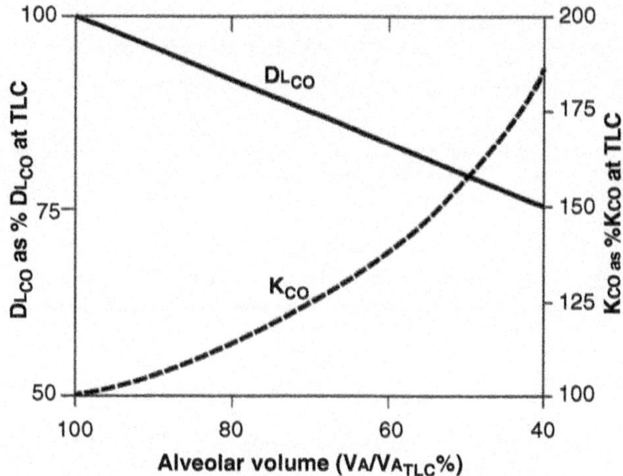

Fig. 12.2 Effect of voluntary changes in lung volume on DL_{CO} and K_{CO}, plotted as the percentage of value at full inflation (approximately TLC) against alveolar expansion expressed as alveolar volume at percent maximum (V_{ATLC} z~93.5% TLC). $K_{CO}/K_{CO}TLC$ at various values of V_A/V_ATLC was calculated from eq. 12.7 [$K_{CO}/K_{CO}TLC = 0.43+0.57/(V_A/V_{ATLC})$], and $DL_{CO}/DL_{CO}(TLC)$ as $K_{CO}/K_{CO}TLC \times V_A/V_ATLC$. Definitions of abbreviations: DL_{CO}: CO diffusing capacity; K_{CO}: rate constant for CO uptake per unit barometric pressure (kCO/Pb* z~DL_{CO}/V_A); V_A; alveolar volume

show a linear association with V_A/V_{ATLC}, which means that DLco/V_A is not DLco corrected for lung volume (Fig. 12.2). Moreover, DLco/V_A does not include the values of V_A. Therefore, Hughes and Pride referred Kco as essentially the rate constant for alveolar CO uptake [2], in contrast to the belief that Kco is of no clinical value because Kco does not correct DLco by lung volume.

2.4 *Kco in Normal Lungs*

Figure 12.2 shows that as $V_A/V_{A\ TLC}$ % decreases, DLco decreases while Kco increases [5, 6]. Due to this compensation, DLco does not fall to 50% of the TLC value at 50% V_A, where Kco increases to 158%.

Roughton and Foester [7] have established the methodology of estimating single-breath diffusing capacity for CO and determined three components that affect DLco. Their research substantially enhanced the clinical applicability of the assessment of CO transfer in the lungs. The Roughton–Foester equation. $1/DL = 1/DM + 1/\Theta^*Vc$ explains the increase in Kco with the decrease in the expansion of the lungs, as follows

$$V_A / DLco = 1 / K_{CO} = V_A / DMco + V_A / \Theta bLco * Vc \qquad (12.4)$$

- DMco is the membrane diffusing capacity.
- ΘbLco is the rate of reaction of CO with blood, adjusted to standard hemoglobin (Hb) concentration.
- Vc is the pulmonary capillary volume.

Fall in V_A/DLco (increase in Kco) is explained by a decrease in V_A/Vc, i.e., Vc/V_A increases because V_A/DMco remains almost constant.

Anemia, alveolar PO_2, and Pco are the factors affecting Kco measurements. Low Kco is found in subjects with anemia where decreased Hb levels lead to a decrease in ΘbLco.

ΘbLco decreases as P_AO2 increases, resulting in decreased Kco. Recent smoking or multiple measurements of DLco increases Pco in plasma, which interferes with the further measurements of DLco.

Age, sex, and height can predict DLco [8], while sex and height affect the V_A. Meanwhile, Kco is inversely related to age and height [9]. The maximum value of Kco occurs just before puberty. This phenomenon may imply that the development of pulmonary capillary vessels precedes the formation of alveoli. The effect of mismatch in V/Q in the apex of the lungs on Kco is more apparent in taller subjects in the upright position due to gravity. Moreover, substantial variation exists in the reference equations; however, no consensus on the reference value is available.

2.5 *Kco and V_A in Diseases*

2.5.1 Clinical Causes of Increase or Decrease in Kco

Alveolar and/or microvascular damage and destruction can reduce the rate of CO uptake via decreased DM and Vc. Emphysema and interstitial lung diseases can cause alveolar damage, while bronchiolitis obliterans and severe chronic heart failure can lead to microvascular damage. Incomplete alveolar expansion in alveoli with a preserved structure can elevate Kco because of high Vc/V_A, i.e., increased pulmonary blood flow (Fig. 12.3). In cases of the left-to-right shunt, Vc/V_A is evident in the whole lung, while after pneumonectomy, high Vc/V_A is evident in the remaining part of the lung.

Redistribution of blood flow to the normal part of the lung can lead to increase in Kco in the early stages of interstitial pulmonary fibrosis and sarcoidosis, where the damage of the interstitial part or vasculature has not extended to the entire lung. Regarding high Kco, increased perfusion has been reported in the apices of the lungs in patients with asthma. This could also be attributed to the raised capillary volume and decreased DM in some obese subjects. Further, pulmonary vascular congestion has been reported in severe chronic heart failure.

Increase in Kco after pneumonectomy is calculated based on the following equation [10],

$$\Delta \text{Kco}(\%\text{predicted}) = 0.4x + 2.1 \tag{12.5}$$

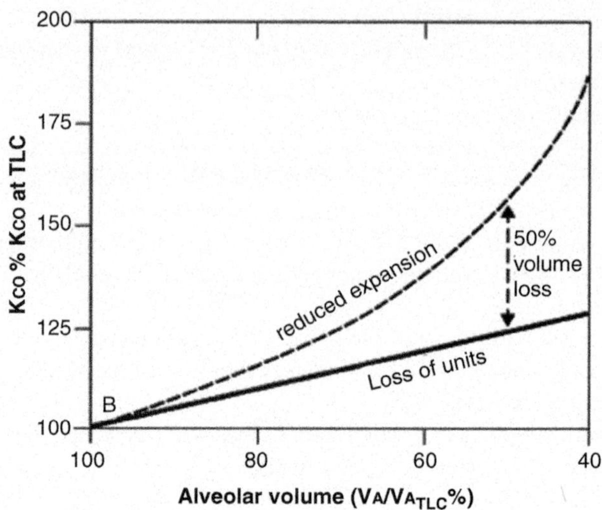

Fig. 12.3 $K_{CO}/K_{CO}TLC$ plotted against volume loss (V_A/V_ATLC) from two causes: (**a**) reduced alveolar expansion (e.g., FRC vs.TLC) and (**b**) loss of units as in lobectomy or pneumonectomy. The calculations of reduced expansion are shown in Fig. 12.2. Loss of units can be calculated from eq. 12.5 (K_{CO} [% predicted] = 0.4x+2.1) assuming that "x" (percentage of blood flow to the lung to be resected) reflects the percentage of lung volume to be resected; DL_{CO}/DL_{CO} (TLC) calculated as $K_{CO}/K_{CO}TLC \times V_A/V_ATLC$. Note that the difference between DL_{CO}/DL_{CO} (TLC) and $K_{CO}/K_{CO}TLC$ is for 50% loss in volume according to the mechanism of the volume deficit

x: percentage of blood flow to the resected lung.

The reason for the increase in K_{CO} is the expected doubling of blood flow per unit volume in the remaining lung.

Additional CO transfer to the extravascular part can raise Kco in pulmonary hemorrhage; although, alveolar capillary uptake is not increased. Kco is more sensitive to pulmonary hemorrhage than DLco, change in which is masked by a slight decrease in V_A.

2.5.2 Clinical Causes of Decreased Accessible Alveolar Volume

Hughes and Pride have suggested three categories of low V_A: associated with an increase in Kco, normal Kco, and decrease in Kco (Table 12.1).

Low V_A with an Increase in Kco

- Incomplete alveolar expansion, i.e., inspiratory muscle weakness, chest wall or pleural restriction, and inadequate expansion to TLC. (Fig. 12.3).
- Poor mixing affecting normal alveolar function, i.e., asthma.

Low V_A with Normal Kco

- Localized loss of units, i.e., Pneumonectomy, local destructive or infiltrative pathology (Fig. 12.3).
- Poor mixing and localized loss of units, i.e., bronchiectasis.

Table 12.1 Causes of low V_A and typical Kco findings

Normal or near-normal V_A			
Low Kco (Low DLco)		High Kco (High DLco)	
Mechanism	Disorders	Mechanism	Disorders
Microvascular destruction	Idiopathic pulmonary hypertension [11]	Increased pulmonary blood flow or redistribution	Left-to-right intracardiac shunts or asthma
Microvascular remodeling and dilation	Hepatopulmonary syndrome Pulmonary arteriovenous malformation		
With reduced VA			
Low Kco (Low DLco)		High Kco (Low~high DLco)	
Mechanism	Disorders	Mechanism	Disorders
Alveolar destruction	Emphysema	Incomplete alveolar expansion to TLC	Inspiratory muscle weakness/chest wall restriction
Alveolar destruction	Idiopathic pulmonary fibrosis,	Increased pulmonary blood flow to residual	Pneumonectomy
Microvascular destruction	Bronchitis obliterans/ chronic heart failure	Lungs Microvascular congestion/dilation Alveolar hemorrhage	Obesity Anti-GBM disease, SLE

Anti-GBM disease: Anti-glomerular basement membrane disease
SLE systemic lupus erythematosus

Low V_A with the Decrease in Kco

- Diffuse loss of units, i.e., interstitial lung disease.
- Poor mixing and some alveolar loss/disorganization, i.e., bronchiolitis obliterans.
- Poor mixing and diffuse alveolar disorganization, i.e., chronic obstructive pulmonary disease (COPD).

Airflow limitation can cause ventilation heterogeneity, which can lead to decreased V_A compared to TLC measured with plethysmography or multiple breath methods. Kco with a decreased V_A caused due to poor mixing can vary according to the pathology. When the diseased unit is localized, redistribution of capillary flow to normal units compensates for the value of Kco, while diffuse lung diseases preclude the compensation for Kco.

3 The Clinical Implication of DLco, Uncorrected Kco

3.1 Asthma with or Without Smoking and COPD

Recently, the overlap between features of asthma and COPD has been studied from the various points of views. Regarding the diffusing capacity of CO in the lungs, a document issued by the joint collaboration of the Global Initiative for Asthma

(GINA) and the Global Initiative for Chronic Obstructive Lung Disease stated that DLco is normal (or slightly elevated) in patients with asthma and often is reduced in patients with COPD [12, 13]. The change in Kco levels, however, has not been elaborated in the document. Moreover, few studies have studied the changes in DLco and Kco levels in smokers with asthma, while several studies have shown normal or increased DLco levels in nonsmokers with asthma.

The clinical role of DLco and Kco is further elaborated in the following sections [14].

3.1.1 DLco and Kco in Nonsmokers and Smokers with Asthma

The relationship between airflow limitation and DLco or Kco was studied. Nonsmokers with asthma ($N = 143$) were predominantly females, while smokers with asthma ($N = 77$) were predominantly males. Nonsmokers with asthma had significantly higher % FEV_1, FEV_1/Forced vital capacity (FVC), and % Kco values, and lower % Residual Volume values than smokers with asthma. (Table 12.2).

% DLco did not show any significant correlation with % FEV_1 in both nonsmokers and smokers with asthma (Fig. 12.4a, c). Meanwhile, as % FEV_1 decreased, % Kco increased significantly in nonsmokers with asthma (Fig. 12.4b), whereas it decreased significantly, though, to a lesser extent (Fig. 12.4d), in smokers with asthma (Fig. 12.4b).

Table 12.2 Characteristics of nonsmokers or smokers with asthma

	Nonsmokers with asthma	Smokers with asthma	*p*-value
Patients, *n*	143	77	N.S.
Sex, male, %	21.0	70.1	<0.001
Age, years	58.7 ± 15.0	61.4 ± 11.5	N.S.
Body mass index, kg/m^2	24.8 ± 4.6	25.2 ± 5.6	N.S.
Severe asthma, %	24.5	33.8	N.S.
%FVC, %	109.4 ± 15.7	108.8 ± 16.9	N.S.
%FEV_1, %	90.1 ± 20.1	81.9 ± 19.6	0.004
FEV_1/FVC, %	67.5 ± 12.6	62.2 ± 12.8	0.004
%RV, %	108.2 ± 21.2	115.5 ± 22.7	0.03
RV/TLC, %	35.5 ± 6.9	36.6 ± 6.8	N.S.
%TLC, %	112.2 ± 13.7	112.2 ± 14.2	N.S.
%DLco, %	102.8 ± 19.7	104.0 ± 24.9	N.S.
%Kco, %	114.3 ± 21.0	99.4 ± 20.7	<0.001
V_A, (male/female), L	4.4 ± 0.7/3.2 ± 0.6	4.6 ± 0.6/3.1 ± 0.6	N.S.

Data are shown as means _ SD. As-NS, nonsmokers; As-Sm, smokers with asthma; DLco, diffusing capacity of the lung for carbon monoxide; FEV_1, forced expiratory volume in 1 s; FVC, forced vital capacity; FEV1/FVC, forced expiratory volume in 1 s/forced vital capacity (absolute %value); Kco, transfer coefficient; NS, not significant; RV, residual volume; TLC, total lung volume; V_A, alveolar volume. *P* values were obtained using a chi-square test, *t*-test, or Wilcoxon test as appropriate

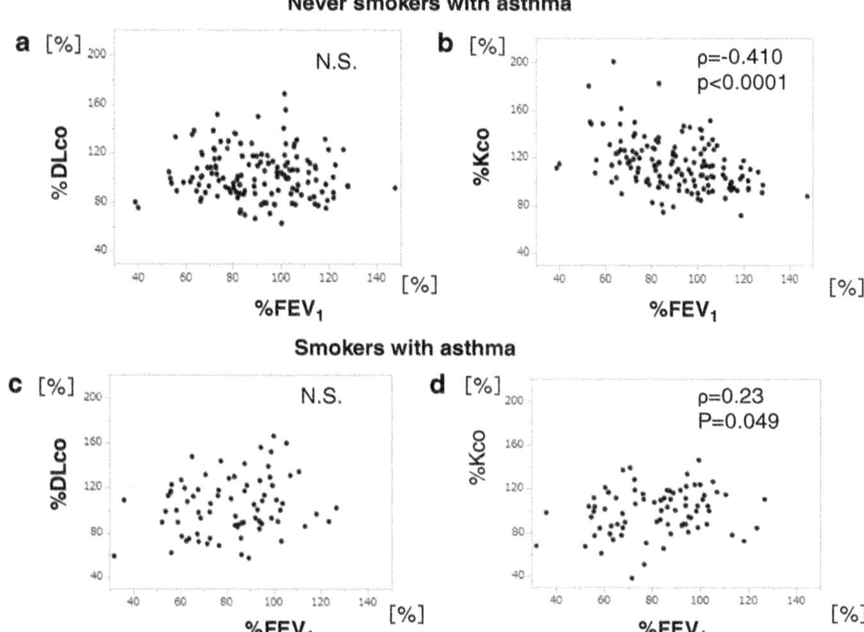

Fig. 12.4 The relationship between %FEV_1 and %DLco or %Kco in non-smokers and smokers with asthma. %FEV_1 decreased, %Kco increased in non-smokers with asthma (**b**), decreased in smokers with asthma (**d**). Such a unique and contrasting correlation relationships were not evident in the case of % DLco. (**a**, **c**) % FEV_1: forced expiratory volume in one s; % DLco: carbon monoxide diffusion capacity; % Kco: transfer coefficient

Vasoconstriction may divert flow and volume of blood from the gas-exchanging units to the "accessible" units, which leads to increased Kco. V_A is calculated considering these measurements based on the assumption that helium gas is in equilibrium in all parts of the whole lungs. Thus, if ventilation were unequal in the lungs, the calculated V_A would be smaller consequently compromising % DLco both in nonsmokers with asthma and smokers with asthma. n smokers with asthma, emphysematous changes, however, prevent the compensatory increase in Kco. Hence, % DLco will be normal or decreased depending on the values of V_A.

These results may suggest that % Kco is a more sensitive parameter compared with % DLco in patients with asthma, which could reflect their pathophysiology more precisely either way.

3.1.2 DLco and Kco in Obstructive Lung Diseases

A study comparing smokers (pack-years>10)(N = 69) and nonsmokers (pack-years<10)(N = 53) with asthma and pre-bronchodilator $FEV_1/FVC < 0.7$ and the COPD(N = 34) was conducted. Nonsmokers with asthma were predominantly females, whereas the other two groups comprised predominantly males. No significant differences were found in age among the three groups. Patients with COPD showed the highest pack-year value, followed by smokers with asthma and nonsmokers with asthma. There were no significant differences in % FVC, % FEV_1, FEV_1/FVC, RV/TLC, and %TLC among the three groups. (Table 12.3).

% DLco was significantly lower in patients with COPD compared to that in nonsmokers and smokers with asthma, despite the similar levels of % FEV_1, FEV_1/FVC (Fig. 12.5). No significant differences were found in % DLco between two asthmatic groups (Fig. 12.5a). Nonsmokers with asthma showed a significantly higher % Kco compared with smokers with asthma. As expected, % Kco was lowest in patients with COPD (Fig. 12.5b).

Collectively, % Kco was significantly associated with % FEV_1 and may have a role in differentiating between the three groups of nonsmokers with asthma, smokers with asthma, and patients with COPD.

Shimizu et al. performed morphometrical assessment of emphysema and small vessels in the lungs using computed tomography, by analysing the low attenuation volume (LAV) [15] and percentage cross-sectional area of small pulmonary vessels (%$CSA_{<5}$)) [16].

Patients with COPD showed the highest % LAV, followed by smokers with asthma, and nonsmokers with asthma. Conversely, patients with COPD had the lowest % $CSA_{<5}$, while no significant differences were observed between the asthmatic groups in %$CSA_{<5}$. Both % LAV and % $CSA_{<5}$ had stronger correlations with % Kco (Fig. 12.6b, d) than with % DLco (Fig. 12.6a, c).

Table 12.3 Characteristics of nonsmokers or smokers with asthma and patients with COPD

	Nonsmoker with asthma	Smokers with asthma	COPD	*p*-value
Patients, *n*	69	53	34	
Male/female, *n*	16/53	43/10	29/5	<0.001
Age, years	65.4 ± 10.6	64.9 ± 9.1	69.7 ± 8.0	N.S.
Pack-years	1.1 ± 2.2	34.6 ± 21.7	62.0 ± 26.1	<0.001
%FVC, %	108.7 ± 17.2	109.6 ± 17.2	109.9 ± 19.3	N.S.
%FEV_1, %	78.5 ± 16.1	75.9 ± 17.9	78.0 ± 17.9	N.S.
FEV_1/FVC, %	58.0 ± 7.2	56.6 ± 10.4	57.0 ± 7.3	N.S.
%RV, %	110.9 ± 21.5	119.3 ± 23.3	112.3 ± 26.3	N.S.
RV/TLC, %	38.5 ± 6.2	38.0 ± 6.6	37.7 ± 7.6	N.S.
%TLC, %	112.2 ± 13.7	112.2 ± 14.2	112.2 ± 14.2	N.S.
V_A, (male/female), L	4.2 ± 0.6/3.1 ± 0.5	4.5 ± 0.7/ 3.2 ± 0.5	4.3 ± 0.8 /3.7 ± 0.6	N.S.

Data are shown as means _ SD. As-NS, nonsmokers; As-Sm, smokers with asthma; DLco, diffusing capacity of the lung for carbon monoxide; FEV_1, forced expiratory volume in 1 s; FVC, forced vital capacity; FEV1/FVC, forced expiratory volume in 1 s/forced vital capacity (absolute %value); Kco, transfer coefficient; NS, not significant; RV, residual volume; TLC, total lung volume; V_A, alveolar volume. *P* values were obtained using a chi-square test, *t*-test, or Wilcoxon test as appropriate

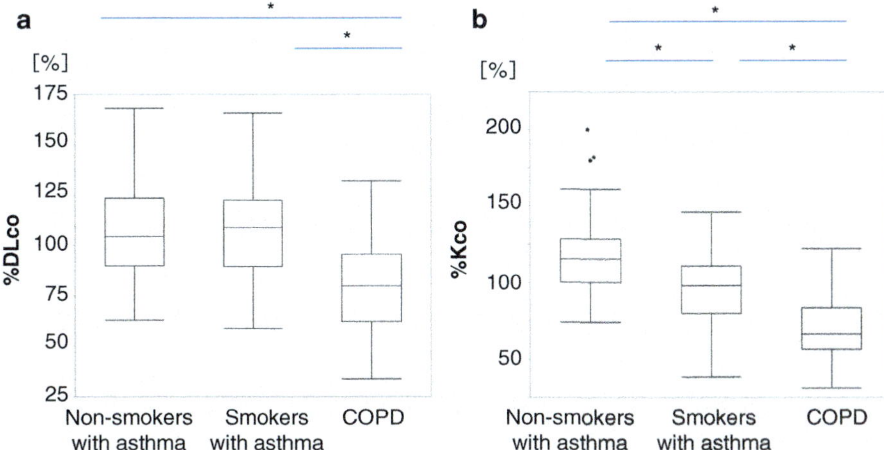

Fig. 12.5 % DLco or % Kco in nonsmokers with asthma, smokers with asthma with, FEV1/FVC < 0.7, and COPD with similar airflow limitations. (**a**) % DLco was the lowest in patients with COPD compared to that in nonsmokers with asthma, or in smokers with asthma with despite similar airflow limitations. There was no significant difference between nonsmokers with asthma and smokers with asthma in % DLco, regardless of the smoking status. (**b**) % Kco was significantly lower in smokers with asthma compared to that in nonsmokers with asthma and it was the lowest in patients with COPD. % DLco: carbon monoxide diffusion capacity; % Kco: transfer coefficient; COPD: chronic obstructive pulmonary disease; $FEV_1/FVC < 0.7$: the ratio of forced expiratory volume in one sec to forced vital capacity <0.7

Presumably, % Kco is sensitive to increased (often secondary to an increase in blood flow) or decreased blood volume in the entire lungs and/or greater blood flow and volume redistribution to the efficiently ventilated areas [17, 18] in nonsmokers with asthma, led to an increase in % Kco, especially in cases where airflow limitation was severe. However, DLco is influenced by both Kco and V_A. Ventilation inequality should concurrently occur in asthmatic lungs.

In smokers with asthma, emphysematous changes prevent the compensatory increase in Kco, which occur in nonsmokers with asthma with reduced % FEV_1, and hence, % DLco could be normal or decreased depending on the values of V_A. In patients with COPD, Kco typically showed a stronger correlation than DLco with emphysema scores. These findings suggest that Kco may be more sensitive than DLco for the detection of emphysematous changes. In this study, we extended these observations to smokers and nonsmokers with asthma.

Therefore, these results show that % Kco is a more sensitive parameter than % DLco, because we found a significant difference between smokers and nonsmokers with asthma in % Kco, and not in % DLco, despite the similar level of airflow limitation.

Kco is essentially the rate constant for alveolar uptake of CO measured while holding breathing in a single breath and independent of V_A. In contrast, Kco and V_A, both of which are mutually independent, affect DLco, at the suggested V_A, Kco may

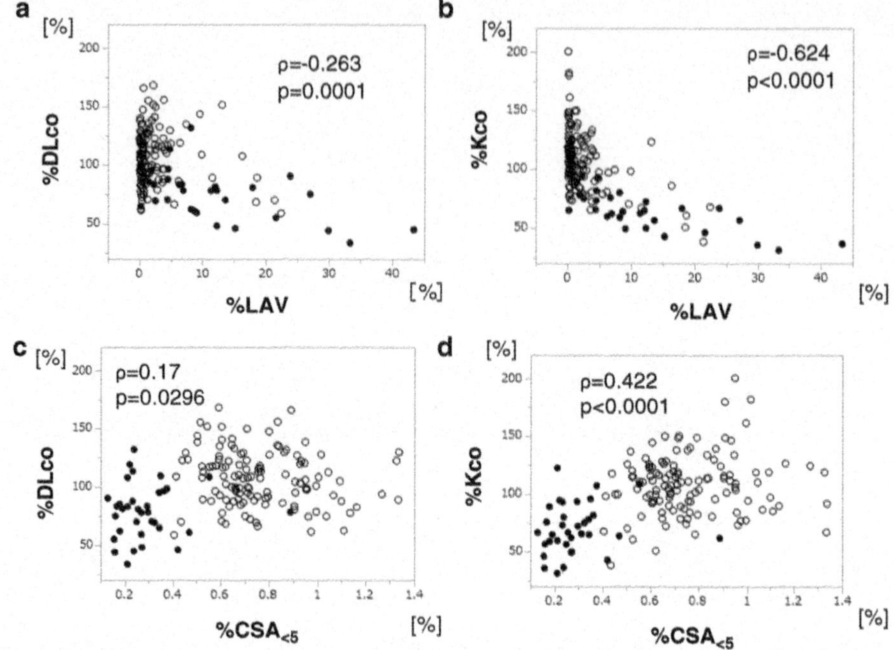

Fig. 12.6 Correlation between the CT parameters of %LAV or $\%CSA_{<5}$, and %DLco or %Kco. %Kco showed the stronger correlations with %LAV (**b**) or $\%CSA_{<5}$ (**d**) than did %DLco. (**a**, **c**)

vary according to the pathogenesis of the disease. In other words, uncorrected Kco can provide information regarding the physiological alterations in individual lungs. In conclusion, for the value of uncorrected Kco, VA and DLco make the best use of single breath, instead of correcting Kco or single use of DLco.

3.1.3 Further Analyses Using the Alternative Equation for Diffusion Capacity Indices

Burrows equation was used in the main assessment [19]. Overall, alternative equations for DLco and/or Kco led to similar results. We found significant positive correlations between % FEV_1 and % DLco in smokers with asthma and not in nonsmokers with asthma; these results were contradictory to the results obtained from the Burrows equation. However, the other results were reconfirmed using the other two equations. When Nishida [20] and Crapo's equations [21] were used, % FEV_1 and % Kco were negatively correlated in nonsmokers with asthma, while no significant correlation was observed in smokers with asthma.

% Kco of nonsmokers with asthma with $FEV_1/FVC < 0.7$ was higher compared to the % Kco of smokers with asthma with $FEV_1/FVC < 0.7$, while % DLco was not significantly different between the two asthma groups.

3.2 *DLco in Interstitial Lung Diseases*

Compared with DLco, Kco has been claimed to be of no use and a misleading parameter in interstitial lung disease (ILD) [5, 22] because decreased lung volume causes a much larger effect (predominantly an increase), subsequently rendering Kco a less sensitive parameter.

3.3 *The Implication of Corrected Kco*

DLco/V_A is not an adjusted or corrected value of DLco, as shown in Fig. 12.2, which demonstrates that DLco/V_A, i.e., Kco is not constant to V_A. This discrepancy led to the belief that DLco/V_A is an arithmetically flawed index with no clinical value. Therefore, some corrections had been suggested. The typical relationships in normal subjects between changes in lung volume and DL_{CO}/V_A (K_{CO}) are shown [5].

$$\mathrm{DLco} / \mathrm{DLco}_{\mathrm{TLC}} = 0.58 + 0.42 * \left(\mathrm{V_A} / \mathrm{V_{ATLC}}\right) \text{where } \mathrm{DL_{CO}TLC} \tag{12.6}$$

and V_ATLC are expected values for DL_{CO} and V_A at a normal predicted TLC.

$$\mathrm{Kco} / \mathrm{Kco}_{\mathrm{TLC}} = 0.43 + 0.57 / \left(\mathrm{V_A} / \mathrm{V_{ATLC}}\right) \tag{12.7}$$

Johnson proposed the formula for the conversion of % Kco to % DLco for patients with asthma, emphysema, extrapulmonary restriction, interstitial lung disease, and lung resection [5]. After adjusting the value of K_{CO} for a low V_A, DL_{CO} and K_{CO} values tended to converge, where uncorrected %Kco exceeded more than 50% compared to % DLco in several diseases. Frans and colleagues [23] also adjusted the DL_{CO} and K_{CO} values for the low V_A in diffuse interstitial lung disease. Returning to the high % Kco in extrapulmonary restriction to 90–100% of the predicted range and in interstitial lung disease to 50–75% of the predicted range, was more in line with the uncorrected DL_{CO}, to prevent us from having the false impression of normal because the volume restriction had "misleadingly" increased the Kco.

3.4 *DLco, Kco in Combined Pulmonary Fibrosis and Emphysema (CPFE)*

Both emphysema and fibrosis are factors that reduce DLco. Thus, patients with CPFE show remarkably low DLco. Assessment of DLco is of great importance because values of spirometry may be normal due to the presence of opposite physiology in the lungs; obstructive in emphysema and restrictive in fibrosis [24]. Combined fibrotic changes may be suspected in patients with COPD when FEV_1/FVC has been normalized despite an additional decrease in DLco.

3.5 DLco in Pulmonary Hypertension

Treatment for pulmonary arterial hypertension (PAH) has improved dramatically in recent years, and hence, early and accurate diagnosis of PAH has been in great need. For the diagnosis of pulmonary hypertension, right heart catheterization (RHC) is mandatory. RHC is, however, highly invasive and is recommended to be performed by well-trained physicians in high-volume PH centers, according to the current guidelines for PH.

Thus, noninvasive assessment methods have been investigated for the screening of PH. Pulmonary function tests are noninvasive and values of DL_{CO} might be abnormal in patients with pulmonary hypertension. Meyrick and Reid [25] suggested that muscularization of smaller, more peripheral pulmonary arteries, medial and intimal thickening of the muscular arteries, and a reduction in peripheral vascular bed may be involved in the mechanism of decreasing DLco. Low DLco has been reported to be associated with a poor prognosis in PH [9]. Patients with pulmonary veno-occlusive disease (PVOD)/pulmonary capillary hemangiomatosis (PCH) show remarkably low values of DLco.

3.5.1 Pulmonary Arterial Hypertension

Three-quarters of IPAH patients have decreased DLco with a mean value across studies of 59–71% of the predicted values [26]. Even after the adjustments for smoking, values of DL_{CO} remained abnormal in patients with idiopathic PAH (IPAH). V_A/TLC values are preserved in PAH, so that this reduction in values of DL_{CO} may not be completely explained by the ventilation perfusion mismatch. Further analysis revealed that both pulmonary membrane diffusion capacity and the pulmonary capillary blood volume contributing to alveolar gas exchange were reduced [27]. Reductions in Dm may be caused by thickening of the alveolar capillary membrane due to endothelial cell proliferation, while reductions in Vc may be the result of increased pulmonary vascular resistance, reduced cardiac output, and local thrombosis.

3.5.2 Connective Tissue Disease-Associated Pulmonary Arterial Hypertension

DL_{CO} is reduced in systemic sclerosis in conjunction with PAH, even in the absence of interstitial lung disease [28]. Isolated, as well as disproportionate reductions in DL_{CO} in relation to other lung function parameters, can represent the presence of associated PAH [29]. Increased values of % VC/% DLco, ranging from 1.5 to 2.0, have been reported to be a diagnostic marker of combined PAH.

3.5.3 Pulmonary Hypertension Due to Chronic Lung Disease

PH is more prevalent in patients with severe COPD and ILD, compared with non-severe diseases [30]. However, the severity of PH does not parallel with the severity of the lung condition [31]. Moreover severe PH can be found in patients, who have disproportionate or relatively preserved lung function [32]. This is also a frequent finding in patients with combined pulmonary fibrosis and emphysema. Moreover, the development of PH in patients with chronic lung disease is not always accompanied by worsening of the physiology in respiratory system Thus, regarding associated PH, patients without severe lung diseases should also be managed with caution PH should be suspected in patients with chronic lung disease, where the DL_{CO} is disproportionately low in relation to other lung function values.

3.5.4 Chronic Thromboembolic Pulmonary Hypertension (CTHPH)

DL_{CO} is reported to be reduced to 40–60% in patients with CTEPH, though this has been reported in a case series with small numbers of patients. This reduction in gas transfer is due to a proportional reduction in pulmonary capillary blood. Patients with low Kco in CTEPH reveal poor prognosis even after 5 years of treatment.

4 Conclusions

DLco is the product of the rate of carbon monoxide uptake from alveolar gas to pulmonary capillary blood and V_A, accessible alveolar volume. Kco and V_A are mutually independent and change due to their own underlying pathophysiology. Several causes can reduce $V_{A,}$ and hence, universal adjustment might not be adequate for practical use. The precise interpretation of DLco, Kco, and V_A is the best use of single-breath DLco assessment. DLco and Kco may have a clinically useful role in ACO and PH on diagnosis, chronic lung diseases such as COPD, ILD on the screening of associated PH, together with spirometry and lung volume measurements.

References

1. Krogh M. The diffusion of gases through the lungs of man. J Physiol. 1915;49:271–96.
2. Hughes JMB, Pride NB. In defence of the carbon monoxide transfer coefficient KCO (TL/VA). Eur Respir J. 2001;17:168–74.

3. Hughes JM, Pride NB. Examination of the carbon monoxide diffusing capacity (DL(CO)) in relation to its KCO and VA components. Am J Respir Crit Care Med. 2012 Jul 15;186(2):132–9.
4. Ogilvie CM, Forster RE, Blakemore WS, Morton JW. A standardized breath holding technique for the clinical measurement of the diffusing capacity of the lung for carbon monoxide. J Clin Invest. 1957;36:1–17.
5. Johnson DC. Importance of adjusting carbon monoxide diffusing capacity (DLCO) and carbon monoxide transfer coefficient (K CO) for alveolar volume. Respir Med. 2000;94:28–37.
6. McGrath M, Thomson ML. The effect of age, body size and lung volume change on alveolar–capillary permeability and diffusing capacity in man. J Physiol. 1959;146:572–82.
7. Roughton FJW, Forster RE. Relative importance of diffusion and chemical reaction in determining rate of exchange of gases in the human lung. J Appl Physiol. 1957;11:290–302.
8. Cotes JE, Chinn DJ, Quanjer PH, Roca J, Yernault J-C. Standardization of the measurement of transfer factor (diffusing capacity). Eur Respir J. 1993;6:41–52.
9. Hughes JMB, Pride NB. Carbon monoxide transfer coefficient (transfer factor/alveolar volume) in females versus males. Eur Respir J. 2003;22:186–9. [Published erratum appears in Eur Respir J 22:570.]
10. Corris PA, Ellis DA, Hawkins T, Gibson GJ. Use of radionuclide screening in the preoperative estimation of pulmonary function after pneumonectomy. Thorax. 1987;42:285–91.
11. Steenhuis LH, Groen HJM, Koëter GH, van der Mark Th W. Diffusion capacity and haemodynamics in primary and chronic thromboembolic pulmonary hypertension. Eur Respir J. 2000;16:276–81.
12. American Thoracic Society. Proceedings of the ATS workshop on refractory asthma: current understanding, recommendations, and unanswered questions. Am J Respir Crit Care Med. 2000;162:2341–51.
13. MacIntyre N, Crapo RO, Viegi G, Johnson DC, van der Grinten CP, Brusasco V, Burgos F, Casaburi R, Coates A, Enright P, Gustafsson P, Hankinson J, Jensen R, McKay R, Miller MR, Navajas D, Pedersen OF, Pellegrino R, Wanger J. Standardisation of the single-breath determination of carbon monoxide uptake in the lung. Eur Respir J. 2005;26:720–35.
14. Shimizu K, Konno S, Makita H, Kimura H, Kimura H, Suzuki M, Nishimura M. Transfer coefficients better reflect emphysematous changes than carbon monoxide diffusing capacity in obstructive lung diseases. J Appl Physiol (1985). 2018;125(1):183–9.
15. Hasegawa M, Nasuhara Y, Onodera Y, Makita H, Nagai K, Fuke S, Ito Y, Betsuyaku T, Nishimura M. Airflow limitation and airway dimensions in chronic obstructive pulmonary disease. Am J Respir Crit Care Med. 2006;173:1309–15.
16. Matsuoka S, Washko GR, Yamashiro T, Estepar RS, Diaz A, SilvermanEK HE, Fessler HE, Criner GJ, Marchetti N, Scharf SM, Martinez FJ, Reilly JJ, Hatabu H, National Emphysema Treatment Trial Research Group. Pulmonary hypertension and computed tomography measurement of small pulmonary vessels in severe emphysema. Am J Respir Crit Care Med. 2010;181:218–25.
17. Collard P, Njinou B, Nejadnik B, Keyeux A, Frans A. Single breath diffusing capacity for carbon monoxide in stable asthma. Chest. 1994;105:1426–9.
18. Stewart RI. Carbon monoxide diffusing capacity in asthmatic patients with mild airflow limitation. Chest. 1988;94:332–6.
19. Burrows B, Kasik JE, Niden AH, Barclay WR. Clinical usefulness of the single-breath pulmonucy diffusing capacity test. Am Rev Respir Dis. 1961;84:789–806.
20. Nishida O, Sewake N, Kambe M, Okamoto T, Takano M. Pulmonary function in healthy subjects and its prediction. 4. Subdivisions of lung volume in adults (author's transl). Rinsho Byori. 1976;24:837–41.
21. Crapo RO, Morris AH. Standardized single breath normal values for carbon monoxide diffusing capacity. Am Rev Respir Dis. 1981;123:185–9.
22. Stam H, Kreuzer FJ, Versprille A. Effect of lung volume and positional changes on pulmonary diffusing capacity and its components. J Appl Physiol (1985). 1991;71:1477–88.

23. Frans A, Nemery B, Veriter C, Lacquet L, Francis C. Effect of alveolar volume on the interpretation of the single breath DLCO. Respir Med. 1997;91:263–73.
24. Cottin V. The impact of emphysema in pulmonary fibrosis. Eur Respir Rev. 2013;22(128):153e157.
25. Meyrick B, Reid L. Pulmonary hypertension. Anatomic and physiologic correlates. Clin Chest Med. 1983;4(2):199–217.
26. Low AT, Medford AR, Millar AB, Tulloh RM. Lung function in pulmonary hypertension. Respir Med. 2015;109(10):1244–9.
27. Farha S, Laskowski D, George D, Park MM, Tang WHW, Dweik RA, Erzurum SC. Loss of alveolar membrane diffusing capacity and pulmonary capillary blood volume in pulmonary arterial hypertension. Respir Res. 2013;22(14):6.
28. Stupi AM, Steen VD, Owens GR, Barnes EL, Rodnan GP, Medsger TA. Pulmonary-hypertension in the crest syndrome variant of systemic-sclerosis. Arthritis Rheum. 1986;29:515e524.
29. Wells AU. Pulmonary function tests in connective tissue disease. Semin Respir Crit Care Med. 2007;28:379e388.
30. Seeger W, Adir Y, Barbera JA, Champion H, Coghlan JG, Cottin V, De Marco T, Galie NGS, Gibbs S, Martinez FJ, Semigran MJ, Simonneau G, Wells, Vachier J-L. J Am Coll Cardiol. 2013;62:D109–16.
31. Hurdman J, Condliffe R, Elliot CA, Swift A, Rajaram S, Davies C, Hill C, Hamilton N, Armstrong IJ, Billings C, Pollard L, Wild JM, Lawrie A, Lawson R, Sabroe I, Kiely DG. Pulmonary hypertension in COPD: results from the aspire registry. Eur Respir J. 2013;41:1292–301.
32. Andersen KH, Iversen M, Kjaergaard J, Mortensen J, Nielsen-Kudsk JE, Bendstrup E, Videbaek R, Carlsen J. Prevalence, predictors, and survival in pulmonary hypertension related to end-stage chronic obstructive pulmonary disease. J Heart Lung Transpl. 2012;31:373–80.

Chapter 13
Basic Perspective of Simultaneous Measurement of D_{LCO} and D_{LNO}: What Are the Most Legitimate Assumptions When Estimating D_{LCO} and D_{LNO}?

Anh Tuan Dinh-Xuan

Abstract It is technically feasible to use two different hemoglobin-reactive gases, namely nitric oxide (NO) and carbon monoxide (CO) inhaled from the same reservoir in a single breath-holding maneuver to measure lung diffusing capacity to both NO (D_{LNO}) and CO (D_{LCO}). Although it is tempting to apply the equation published in 1957 by Roughton and Forster to decipher alveolar-capillary membrane conductance and pulmonary capillary volume, this might not be advisable in the case of D_{LNO}–D_{LCO} simultaneous measurement as there remain unknowns and controversies regarding the values of the specific transfer conductance of blood for NO (ΘNO) and CO (ΘCO). Instead, it is now proposed to discuss the interpretation of either D_{LCO} and D_{LNO} separately with D_{LNO} being weighted by the membrane gas conductance, while the D_{LCO} is dominated by ΘCO. Alternatively the D_{LNO}/D_{LCO} could be an acceptable ratio to explore lung gas transfer conductance or disturbances.

Keywords Lung gas diffusion capacity · Nitric oxide · Carbon monoxide
Pulmonary capillary volume · Alveolar-capillary membrane conductance

A. T. Dinh-Xuan (✉)
Respiratory Physiology Unit, Department of Thoracic Medicine and Surgery,
Cochin Hospital, University of Paris, Paris, France
e-mail: anh-tuan.dinh-xuan@aphp.fr

K. Yamaguchi (ed.), *Structure-Function Relationships in Various Respiratory Systems*, Respiratory Disease Series: Diagnostic Tools and Disease Managements, https://doi.org/10.1007/978-981-15-5596-1_13

1 Introduction

Respiration and gas exchanges are vital processes underlying the status of health of all living organisms, including humans. From its origin in the inspired ambient air, gas molecules must undergo different steps before they eventually reach hemoglobin (Hb) inside the red blood cells which are continuously flowing through pulmonary capillaries of perfused ventilated lung areas. The first step is a hydrodynamic convection process enabling the gas to flow down from the upper airways to the acini entrances. After entering the alveolar region, gas molecules are transported by both hydrodynamic convection and diffusion inside the breathing acini. The following step of gas transfer involves its extremely rapid (almost instantaneous) dissolution in the alveolar membrane. Once dissolved, gas molecules diffuse across the membrane and capillary plasma to reach the red blood cells. The gas molecules then have to cross the very thin red blood cell membrane to enter the inner cytoplasm in which they diffuse to reach, and subsequently react with, Hb to form the combined Hb-gas whose rate of reaction will depend on the nature of the binding gas molecule. More than 60 years ago, Roughton and Forster [1] have proposed a model describing the transfer of carbon monoxide (CO) or any diffusible gases, including oxygen (O_2) and nitric oxide (NO), from air to blood as a two-step process in which the gas has to overcome two resistances in series according to eq. (13.1) (Fig. 13.1):

$$1/DL = 1/DM + 1/\Theta \cdot VC \tag{13.1}$$

where

- 1/DL is the total resistance of the blood–gas barrier to gas transfer ($mmol^{-1} \cdot min \cdot kPa$ in SI units or $mL^{-1} \cdot min \cdot mmHg$ in traditional units),
- DM is the membrane diffusing capacity, whereas 1/DM is the resistance to passive diffusion across the alveolar-capillary membrane and intracapillary plasma,
- Θ is the specific transfer conductance of blood (measured in vitro) for a specified gas, VC is the pulmonary capillary blood volume (measured in mL), whereas $1/\Theta \cdot VC$ is the resistance to gas transfer of the red blood cell for reactive gases such as CO, NO, or oxygen.

From the above, Roughton and Forster viewed lung gas transfer as a journey that a given gas molecule has to undertake by flowing through two successive resistances whose inverse values are known as conductance. The first is membrane conductance, related to the diffusing capacity of alveolar-capillary membrane. This conductance is the result of passive diffusion through a very thin tissue barrier, consisting of the alveolar epithelial type 1 cell, a basement membrane, and a capillary endothelial cell (Fig. 13.1). Additionally, there is an intracapillary component consisting of a plasma layer of variable thickness between the endothelium and the red cell membrane. The second is lung capillary blood conductance which depends on both the reactivity of the gas with Hb and the mass of Hb in the lung capillaries which in turn depends on lung capillary blood volume (VC).

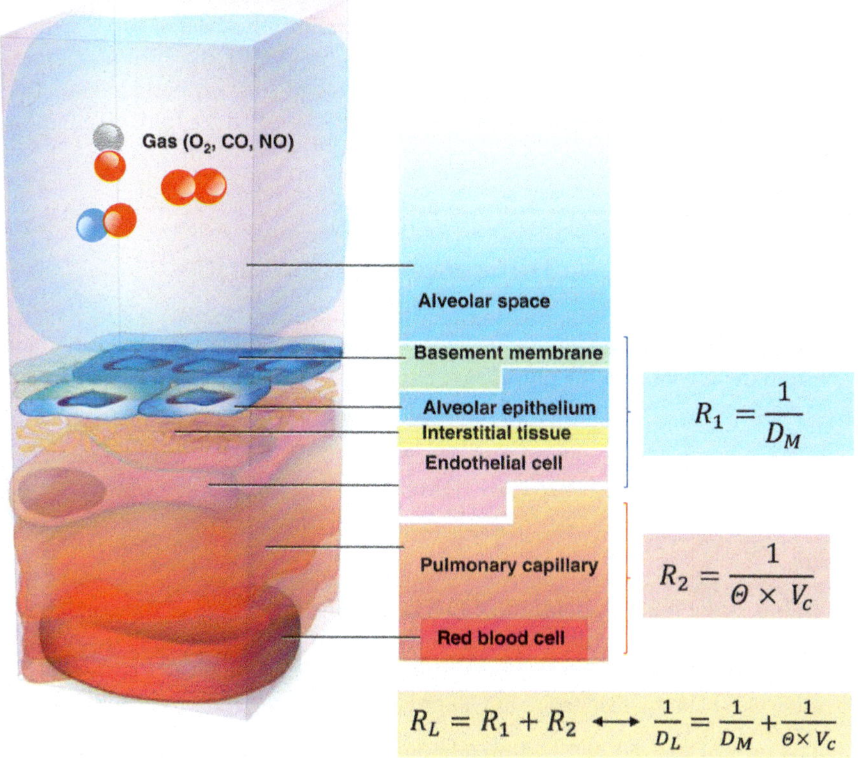

Fig. 13.1 Transport of gas from alveolar space to red blood cell

In the clinical setting, the most commonly used technique for measuring gas transfer from the alveolus to the pulmonary capillary blood is the single-breath diffusing capacity, or transfer factor, of the lung for CO (D_{LCO} or TLCO, respectively) [2, 3]. The breath-holding technique for the clinical measurement of D_{LCO} has undergone little change since its original description in 1957 by Ogilvie et al. [4]. In the 1980s, the single breath measurement of diffusing capacity of the lung using different hemoglobin (Hb)-reactive gases, namely nitric oxide (NO), was almost simultaneously, but independently, introduced by two European groups of investigators based in Bordeaux (France) [5] and Cambridge (UK) [6, 7]. Both groups of investigators applied the same single breath technique, whereby both tracer gases (CO and NO) were inhaled from the same reservoir together with an inert volume-marker gas (usually helium). Using two different Hb-reactive gases, namely CO and NO, would allow solving the equation established more than 60 years ago by Roughton and Forster [1].

Although questioned by a French group of investigators who criticized the Roughton and Forster model for its apparent oversimplification of supposedly complex phenomena related to gas transport and capture [8, 9], the current chapter will focus on recent evidence deciphering specific blood transfer conductance for NO and CO from their membrane diffusing capacities based on the initial Roughton and Forster equation [1].

2 D_{LNO} Versus D_{LCO}: Differences and Similarities

As previously mentioned, according to Roughton and Forster [1] the conductance of CO from the alveolus to the pulmonary capillary blood can be described by the classical equation:

$$1/D_{\text{LCO}} = 1/\text{DMCO} + 1/\Theta CO \cdot \text{VC} \tag{13.2}$$

where

- $1/D_{LCO}$ is the lung resistance to CO transfer (the reciprocal of the conductance),
- 1/DMCO (the molecular diffusion resistance of the lung membranes—from the surfactant lining layer to the red cell membrane, and
- $1/\Theta CO \cdot VC$ is the resistance to CO transfer within the red blood cell.

The sum of the red cell and membrane resistances, which are in series, are the components of the overall resistance, $1/D_{LCO}$. The same equation applies to NO uptake:

$$1/D_{\text{LNO}} = 1/\text{DMNO} + 1/\Theta NO \cdot \text{VC} \tag{13.3}$$

2.1 Differences Between NO and CO Blood Diffusivity and Hb Reactivity

Although NO and CO follow the same path while crossing the blood–gas barrier, both gases are differently handled by the lung tissues and pulmonary capillary blood as they do not share the same chemical properties and behaviors with regard to Hb capillary blood. NO diffusivity (solubility/MW^2) in plasma is 1.97 times faster than that of CO, and its specific conductance (Θ), i.e., the rate of NO uptake per mmHg of NO tension per mL of blood is 5.75 times faster than the uptake of CO at a *P*O2 of 100 mm Hg [10]. The chemical reactions of NO and CO with Hb also differ. NO reacts directly with the oxygen of oxyhemoglobin to form nitrate (NO_3) plus a deoxygenated form of Hb called methemoglobin (metHb) in which the iron atoms of the heme ring are oxidized from the ferrous (Fe^{2+}) to the ferric (Fe^{3+}) forms [11]:

$$NO + Hb(Fe^{2+})O_2 \rightarrow metHb(Fe^{3+})O_2 + NO_3^- \quad (13.4)$$

Unlike NO, CO does not react with O_2. Rather, CO competes with O_2 to occupy the Fe^{2+} site on the heme ring:

$$CO + Hb(Fe^{2+})O_2 \rightarrow Hb(Fe^{2+})CO + Hb(Fe^{2+})O_2 \quad (13.5)$$

The increased affinity of CO for Hb is due to the different angles of attachment of CO and O_2 to the heme ring [12]. Both NO and CO are tightly bound to Hb through their extremely slow dissociation constants but unlike NO, the rate of reaction of CO with oxyhemoglobin is PO_2 dependent, as shown by the linear relationship between PO_2 and $1/\Theta_{CO}$, the specific resistance reaction rate. The higher the PO_2 the greater the resistance rate to CO transfer within the red blood cell. D_{LNO}, on the other hand, is independent of the level of alveolar PO_2 (PAO_2) [13] because NO reacts readily with hemoglobin (eq. 13.4) instead of competing with oxygen for Hb binding sites as does CO (eq. 13.5).

From the above, one might question the existence of a significant blood resistance to NO uptake, assuming 1/ΘNO negligible and consequently consider from eq. (13.3) that $1/D_{LNO}$ might approximately equate to 1/DMNO and conclude that D_{LNO} reflects lung membrane diffusing capacity [5]. Experimental evidence however suggests that resistance to NO uptake, which does not relate to the extremely fast chemical combination of NO with Hb, might stem in the red blood cell membrane or its vicinity, including the stagnant layer of plasma immediately surrounding the cell. Consistent with this view is the demonstration that red blood cell lysis (by the addition of water to blood in a membrane oxygenator model of NO and CO transfer [14]), or red blood cell substitution, in anaesthetized dogs, with cell-free heme-based oxyglobin [15] markedly increased D_{LNO} without affecting D_{LCO}.

2.2 *ΘNO and ΘCO and Absolute Values of DMCO and VC*

Accurate evaluation of ΘNO and ΘCO is essential for the calculation of the membrane and red blood cell conductance for these gases from D_{LNO} and D_{LCO} measurements [16]. There are however ongoing debates regarding the choice of ΘNO and ΘCO values that should be used when computing DMCO and VC according to the initial concept proposed by Roughton and Forster in 1957 (eq. 13.1) [1]. In 2017, members of a taskforce endorsed by the European Respiratory Society have published a report [17] recommending the use of a ΘNO of 4.5 mLNO/min/mmHg/mL blood, a DMNO to DMCO ratio (DMNO/DMCO) of 1.97 and values of ΘCO derived from the study published by Guénard et al. in 2016 [18]. This recommendation is not unanimously followed as investigators from the Mayo Clinic (Rochester, Minnesota, USA) [19, 20] prefer to calculate DMCO and VC using an infinite ΘNO, a DMNO to DMCO ratio of 2.26, and ΘCO values derived from

Reeves and Park's formula published in 1992 [21]. To justify their choice, these investigators [19, 20] have argued that the combination of an infinite ΘNO, an empirical value for DMNO/DMCO (>2.0) and Reeves and Park's ΘCO gives a value of DMCO (using a combined D_{LNO}–D_{LCO} analysis) which agrees with the DMCO value calculated separately by the classical two-stage oxygen technique of Roughton and Forster [1]. To reconcile those two opposite views, Borland et al. have elegantly put forward a plausible hypothesis based on the fundamental rules of combined diffusion and chemical reaction to a red cell to explain why ΘNO could be finite in vitro but effectively infinite in vivo [22]. If this hypothesis holds true $1/D_{LNO}$ would mainly reflect the resistance of the alveolar-capillary membrane (~60%) with a smaller contribution from plasma and minimal contribution from the outermost layers of the red cell (Fig. 13.2) [17]. But if this hypothesis is

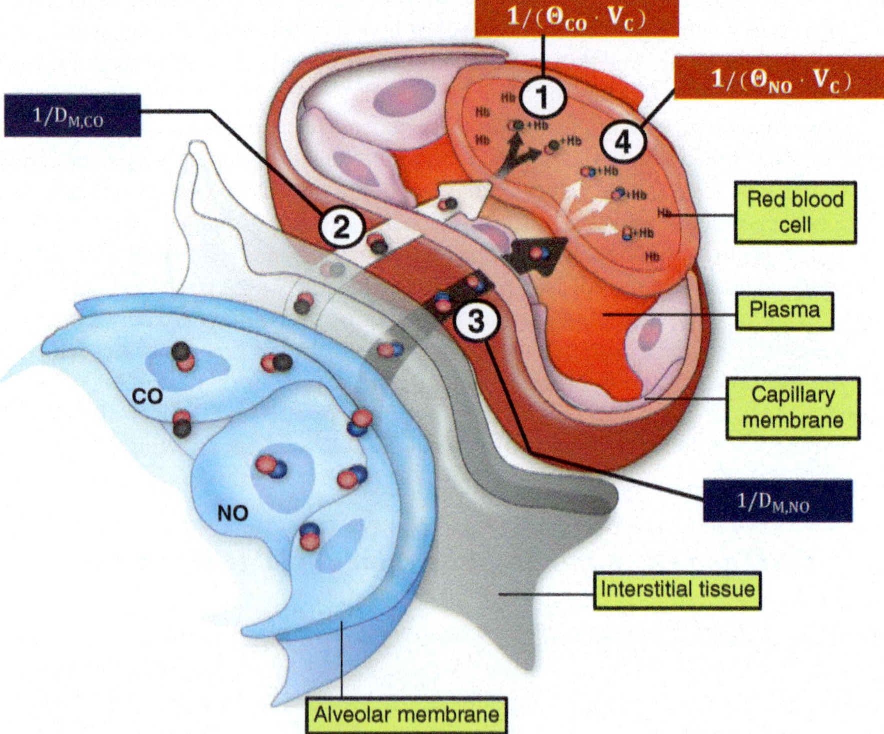

Fig. 13.2 Diagram of the uptake pathways for inhaled nitric oxide (NO) and carbon monoxide (CO) from the alveolar membrane to their combination with hemoglobin (Hb) within the red blood cell, in terms of the Roughton and Forster equation [1], 1/DL=1/DM+1/(Θ·VC), where 1/DL is the total resistance to NO or CO uptake, 1/DM is the resistance from the alveolar membrane to the red cell membrane (membrane resistance) and 1/(Θ·VC) is the diffusion and chemical combination resistance (red cell resistance) within the erythrocyte (1). The chief barrier to CO uptake is within the red cell (~70-80%); the ~25% remaining resistance to CO diffusion is located in the alveolar membrane (2). The main resistance (~60%) barrier for NO lies within the alveolar-capillary membranes (3), while the red cell resistance (4) only accounts for less than 40% of the resistance to NO diffusion [15]. Reproduced and adapted from [16] with permission from the publisher

correct, this would also imply that one cannot accurately calculate DMCO and VC in a combined D_{LNO}–D_{LCO} maneuver since these variables would differ for NO and CO [22].

3 Conclusion

In clinical studies, the fact that D_{LNO}, unlike D_{LCO}, was relatively independent of changes in the inspired oxygen concentration, and thus alveolar oxygen pressure [13, 18] and hematocrit [23], which operate through variations in the Θ value for blood, seemed to support the original notion that ΘNO was "effectively" infinite, and that D_{LNO} is a surrogate for the alveolar membrane diffusing capacity, i.e., D_{LNO} = DMNO = 1.97·D_{LCO}; this view is still held by some investigators [19, 20]. But the current consensus is that ΘNO could be finite in vitro but effectively infinite in vivo [22] and that D_{LNO} is weighted, but not dominated, by the membrane gas conductance, while the D_{LCO} is dominated by ΘCO (Fig. 13.2) [17]. The D_{LNO}/D_{LCO} ratio has been studied in several clinical situations [24, 25] and will be reviewed in Chap. 15.

References

1. Roughton FJ, Forster RE. Relative importance of diffusion and chemical reaction rates in determining rate of exchange of gases in the human lung, with special reference to true diffusing capacity of pulmonary membrane and volume of blood in the lung capillaries. J Appl Physiol. 1957;11:290–302.
2. Hughes JM, Bates DV. Historical review: the carbon monoxide diffusing capacity (DLCO) and its membrane (DM) and red cell (Θ·VC) components. Respir Physiol Neurobiol. 2003;138:115–42.
3. Hughes JM, Pride NB. Examination of the carbon monoxide diffusing capacity (DLCO) in relation to its KCO and VA components. Am J Respir Crit Care Med. 2012;186:132–9. https://doi.org/10.1164/rccm.201112-2160CI.
4. Ogilvie CM, Forster RE, Blakemore WS, Morton JW. A standardized breath holding technique for the clinical measurement of the diffusing capacity of the lung for carbon monoxide. J Clin Invest. 1957;36:1–17. https://doi.org/10.1172/JCI103402.
5. Guénard H, Varène N, Vaida P. Determination of lung capillary blood volume and membrane diffusing capacity in man by the measurements of NO and CO transfer. Respir Physiol. 1987;70:113–20.
6. Borland CD, Cracknell N, Higenbottam TW. Is the measurement of "DLNO" a true measure of membrane diffusing capacity? Clin Sci. 1984;67:41P.
7. Borland CD, Higenbottam TW. A simultaneous single breath measurement of pulmonary diffusing capacity with nitric oxide and carbon monoxide. Eur Respir J. 1989;2:56–63.
8. Kang MY, Sapoval B. Time-based understanding of DLCO and DLNO. Respir Physiol Neurobiol. 2016;225:48–59. https://doi.org/10.1016/j.resp.2016.01.008.
9. Kang MY, Grebenkov D, Guénard H, Katz I, Sapoval B. The Roughton-Forster equation for DLCO and DLNO re-examined. Respir Physiol Neurobiol. 2017;241:62–71. https://doi.org/10.1016/j.resp.2016.12.014.

10. Carlsen E, Comroe JH Jr. The rate of uptake of carbon monoxide and of nitric oxide by normal human erythrocytes and experimentally produced spherocytes. J Gen Physiol. 1958;42:83–107.
11. Umbreit J. Methemoglobin—it's not just blue: a concise review. Am J Hematol. 2007;82:134–44.
12. Haab P, Dyrand-Arczynska W. Carbon monoxide effects on oxygen transport. In: Crystal RG, editor. The lung: Scientific Foundations. 2nd ed. Philadelphia: Lippincott Williams & Wilkins; 1997. p. 1673–80.
13. Borland CD, Cox Y. Effect of varying alveolar oxygen partial pressure on diffusing capacity for nitric oxide and carbon monoxide, membrane diffusing capacity and lung capillary blood volume. Clin Sci (Lond). 1991;81:759–65.
14. Borland CD, Dunningham H, Bottrill F, Vuylsteke A. Can a membrane oxygenator be a model for lung NO and CO transfer? J Appl Physiol (1985). 2006;100:1527–38.
15. Borland CD, Dunningham H, Bottrill F, Vuylsteke A, Yilmaz C, Dane DM, Hsia CC. Significant blood resistance to nitric oxide transfer in the lung. J Appl Physiol (1985). 2010;108:1052–60. https://doi.org/10.1152/japplphysiol.00904.2009.
16. Martinot JB, Guénard H, Dinh-Xuan AT, Gin H, Dromer C. Nitrogen monoxide and carbon monoxide transfer interpretation: state of the art. Clin Physiol Funct Imaging. 2017;37:357–65. https://doi.org/10.1111/cpf.12316.
17. Zavorsky GS, Hsia CC, Hughes JM, Borland CD, Guénard H, van der Lee I, Steenbruggen I, Naeije R, Cao J, Dinh-Xuan AT. Standardisation and application of the single-breath determination of nitric oxide uptake in the lung. Eur Respir J. 2017;49:1600962. https://doi.org/10.1183/13993003.00962-2016.
18. Guénard H, Martinot JB, Martin S, Maury B, Lalande S, Kays C. In vivo estimates of NO and CO conductance for haemoglobin and for lung transfer in humans. Respir Physiol Neurobiol. 2016;228:1–8. https://doi.org/10.1016/j.resp.2016.03.003.
19. Taylor BJ, Coffman KE, Summerfield DT, Issa AN, Kasak AJ, Johnson BD. Pulmonary capillary reserve and exercise capacity at high altitude in healthy humans. Eur J Appl Physiol. 2016;116:427–37. https://doi.org/10.1007/s00421-015-3299-1.
20. Coffman KE, Chase SC, Taylor BJ, Johnson BD. The blood transfer conductance for nitric oxide: infinite vs. finite ΘNO. Respir Physiol Neurobiol. 2017;241:45–52. https://doi.org/10.1016/j.resp.2016.12.007.
21. Reeves RB, Park HK. CO uptake kinetics of red cells and CO diffusing capacity. Respir Physiol. 1992;88:1–21.
22. Borland C, Patel S, Zhu Q, Vuylsteke A. Hypothesis: why ΘNO could be finite in vitro but infinite in vivo. Respir Physiol Neurobiol. 2017;241:58–61. https://doi.org/10.1016/j.resp.2017.02.013.
23. van der Lee I, Zanen P, Biesma DH, van den Bosch JM. The effect of red cell transfusion on nitric oxide diffusing capacity. Respiration. 2005;72:512–6.
24. Hughes JM, van der Lee I. The TL,NO/TL,CO ratio in pulmonary function test interpretation. Eur Respir J. 2013;41:453–61. https://doi.org/10.1183/09031936.00082112.
25. Hughes JMB, Dinh-Xuan AT. The DLNO/DLCO ratio: physiological significance and clinical implications. Respir Physiol Neurobiol. 2017;241:17–22. https://doi.org/10.1016/j.resp.2017.01.002.

Chapter 14
Reference Equations for Simultaneously Measured D_{LCO} and D_{LNO}: How Are Reference Equations Influenced by Age, Sex, and Anthropometric Variables?

Mathias Munkholm and Jann Mortensen

Abstract To this date, only a few reference equations for the simultaneously measured D_{LCO} and D_{LNO} ($D_{L,CO,NO}$) method have been published in adults. For children, the number is even less. However, the different reference equations are largely alike. Independent variables for most of the outcomes for the $D_{L,CO,NO}$ method are sex, height, age, and/or age squared. Outcomes increase with increasing height and are generally higher in males than in females. In adults, there seems to be an accelerated loss of diffusing capacity with increasing age. Weight is not an independent variable for any of the outcomes. The possible influence of ethnicity on the outcomes has yet not been investigated.

Keywords Reference equations · $D_{L,CO,NO}$ · Independent variables · D_{LCO} · D_{LNO}

1 Introduction

To this date, a rather limited number of reference values for the simultaneously measured D_{LCO} and D_{LNO} ($D_{L,CO,NO}$) method have been published. For adults, some of these studies include only a small number of subjects (10–70 subjects) [1]. However, 2 studies include around 130 subjects each [2, 3], while 2 other studies include around 300 subjects each [4, 5]. The latest of these has a rather large age span ranging from 18 to 97 years [5]. In addition, one study has combined and

M. Munkholm
Department of Internal Medicine, Odense University Hospital, Svendborg, Denmark

J. Mortensen (✉)
Department of Clinical Physiology, Nuclear Medicine and PET, Diagnostic Center, Rigshospitalet, Copenhagen University Hospital, Copenhagen, Denmark
e-mail: jann.mortensen@regionh.dk

K. Yamaguchi (ed.), *Structure-Function Relationships in Various Respiratory Systems*, Respiratory Disease Series: Diagnostic Tools and Disease Managements, https://doi.org/10.1007/978-981-15-5596-1_14

reanalyzed data from three of the before mentioned studies to achieve one combined set of reference equations [6]. This study has an age span from 18 to 93 years.

When it comes to children, one larger study exists. It comprises 312 children being 5–17 years of age [7]. Apart from this, only a few smaller studies on children exist [8, 9].

2 Independent Variables in the Reference Equations

Since the $D_{L,CO,NO}$ method has many different outcomes such as $D_{L,CO}$, $D_{L,NO}$, Dm, Vc, K_{CO}, K_{NO}, and V_A, naturally many different reference equations have to be established for this method. As one could expect, the independent variables vary somewhat between the different reference equations. However, in general the reference equations include sex, height, age, and/or age squared.

Age: All outcomes decrease with age in adults. Furthermore, there seems to be an accelerated loss of diffusing capacity with age, which has been observed in all studies including a relatively high number of old people (Fig. 14.1). This accelerated loss can be taken into account by including age squared in the reference equation.

Sex: Generally, outcomes are higher in men than in women (Fig. 14.1). However, as suggested by two of the largest studies, the opposite seems to be the case when looking at Vc/VA [5, 6].

Height: When it comes to $D_{L,CO}$, $D_{L,NO}$, Dm, and Vc, increasing height renders higher outcomes.

Weight: No studies have found weight to be a significant independent variable.

Ethnicity: To this date, no studies have explored the potential significance of ethnicity on the outcomes of the $D_{L,CO,NO}$ method. Most studies have been performed on Caucasians, while one study has examined North African children [8].

3 Children

Not surprisingly, reference equations for the $D_{L,CO,NO}$ method in children are more limited than for adults. Two smaller studies have been published. One involving 85 children and the other involving 50 children over 8 years of age [8, 9]. In addition, in 2014 a larger study comprising 312 children in the age of 5–17 years was published [7]. However, because these reference equations are presented in a rather complex way including exponential functions, direct comparison to the other reference equations is difficult. In general, increasing height renders higher outcomes in children.

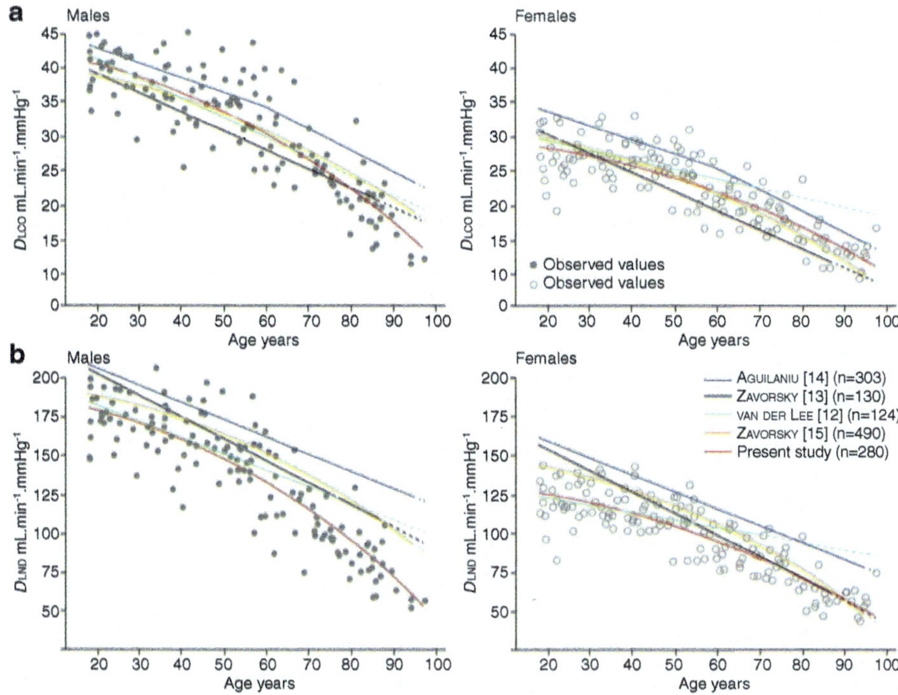

Fig. 14.1 Reference equations for diffusion capacity for CO (**a**) and NO (**b**) from five publications in adults. We acknowledge the European Respiratory Society (ERS) for letting us reproduce this figure from: Munkholm M, Marott JL, Bjerre-Kristensen L, et al. Reference equations for pulmonary diffusing capacity of carbon monoxide and nitric oxide in adult Caucasians. Eur Respir J 2018; 52: 1500677. [https://doi.org/10.1183/13993003.00677-2015]

4 Conclusion

Independent variables for most of the outcomes for the $D_{L,\,CO,\,NO}$ method are sex, height, age, and/or age squared. Outcomes increase with increasing height and are generally higher in males than in females. In adults, there seems to be an accelerated loss of diffusing capacity with increasing age. Weight is not an independent variable for any of the outcomes. The possible influence of ethnicity on the outcomes has yet not been investigated.

References

1. Hughes JM, van der Lee I. The TL,NO/TL,CO ratio in pulmonary function test interpretation. Eur Respir J. 2013;41(2):453–61. https://doi.org/10.1183/09031936.00082112.
2. Zavorsky GS, Cao J, Murias JM. Reference values of pulmonary diffusing capacity for nitric oxide in an adult population. Nitric Oxide Biol Chem. 2008;18(1):70–9. https://doi.org/10.1016/j.niox.2007.10.002.
3. van der Lee I, Zanen P, Stigter N, van den Bosch JM, Lammers JW. Diffusing capacity for nitric oxide: reference values and dependence on alveolar volume. Respir Med. 2007;101(7):1579–84. https://doi.org/10.1016/j.rmed.2006.12.001.
4. Aguilaniu B, Maitre J, Glenet S, Gegout-Petit A, Guenard H. European reference equations for CO and NO lung transfer. Eur Respir J. 2008;31(5):1091–7. https://doi.org/10.1183/09031936.00063207.
5. Munkholm M, Marott JL, Bjerre-Kristensen L, Madsen F, Pedersen OF, Lange P, et al. Reference equations for pulmonary diffusing capacity of carbon monoxide and nitric oxide in adult Caucasians. Eur Respir J. 2018;52(1):1500677. https://doi.org/10.1183/13993003.00677-2015.
6. Zavorsky GS, Hsia CC, Hughes JM, Borland CD, Guenard H, van der Lee I, et al. Standardisation and application of the single-breath determination of nitric oxide uptake in the lung. Eur Respir J. 2017;49(2):1600962. https://doi.org/10.1183/13993003.00962-2016.
7. Thomas A, Hanel B, Marott JL, Buchvald F, Mortensen J, Nielsen KG. The single-breath diffusing capacity of CO and NO in healthy children of European descent. PLoS One. 2014;9(12):e113177. https://doi.org/10.1371/journal.pone.0113177.
8. Rouatbi SKM, Garrouch A, Ben SH. Reference values of capillary blood volume and pulmonary membrane diffusing capacity in North African boys aged 8 to 16 years. Egyptian J Chest Dis Tuberculosis. 2014;63(3):705–15.
9. Rouatbi S, Ouahchi YF, Ben Salah C, Ben Saad H, Harrabi I, Tabka Z, et al. Physiological factors influencing pulmonary capillary volume and membrane diffusion. Rev Mal Respir. 2006;23(3 Pt 1):211–8.

Chapter 15
Differential Diagnosis Based on the Newly Developed Indicator of D_{LNO}/D_{LCO}: What Are Pathophysiological Backgrounds on D_{LNO}/D_{LCO}?

Anh Tuan Dinh-Xuan

Abstract The diffusing capacity of the lung for nitric oxide (D_{LNO}) can now be measured simultaneously with the diffusing capacity of the lung for carbon monoxide (D_{LCO}) in a single breath-holding maneuver. The D_{LNO}/D_{LCO} ratio has the advantage of remaining stable over time in a given healthy individual, unlike the well-known decline in D_{LCO} and D_{LNO} values with aging, making the D_{LNO}/D_{LCO} ratio an interesting parameter to consider when assessing alveolar-capillary gas exchange in the elderly. In clinical studies, three patterns emerged for the D_{LNO}/D_{LCO} ratio. A high D_{LNO}/D_{LCO} ratio is associated with pulmonary vascular disease, a normal ratio is associated with mild to moderate COPD, chronic heart failure, and morbid obesity, and a low D_{LNO}/D_{LCO} ratio is seen in lung disease with alveolar destruction (severe COPD with emphysema), or interstitial fibrosis including cystic fibrosis and interstitial fibrotic lung disease.

Keywords Lung gas diffusion capacity · Nitric oxide · Carbon monoxide Pulmonary capillary volume · Alveolar-capillary membrane conductance

1 Introduction

More than 60 years ago, Roughton and Forster [1] have proposed a model describing the transfer of carbon monoxide (CO) or any diffusible gases, including oxygen (O_2) and nitric oxide (NO), from air to blood as a two-step process in which the gas has to overcome two resistances in series according to eq. 15.1 (Fig. 15.1):

A. T. Dinh-Xuan (✉)
Respiratory Physiology Unit, Department of Thoracic Medicine and Surgery, Cochin Hospital, University of Paris, Paris, France
e-mail: anh-tuan.dinh-xuan@aphp.fr

K. Yamaguchi (ed.), *Structure-Function Relationships in Various Respiratory Systems*, Respiratory Disease Series: Diagnostic Tools and Disease Managements, https://doi.org/10.1007/978-981-15-5596-1_15

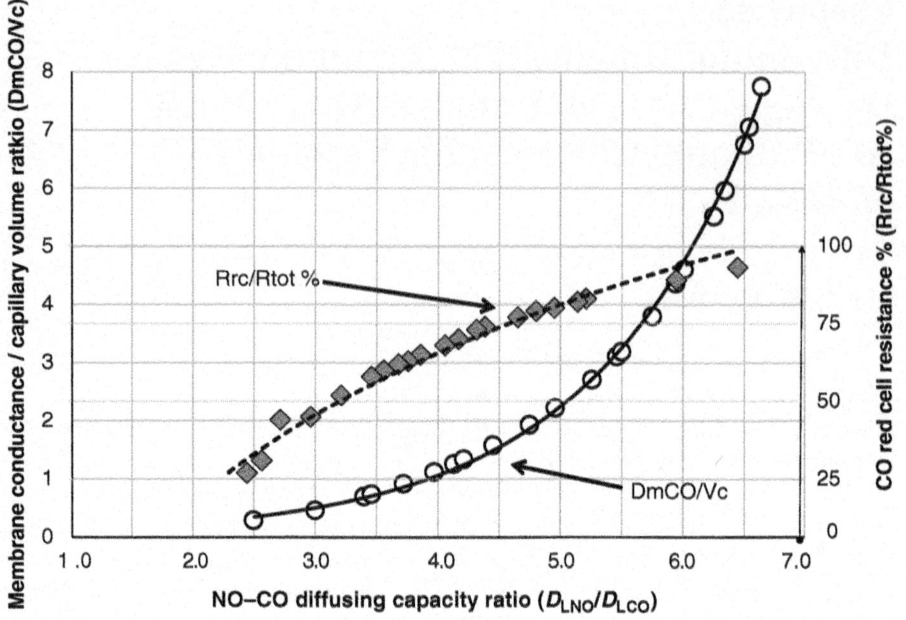

Fig. 15.1 Ratio of carbon monoxide (CO) membrane diffusing capacity (DMCO) to pulmonary capillary volume (VC), DMCO/VC, and red cell diffusion resistance (Rrc) for CO as a percentage of total (Rtot) CO diffusion resistance, Rrc/Rtot%, plotted against ratio of diffusing capacities for nitric oxide (NO) and CO, D_{LNO}/D_{LCO}, for normal male, aged 45 years, height 175 cm. Individual points calculated using the computer program in the online supplement of Zavorsky et al. [14]. The relationship is given by the equation DMCO/VC = $0.0536 \cdot \exp^{0.7465 \cdot DLNO/DLCO}$ and Rrc/Rtot% = 69.805 $\ln(D_{LNO}/D_{LCO}) - 31.963$. Reproduced and adapted from [10] with permission from the publisher

$$1/\mathrm{DL} = 1/\mathrm{DM} + 1/\Theta \cdot \mathrm{VC} \tag{15.1}$$

where

- 1/DL is the total resistance of the blood–gas barrier to gas transfer ($\mathrm{mmol}^{-1} \cdot \mathrm{min} \cdot \mathrm{kPa}$ in SI units or $\mathrm{mL}^{-1} \cdot \mathrm{min} \cdot \mathrm{mm\ Hg}$ in traditional units),
- DM is the membrane diffusing capacity, whereas 1/DM is the resistance to passive diffusion across the alveolar-capillary membrane and intracapillary plasma,
- Θ is the specific transfer conductance of blood (measured in vitro) for a specified gas, VC is the pulmonary capillary blood volume (measured in mL), whereas $1/\Theta \cdot \mathrm{VC}$ is the resistance to gas transfer of the red blood cell for reactive gases such as CO, NO, or oxygen.

In the 1960s and 1970s, the Roughton and Forster's approach (eq. 15.1) was applied by means of repeated measurements of diffusing capacity of the lung for CO (D_{LCO}) using two different partial pressure of oxygen (PO_2) to derive DM and VC [2, 3]. This approach was time consuming and was therefore not very often applied. As a result, the single breath D_{LCO} method without computing DM and VC remained the most commonly used technique for measuring gas transfer from the alveolus to the pulmonary capillary blood in the clinical setting [4, 5]. In the 1980s, two European groups of investigators based in Bordeaux (France) [6] and Cambridge (UK) [7, 8] had the idea of mixing a second hemoglobin (Hb)-reactive gas, namely nitric oxide (NO), together with the already used gas CO in the same reservoir from which the patient can inhale both tracer gases (NO and CO) together with an inert volume-marker gas (usually helium), a maneuver is known as the simultaneous single breath D_{LNO}–D_{LCO} technique [8]. However, as discussed in Chap. 13, while VC alveolar-capillary membrane diffusing capacity for carbon monoxide (DMCO) may be calculated from the Roughton and Forster technique (eq. 15.1) using two different PO_2, there remain unknowns and controversies regarding the choice of ΘNO and ΘCO values that should be used when computing DMCO and VC according to the initial concept proposed by Roughton and Forster in 1957 and it is now proposed to focus the D_{LNO}/D_{LCO} ratio [9, 10] or to discuss on the interpretation of either D_{LCO} and D_{LNO} separately (see Chap. 13). The present chapter will review evidence from the literature on the physiological significance and clinical implications of the D_{LNO}/D_{LCO} ratio.

2 D_{LNO}/D_{LCO} Ratio: Physiological Interpretation

With the classical analysis proposed by Roughton and Forster [1] and from eq. (15.1), the two resistances (1/DMCO and 1/ΘCO·VC) are approximately equal, with a 50:50 split for CO [11]. For NO, the total resistance to alveolar-capillary diffusion ($1/D_{LNO}$) is much less, 20–25% of that for CO, thus D_{LNO} is four to five times greater than D_{LCO} [9]. On clinical grounds, the D_{LCO} data from pulmonary vascular disease and anemia seemed to favor a 25:75 partition between the membrane resistance, 1/DMCO (25%) and the red cell resistance, 1/ΘCO·VC (75%), and this was supported by calculations of DMCO and VC, using more recent estimates [12] of the 1/ΘCO-PO_2 relationship [13]. Recent work using the D_{LNO}–D_{LCO} method with finite values for ΘNO and ΘCO [14] suggests that the red cell resistance percentage for $1/D_{LNO}$ is much less (37%) and that $1/D_{LNO}$ is weighted, but not dominated, by the membrane gas conductance (63%) [14, 15]. When computing theoretical D_{LNO}, D_{LCO} values and the resulting D_{LNO}/D_{LCO} ratios using the generally accepted value for ΘNO [16] and the newly published equation for the 1/ΘCO-PO_2 relationship [13], we found a curvilinear relationship between DMCO/VC and D_{LNO}/D_{LCO}, which is independent of the absolute values of D_{LNO} and D_{LCO} (Fig. 15.1). A comparable, but not superimposable, curvilinear relationship was also found between red cell diffusion resistance (Rrc), expressed as a percentage of the total resistance (Rtot) to CO diffusion (Rrc/Rtot%) (Fig. 15.1) [10]. In other words, $1/D_{LNO}$ represents 1/(DMCO·1.97), and $1/D_{LCO}$ reflects 1/VC. As a result the D_{LNO}/D_{LCO} ratio can be viewed as an expression of two other related ratios, DMCO/VC and the Rrc/Rtot% fraction for CO (Fig. 15.1).

3 D_{LNO}/D_{LCO} Ratio in Health and Disease

3.1 Normal Aging

As DL has two components (VA & K) and since VA is common to D_{LNO} and D_{LCO} the effects of aging on D_{LNO}/D_{LCO} should only depend on the rate of change in the KNO/KCO ratio. As KCO and KNO decline with aging at the same rate [17–19], D_{LNO}/D_{LCO} ratio remains stable over time, unlike the well-known decline in D_{LCO} and D_{LNO} values with increasing age [9, 10, 14]. Thus, age independence is a benefit when D_{LNO}/D_{LCO} ratios in the elderly are being considered.

3.2 Physical Exercise in Healthy Individuals and Patients with Lung Diseases

As cardiac output and oxygen consumption increase during physical exertion, both D_{LNO} and D_{LCO} increase linearly during moderate to heavy (maximum oxygen uptake 46.5 mL min^{-1} kg^{-1}) exercise [14, 20]. With exercise, pulmonary vascular pressures increase and more alveolar surface is available for gas exchange from the opening up of closed capillary units in the alveolar septa (recruitment) and dilatation of already patent vessels. This recruitment and dilation increase DMNO and DMCO. In addition, D_{LCO} will increase as VC increases. As the increase in D_{LNO} will be smaller than that of D_{LCO}, the resulting D_{LNO}/D_{LCO} ratio decreases linearly with increasing power output [21] by about 17–28% from rest to maximum exercise [14]. In patients with pulmonary sarcoidosis and lung fibrosis, the D_{LNO}/D_{LCO} ratio fell similarly from rest to exercise [22].

3.3 Change in Lung Volume

3.3.1 Healthy Individuals

During the single breath test, D_{LNO} and D_{LCO} are measured at maximum lung inflation, i.e., when the subject reaches his total lung capacity (TLC) before holding his breath. The breath-hold lung volume is then measured by inert gas (usually helium) dilution from the expired gas. A subtraction of dead space volume (instrumental plus anatomic dead space) is made and an "alveolar volume" (VA) calculated. This VA at TLC (~VAmax) is about 94% (SD 7%) of a separately measured TLC by multi-breath dilution [23, 24]; this difference (from 100%) is 3–4% greater than expected from the anatomic dead space and reflects incomplete alveolar mixing during the 10 s breath-hold time. Healthy subjects can voluntarily stop the initial inspiration to TLC at a submaximal volume (1/2VAmax to 0.9VAmax) and D_{LNO} and

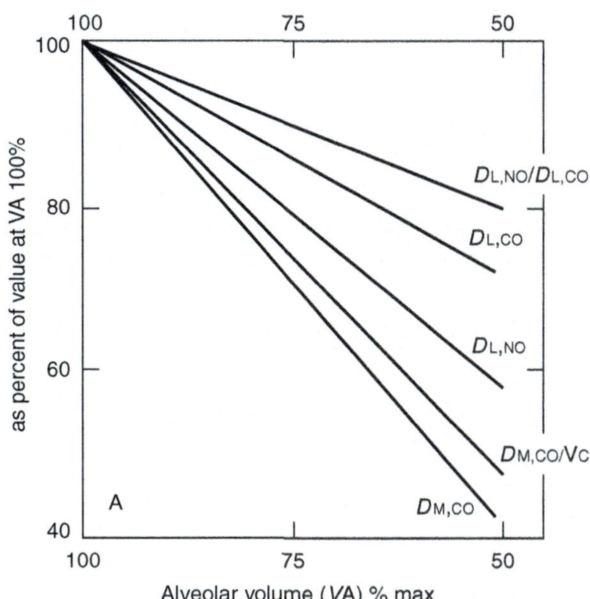

Fig. 15.2 Plot of diffusing capacities for nitric oxide (NO) and carbon monoxide (CO) and their ratio (D_{LNO}/D_{LCO}), membrane diffusing capacity for carbon monoxide (DMCO), and the DMCO/pulmonary capillary volume ratio (DMCO/VC) (on the ordinate) versus alveolar volume (~ lung volume minus anatomic dead space) as % maximum (~ TLC minus dead space) while normal subjects voluntarily changed lung expansion (50% VA max is about FRC). Reproduced and adapted from [14] with permission from the publisher

D_{LCO} and their components measured at different levels of alveolar expansion [24]. In Fig. 15.2, as the alveoli reduce in size (down to 50%VAmax), the reduction in D_{LNO} exceeds that of D_{LCO} and the D_{LNO}/D_{LCO} ratio falls. This is because ΔD_{LNO} is more driven by ΔDMNO than the more VC-weighted D_{LCO}. As the change in DMCO/VC is similar to the change in DMCO itself, VC changes little as the lung becomes smaller. This explains why extrapulmonary restriction which reduces VA to half its maximal value only decreases the D_{LNO}/D_{LCO} ratio by 20% (Fig. 15.2).

3.3.2 Patients with Lung Diseases

In patients with lung diseases or lung involvement from systemic diseases, VA can be reduced by alveolar destruction or filling with fluid or inflammatory tissue. This may be local (e.g., pneumonectomy) or diffuse. A 50% loss of VA in pneumonectomy only results in a 40% (and not 50%) reduction of D_{LCO} of predicted maximal values. Capillary dilatation and recruitment increasing blood flow and alveolar volume in the remaining lung explains explain this relatively smaller than expected fall in D_{LCO} [5]. This "compensatory" effect (mediated by an increase in KCO to 110–120% predicted) might be less for D_{LNO} (though D_{LNO} is also blood volume sensitive), so D_{LNO}/D_{LCO} might theoretically fall; but there is no data to date one way or the other. The third cause of VA reduction is poor distribution of the inhaled marker gases (usually helium); this occurs in airflow obstruction when the separately measured TLC exceeds the single breath VA. The effect on D_{LNO}/D_{LCO} is difficult to predict in those circumstances.

3.4 Altitude-Induced Hypoxia

Several studies [25–29] have looked at the effects of acute and chronic exposure to hypoxia of altitude dwelling (Table 15.1). All values, including those for permanent residents (highlanders) are expressed as percent of that at sea level in the "lowlanders." A fall in D_{LNO}/D_{LCO} ratio is seen in a majority of studies, whereas the rise in VC is unanimously observed by all investigators (Table 15.1). This increase in VC is probably due to an increase in cardiac output and pulmonary capillary recruitment and dilatation. When comparing lowlanders and highlanders (acute versus chronic exposure), highlanders have large increases in D_{LNO}, D_{LCO}, DMCO, and VC, with reductions in D_{LNO}/D_{LCO}. Highlanders have secondary polycythemia, which is particularly marked in those with chronic mountain sickness [26]. The high hemoglobin concentration explains a large part of the high D_{LCO} and VC.

In summary, altitude-induced hypoxia causes the following changes:

- An increase in pulmonary blood flow (as a result of increased cardiac output) on acute exposure thus explaining high D_{LNO}, D_{LCO}, and VC.
- Polycythemia (as a result of chronic hypoxia) thus accounting for increased D_{LCO} and VC through an expansion of the alveolar membrane surface (↑DMCO).
- Capillary recruitment (↑D_{LCO} and ↑VC) also likely occurs in lowlanders after 2/3 day exposure to altitude-induced hypoxia [27, 28].

Table 15.1 Acute and chronic effects of altitude-induced hypoxia in lowlanders and highlanders. All results are as % of sea level controls for each study

Authors	n	Altitude exposure and locations	DL_{NO}/DL_{CO}	DL_{NO} %	DL_{CO} %	Dm_{CO} %	Vc %	Ref
Healthy lowlanders								
de Bisschop et al.	16	4 days at 4000 m. Bolivia.	86	84	97[a]	73	111	[25]
Groepenhoff et al.	15	4 days at 4300 m. Peru.	83	127	155	102	183	[26]
Martinot et al.	25	2–3 days at 4300 m. Peru.	92	121	131	108	138	[27]
Faoro et al.	13	2–4 days at 5150 m. Nepal.	93	103	112[a]	94	120	[28]
Taylor et al.	7	40 days at 5150 m. Nepal.	106	117	110	126	104	[29]
Highlanders								
de Bisschop et al.	8	4000 m. Bolivia.	83	110	133 (Hb 17.1)	129	161 (137[a])	[25]
Groepenhoff et al.	15	4300 m. Peru.	79	132	167 (Hb 18.0)	101	208 (169[a])	[26]
Faoro et al.	28	5150 m. Nepal	82	153	185 (Hb 17.6)	125	222 (196[a])	[28]

[a]Corrected for polycythaemia to the standard hemoglobin level (13.4–14.6 g dL^{-1})

3.5 D_{LNO}/D_{LCO} Ratios in Respiratory Disease

In Table 15.2, there are listed clinical studies [22, 30–39] that have specifically used D_{LNO}/D_{LCO} ratios to distinguish patients from healthy controls and/or to differentiate one group of diseased individuals from another. One important feature is that all studies reported here had their own controls (normal subjects) to report patients' values as "% control." This is important because, at this early stage in the development of the D_{LNO}, the "normal" value of D_{LNO}/D_{LCO} ratios, i.e., the ratios found in healthy control subjects, varies widely (from 3.9 to 5.4) from one study to another. Studies reported in Table 15.2 are also arbitrarily divided into three groups (A, B & C) depending on the values of D_{LNO}/D_{LCO} ratios as compared with controls in a given study.

Studies in Group A [30–32] reporting high (>110% control) D_{LNO}/D_{LCO} ratios are associated with pulmonary vascular disease, either associated with remodeling and dilatation in the pulmonary capillary as seen in the hepatopulmonary syndrome [30] or pulmonary arterial hypertension [31, 32]. In all three studies, D_{LNO} and D_{LCO} values are low, but the DMCO/VC ratio is high as the reduction in VC is greater than the reduction in DMCO.

Table 15.2 Clinical studies of D_{LNO}/D_{LCO} ratios with related values and indices. All values are as percent of study controls

Authors	Diagnosis	*n*	DL_{NO}/DL_{CO}	DL_{NO} %	DL_{CO} %	Ref
Group A						
Degano et al.	HPS	11	111	71	66	[30]
Borland et al.	PAH	12	111	65	62	[31]
van der Lee et al.	PAH	26	114	58	65	[32]
Group B						
van der Lee et al.	ILD	41	105	58	65	[32]
van der Lee et al.	Smokers: GOLD 0	168	107	95	89	[33]
id	COPD: GOLD 1	68	110	95	86	id
id	COPD: GOLD 2	26	97	88	86	id
Magini et al.	CHF		107	82	77	[34]
Farha et al.	PAH	28	100	70	71	[35]
Zavorsky, 2008	Morbid obesity	10	103	108	95	[36]
Group C						
Phansalkar et al.	ILD (sarcoidosis)	25	80	35	43	[22]
Moinard et al.	COPD: GOLD 3–4	10	94	52	56	[37]
Dressell et al.	Cystic fibrosis	21	86	77	87	[38]
Barisione et al	NSIP[9]	30	91	52	58	[39]
id	UIP–ILD	30	89	32	37	id

CHF chronic heart failure, *COPD* chronic obstructive pulmonary disease, *GOLD* Global initiative for chronic obstructive lung disease, *HPS* hepatopulmonary syndrome, *ILD* interstitial lung disease, *NSIP* non-specific interstitial pneumonia, *PAH* pulmonary arterial hypertension, *UIP* Usual interstitial pneumonia. Rows are arranged in three sections (A > 110%, B ≤ 110% ≥ 95%, C < 95%)

Studies in Group B reported normal (110–97% control) D_{LNO}/D_{LCO} ratios [32–36]. Those studies with a reduced D_{LCO} (<80% control) are also associated with pulmonary hypertension (PH), either associated with chronic heart failure (postcapillary PH) [34] or arterial (precapillary) PH [35].

Studies in Group C reported low D_{LNO}/D_{LCO} (<95% control) ratios seen in pathological conditions reducing lung gas exchange surface, e.g., destructive emphysema [37] cystic fibrosis [38] or interstitial lung disease [39], including sarcoidosis [22].

4 Conclusion

When measuring NO transfer from alveolar gas to pulmonary capillary blood one can consider that D_{LNO} mostly reflects alveolar capillary membrane conductance (DM) as NO readily reacts with red cell hemoglobin thus making negligible lung capillary blood resistance to NO diffusion. On the other hand, because CO has a much slower reaction with oxygenated hemoglobin, one can consider that D_{LCO} mainly reflects red cell conductance. As a result, D_{LCO} will be reduced when pulmonary capillary volume (VC) is compromised. According to the Roughton and Forster equation, the D_{LNO}/D_{LCO} ratio is positively related to the DMCO/VC ratio and the CO red cell resistance fraction (Rrc/Rtot) in a curvilinear manner. The relationship between D_{LNO}/D_{LCO} and DMCO/VC or Rrc/Rtot% is independent of the absolute values of D_{LNO} or D_{LCO}. In clinical studies, three patterns emerged for the D_{LNO}/D_{LCO} ratio. A high D_{LNO}/D_{LCO} ratio is associated with pulmonary vascular disease, a normal ratio is associated with mild to moderate COPD, chronic heart failure, and morbid obesity, and a low D_{LNO}/D_{LCO} ratio is seen in lung disease with alveolar destruction (severe COPD with emphysema), or interstitial fibrosis including cystic fibrosis and interstitial fibrotic lung disease.

References

1. Roughton FJ, Forster RE. Relative importance of diffusion and chemical reaction rates in determining rate of exchange of gases in the human lung, with special reference to true diffusing capacity of pulmonary membrane and volume of blood in the lung capillaries. J Appl Physiol. 1957;11:290–302.
2. Bates DV, Varvis CJ, Donevan RE, Christie RV. Variations in the pulmonary capillary blood volume and membrane diffusion component in health and disease. J Clin Invest. 1960;39:1401–12.
3. Rankin J, McNeill RS, Forster RE. The effect of anemia on the alveolar-capillary exchange of carbon monoxide in man. J Clin Invest. 1961;40:1323–30.
4. Hughes JM, Bates DV. Historical review: the carbon monoxide diffusing capacity (DLCO) and its membrane (DM) and red cell (Θ·VC) components. Respir Physiol Neurobiol. 2003;138:115–42.

5. Hughes JM, Pride NB. Examination of the carbon monoxide diffusing capacity (DLCO) in relation to its KCO and VA components. Am J Respir Crit Care Med. 2012;186:132–9. https://doi.org/10.1164/rccm.201112-2160CI.
6. Guénard H, Varène N, Vaida P. Determination of lung capillary blood volume and membrane diffusing capacity in man by the measurements of NO and CO transfer. Respir Physiol. 1987;70:113–20.
7. Borland CD, Cracknell N, Higenbottam TW. Is the measurment of "DLNO" a true measure of membrane diffusing capacity? Clin Sci. 1984;67:41P.
8. Borland CD, Higenbottam TW. A simultaneous single breath measurement of pulmonary diffusing capacity with nitric oxide and carbon monoxide. Eur Respir J. 1989;2:56–63.
9. Hughes JM, van der Lee I. The TL,NO/TL,CO ratio in pulmonary function test interpretation. Eur Respir J. 2013;41:453–61. https://doi.org/10.1183/09031936.00082112.
10. Hughes JMB, Dinh-Xuan AT. The DLNO/DLCO ratio: physiological significance and clinical implications. Respir Physiol Neurobiol. 2017;241:17–22. https://doi.org/10.1016/j.resp.2017.01.002.
11. Hsia CC, McBrayer DG, Ramanathan M. Reference values of pulmonary diffusing capacity during exercise by a rebreathing technique. Am J Respir Crit Care Med. 1995;152:658–65.
12. Forster RE. Diffusion of gases across the alveolar membrane. In: Handbook of physiology. The respiratory system, Section 3, Volume IV, Chapter 5, Gas exchange. Washington, DC: American Physiological Society; 1987. p. 71–88.
13. Guénard H, Martinot JB, Martin S, Maury B, Lalande S, Kays C. In vivo estimates of NO and CO conductance for haemoglobin and for lung transfer in humans. Respir Physiol Neurobiol. 2016;228:1–8. https://doi.org/10.1016/j.resp.2016.03.003.
14. Zavorsky GS, Hsia CC, Hughes JM, Borland CD, Guénard H, van der Lee I, Steenbruggen I, Naeije R, Cao J, Dinh-Xuan AT. Standardisation and application of the single-breath determination of nitric oxide uptake in the lung. Eur Respir J. 2017;49:1600962. https://doi.org/10.1183/13993003.00962-2016.
15. Borland CD, Dunningham H, Bottrill F, Vuylsteke A, Yilmaz C, Dane DM, Hsia CC. Significant blood resistance to nitric oxide transfer in the lung. J Appl Physiol (1985). 2010;108:1052–60. https://doi.org/10.1152/japplphysiol.00904.2009.
16. Carlsen E, Comroe JH Jr. The rate of uptake of carbon monoxide and of nitric oxide by normal human erythrocytes and experimentally produced spherocytes. J Gen Physiol. 1958;42:83–107.
17. Glénet SN, De Bisschop C, Vargas F, Guénard H. Deciphering the nitric oxide to carbon monoxide lung transfer ratio: physiological implications. J Physiol. 2007;582(Pt 2):767–75.
18. Aguilaniu B, Maitre J, Glénet S, Gegout-Petit A, Guénard H. European reference equations for CO and NO lung transfer. Eur Respir J. 2008;31:1091–7. https://doi.org/10.1183/09031936.00063207.
19. Zavorsky GS, Cao J, Murias JM. Reference values of pulmonary diffusing capacity for nitric oxide in an adult population. Nitric Oxide. 2008;18:70–9.
20. Zavorsky GS, Kim DJ, McGregor ER, Starling JM, Gavard JA. Pulmonary diffusing capacity for nitric oxide during exercise in morbid obesity. Obesity (Silver Spring). 2008;16:2431–8. https://doi.org/10.1038/oby.2008.402.
21. Tamhane RM, Johnson RL Jr, Hsia CC. Pulmonary membrane diffusing capacity and capillary blood volume measured during exercise from nitric oxide uptake. Chest. 2001;120:1850–6.
22. Phansalkar AR, Hanson CM, Shakir AR, Johnson RL Jr, Hsia CC. Nitric oxide diffusing capacity and alveolar microvascular recruitment in sarcoidosis. Am J Respir Crit Care Med. 2004;169:1034–40.
23. Roberts CM, MacRae KD, Seed WA. Multi-breath and single breath helium dilution lung volumes as a test of airway obstruction. Eur Respir J. 1990;3:515–20.
24. van der Lee I, Zanen P, Stigter N, van den Bosch JM, Lammers JW. Diffusing capacity for nitric oxide: reference values and dependence on alveolar volume. Respir Med. 2007;101:1579–84.
25. de Bisschop C, Kiger L, Marden MC, Ajata A, Huez S, Faoro V, Martinot JB, Naeije R, Guénard H. Pulmonary capillary blood volume and membrane conductance in Andeans and

lowlanders at high altitude: a cross-sectional study. Nitric Oxide. 2010;23:187–93. https://doi.org/10.1016/j.niox.2010.05.288.
26. Groepenhoff H, Overbeek MJ, Mulè M, van der Plas M, Argiento P, Villafuerte FC, Beloka S, Faoro V, Macarlupu JL, Guenard H, de Bisschop C, Martinot JB, Vanderpool R, Penaloza D, Naeije R. Exercise pathophysiology in patients with chronic mountain sickness exercise in chronic mountain sickness. Chest. 2012;142:877–84. https://doi.org/10.1378/chest.11-2845.
27. Martinot JB, Mulè M, de Bisschop C, Overbeek MJ, Le-Dong NN, Naeije R, Guénard H. Lung membrane conductance and capillary volume derived from the NO and CO transfer in high-altitude newcomers. J Appl Physiol (1985). 2013;115:157–66. https://doi.org/10.1152/japplphysiol.01455.2012.
28. Faoro V, Huez S, Vanderpool R, Groepenhoff H, de Bisschop C, Martinot JB, Lamotte M, Pavelescu A, Guénard H, Naeije R. Pulmonary circulation and gas exchange at exercise in Sherpas at high altitude. J Appl Physiol (1985). 2014;116:919–26. https://doi.org/10.1152/japplphysiol.00236.2013.
29. Taylor BJ, Coffman KE, Summerfield DT, Issa AN, Kasak AJ, Johnson BD. Pulmonary capillary reserve and exercise capacity at high altitude in healthy humans. Eur J Appl Physiol. 2016;116:427–37. https://doi.org/10.1007/s00421-015-3299-1.
30. Degano B, Mittaine M, Guénard H, Rami J, Garcia G, Kamar N, Bureau C, Péron JM, Rostaing L, Rivière D. Nitric oxide and carbon monoxide lung transfer in patients with advanced liver cirrhosis. J Appl Physiol (1985). 2009;107:139–43. https://doi.org/10.1152/japplphysiol.91621.2008.
31. Borland C, Cox Y, Higenbottam T. Reduction of pulmonary capillary blood volume in patients with severe unexplained pulmonary hypertension. Thorax. 1996;51:855–6.
32. van der Lee I, Zanen P, Grutters JC, Snijder RJ, van den Bosch JM. Diffusing capacity for nitric oxide and carbon monoxide in patients with diffuse parenchymal lung disease and pulmonary arterial hypertension. Chest. 2006;129:378–83.
33. van der Lee I, Gietema HA, Zanen P, van Klaveren RJ, Prokop M, Lammers JW, van den Bosch JM. Nitric oxide diffusing capacity versus spirometry in the early diagnosis of emphysema in smokers. Respir Med. 2009;103:1892–7. https://doi.org/10.1016/j.rmed.2009.06.005.
34. Magini A, Apostolo A, Salvioni E, Italiano G, Veglia F, Agostoni P. Alveolar-capillary membrane diffusion measurement by nitric oxide inhalation in heart failure. Eur J Prev Cardiol. 2015;22:206–12. https://doi.org/10.1177/2047487313510397.
35. Farha S, Laskowski D, George D, Park MM, Tang WH, Dweik RA, Erzurum SC. Loss of alveolar membrane diffusing capacity and pulmonary capillary blood volume in pulmonary arterial hypertension. Respir Res. 2013;14:6. https://doi.org/10.1186/1465-9921-14-6.
36. Zavorsky GS, Kim DJ, Sylvestre JL, Christou NV. Alveolar-membrane diffusing capacity improves in the morbidly obese after bariatric surgery. Obes Surg. 2008;18:256–63. https://doi.org/10.1007/s11695-007-9294-9.
37. Moinard J, Guénard H. Determination of lung capillary blood volume and membrane diffusing capacity in patients with COLD using the NO-CO method. Eur Respir J. 1990;3:318–22.
38. Dressel H, Filser L, Fischer R, Marten K, Müller-Lisse U, de la Motte D, Nowak D, Huber RM, Jörres RA. Lung diffusing capacity for nitric oxide and carbon monoxide in relation to morphological changes as assessed by computed tomography in patients with cystic fibrosis. BMC Pulm Med. 2009;9:30. https://doi.org/10.1186/1471-2466-9-30.
39. Barisione G, Brusasco C, Garlaschi A, Baroffio M, Brusasco V. Lung diffusing capacity for nitric oxide as a marker of fibrotic changes in idiopathic interstitial pneumonias. J Appl Physiol (1985). 2016;120:1029–38. https://doi.org/10.1152/japplphysiol.00964.2015.

Chapter 16
Clinical Significance of Simultaneous Measurements of D_{LCO} and D_{LNO}: Can D_{LCO} and D_{LNO} Differentiate Various Kinds of Lung Diseases?

Kazuhiro Yamaguchi

Abstract The advantages and disadvantages of simultaneous measurement of D_{LCO} and D_{LNO} are systematically reviewed. We have learned many important phenomena about abnormal behaviors of D_{LCO} and D_{LNO} in a variety of lung and cardiac diseases from papers published so far. They showed that sensitivity to detecting pathological changes in alveolar septa is higher in D_{LNO} than in D_{LCO}, but it is almost identical in a practical sense. The most attractive point of simultaneous measurement of D_{LCO} and D_{LNO} is to partition D_{LCO} and D_{LNO} into membrane component of D_L (D_{MCO}) and pulmonary capillary blood volume (V_C). However, there are serious problems with this partitioning. Relative Krogh diffusion constants of NO against CO in various diffusion paths, i.e., $\Upsilon_{NO/CO}$ (D_{MNO}/D_{MCO}), and specific gas conductance of NO in blood (θ_{NO}) are not conclusively warranted. Therefore, the author considers that one should wait for partitioning D_{LCO} and D_{LNO} into D_{MCO} and V_C until reliable $\Upsilon_{NO/CO}$ and θ_{NO} values are ensured. Instead, the author advises one to use D_{LNO}/D_{LCO}, which is certified to act as representative of D_{MCO}/V_C under any pathophysiological conditions. Important role of D_{LNO}/D_{LCO} was identified in patients with IPF, NSIP, allogeneic stem cell transplantation, lung transplantation, and so forth, who showed decreasing D_{LNO}/D_{LCO}, reflecting a decrease in D_{MCO} (including heterogeneity of D_{MCO} distribution), which is caused by augmented injury of alveolocapillary tissue barrier. On the other hand, patients with pulmonary arterial hypertension showed increasing D_{LNO}/D_{LCO}, reflecting a decrease in V_C (including heterogeneity of V_C distribution), which is elicited by augmented injury of pulmonary microcirculation. As such, D_{LNO}/D_{LCO} works as a good, clinical indicator for differentiating between pathophysiological abnormalities in the alveolocapillary tissue barrier and those in pulmonary microcirculation without awareness of actual values of D_{MCO} and V_C.

K. Yamaguchi (✉)
Department of Respiratory Medicine, Tokyo Medical University, Tokyo, Japan

Division of Respiratory Medicine, Tohto Clinic, Kenko-Igaku Association, Tokyo, Japan
e-mail: yamaguc@sirius.ocn.ne.jp

K. Yamaguchi (ed.), *Structure-Function Relationships in Various Respiratory Systems*, Respiratory Disease Series: Diagnostic Tools and Disease Managements, https://doi.org/10.1007/978-981-15-5596-1_16

Keywords Diffusing capacity (D_L) for CO or NO (D_{LCO}, D_{LNO}) · Membrane component of D_L (D_{MCO}, D_{MNO}) · Blood component of D_L (D_{BCO}, D_{BNO}) · Pulmonary capillary blood volume (V_C) · Specific gas conductance of CO or NO in blood (θ_{CO}, θ_{NO}) · D_{LNO}/D_{LCO}

1 Introduction

More than 100 years ago, Marie Krogh developed the method to measure the single-breath uptake of carbon monoxide (CO) from alveolar gas to hemoglobin (Hb) packed in the erythrocyte (breath-holing time: 6 ~ 10 s) [1]. Since then, the single-breath diffusing capacity for CO (D_{LCO}) has gradually gained a position as the most clinically useful pulmonary function test after spirometry and lung volume examination. The practical method to measure the single-breath D_{LCO} was established by Ogilvie et al. [2] and has been utilized in the field of clinical practice to date. At the same time, Roughton and Forster [3] proposed a model describing the gas transfer of CO in the lung. They postulated that CO transfer process from alveolar gas to Hb packed in the erythrocyte is prescribed by aqueous-phase diffusion and reaction with Hb but not by capillary perfusion, leading them to simplify the CO transfer process into two distinct processes, i.e., (1) the membrane component of diffusing capacity for CO (D_{MCO}) that reflects the aqueous-phase diffusion limitation across the effective alveolocapillary membrane consisting of alveolocapillary tissue barrier and plasma layer surrounding the erythrocyte, and (2) the blood component of diffusing capacity for CO (D_{BCO}) that is defined as the product of specific gas conductance of CO in the blood (θ_{CO}) and pulmonary capillary blood volume (V_C), i.e., $D_{BCO} = \theta_{CO} \cdot V_C$. The θ_{CO} signifies the diffusive and reactive process across the interior of the erythrocyte (including erythrocyte membrane, if it has any resistance) incorporated with the competitive, replacement reaction of CO with oxyhemoglobin (HbO_2). As the reciprocals of D_{MCO} and D_{BCO} are the gas transfer resistances that are connected in series, the total resistance for CO transfer, $1/D_{LCO}$, is expressed as (Fig. 16.1):

$$1/D_{LCO} = 1/D_{MCO} + 1/(\theta_{CO} \bullet V_C) \tag{16.1}$$

Eq. 16.1 allows one to decide D_{MCO} and V_C from D_{LCO} values measured at two different alveolar partial pressures of O_2 (PO_2) (Roughton–Forster's (R–F) classic two-step PO_2 method [3]).

About 30 years ago, Guénard et al. [4] and Borland and Higenbottamet [5] independently proposed a novel method, in which D_{LCO} and diffusing capacity for nitric oxide (D_{LNO}) are simultaneously measured (simultaneous one-step CO–NO method). Differing from the R–F classic two-step PO_2 method, the simultaneous one-step CO–NO method requires only the single measurement of D_{LCO} and D_{LNO} at a certain

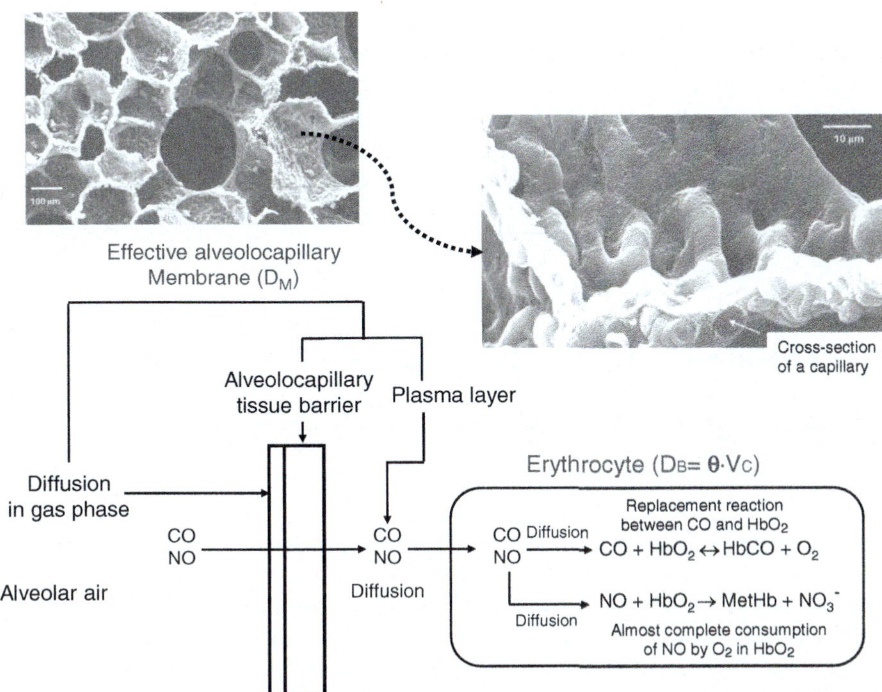

Fig. 16.1 CO and NO transfer from alveolar gas to hemoglobin (Hb) in erythrocyte. Membrane component (D_M) of diffusing capacity is prescribed by diffusion through the effective alveolocapillary membrane that includes gas-phase diffusion (stratified heterogeneity of inspired gas), alveolocapillary tissue barrier, and plasma layer, while blood component ($D_B = \theta \cdot V_C$) is prescribed by diffusion in erythrocytes for NO but diffusion and chemical reaction in erythrocytes for CO. Reaction of CO with oxyhemoglobin (HbO_2) is replacement reaction, while that of NO with HbO_2 is oxidative reaction in which ferrous HbO_2 is converted to ferric methemoglobin (MetHb) that loses the ability to bind to O_2. Oxidative reaction by NO produces NO_3^- simultaneously. Scanning electron microscopic images of alveolar septa are personal provision from Weibel

alveolar PO_2 for determining D_{MCO} (or D_{MNO}) and V_C under some unique assumptions. Replacing CO with NO, Eq. (16.1) can be applied for estimating D_{LNO}, as well. The growing demand for the D_{LNO} measurement in a clinical practice compels the Task Force Panel organized in the European Respiratory Society (ERS) to establish the standardization on the method and interpretation concerning the simultaneous measurement of D_{LCO} and D_{LNO} [6], which is called the 2017-ERS Standards in the current study. If the assumptions adopted by the 2017-ERS Standards are valid, the simultaneous one-step CO–NO method is clinically more convenient than the R–F classic two-step PO_2 method and curtails the time required for D_L measurement. Unfortunately, however, the validity of basic assumptions underlying the simultaneous one-step CO–NO method has not been firmly certified. Furthermore, it has not been well understood about the matter of what should be cautious when interpreting the measures of D_{LCO} and D_{LNO} in a clinical setting. The current study, therefore, shed light on the following issues: (1) physical and chemical properties that build D_{LCO} and D_{LNO}, including physical and chemical properties of CO and NO in various tissue components, including effective alveolocapillary membrane and erythrocytes; (2) what are the most appropriate values of θ_{CO} and θ_{NO} in capillary blood?; (3) considerations for various assumptions introduced by Roughton and Forster for describing diffusing capacity; (4) precautions on D_{LCO} and D_{LNO} measured in clinical practice; and (5) learning from D_{LCO} and D_{LNO} measures in a variety of lung and cardiac diseases.

2 Physical and Chemical Properties Building D_{LCO} and D_{LNO}

2.1 Physical Properties of CO and NO in Different Diffusion Paths

The solubility (α, Bunsen solubility coefficient expressed in mL/mL/atm) in water is 0.0189 for CO and 0.035 for NO (Table 16.1). Although reliable values concerning the solubility of these gases in lung tissue have not been reported, Power [16] identified that the solubility of CO in human lung tissue is depressed by ~5% in comparison with that in water. Differing from CO and NO, the physical properties of O_2 have been well studied under a variety of aqueous conditions. Therefore, the author discusses the general difference in gas solubility in various diffusion paths based on the α_{O2} measured in water, protein solution, erythrocytes, or lung tissue. The α_{O2} in water is 0.0238, which is decreased as the solvent density (ρ, g/cm^3) is increased by the addition of albumin, while the α_{O2} in erythrocytes is 1.2-fold that in water due to the solvent effect of Hb [17, 18]. The α_{O2} in human lung tissue is 0.0213, which is 10% lower than that in water [16]. The evidence harvested for α_{O2} indicates that although the solubility of a certain gas differs depending on the solvent conditions, the difference is tolerably small. These facts suggest that the solubility of a gas in various diffusion paths from the alveolocapillary tissue barrier to the interior of the erythrocyte is nearly identical and approximated by that in water.

On the other hand, the diffusivity (d, cm^2/s) of CO or NO in various diffusion paths has a critical problem. Since trustworthy values of d_{CO} and d_{NO} in plasma layer, erythrocyte, and lung tissue have not been investigated so far, the author thinks over the behavior of d of a certain gas under different solvent conditions based on the well-recognized d_{O2} data. The d_{O2} in water is $2.83 \cdot 10^{-5}$ at 37 °C [19], which is reduced as the viscosity (μ, g/cm/s) of the solvent is increased by the addition of albumin or serum proteins [20, 21]. The d_{O2} in the solution with a normal albumin concentration of 5 g/dL is $2.49 \cdot 10^{-5}$, which is 88% of that in water. Stein et al. [22] demonstrated that the d_{O2} in erythrocytes is greatly reduced as the solvent μ is augmented by a high Hb concentration, resulting in that the d_{O2} at a normal Hb concentration of 33 g/dL packed in erythrocytes is $7.5 \cdot 10^{-6}$, which is only 27% of the d_{O2} in water. These facts surely indicate that differing from the gas solubility, the

Table 16.1 Physical and chemical properties of CO and NO

	CO	NO
Bunsen solubility coefficient in water[a] (α, mL/mL/atm)	0.018	0.035
Molecular weight (g/mol)	28	30
Relative diffusivity in water[a] by Fick/Graham's law	1	0.97
Relative Krogh diffusion constant in water ($\Upsilon_{NO/CO}$)[a] by Fick/Graham's law	1	1.97
Measured diffusivity in water[b] (cm^2/sec)	$3.26 \cdot 10^{-5}$	$4.80 \cdot 10^{-5}$
Relative Krogh diffusion constant in water ($\Upsilon_{NO/CO}$)[b] estimated from measured diffusivity	1	3.35
Association rate constant with Hb[c] (κ': $\mu M^{-1} \cdot s^{-1}$)	6.0	60
Dissociation rate constant with Hb[c] (κ: s^{-1})	0.008	0.00003
Equilibrium constant with Hb[c] (K: μM^{-1})	750	$2 \cdot 10^6$
Relative affinity with Hb[c] in contrast to the affinity of O_2 with Hb	234	$6.25 \cdot 10^5$
PO_2-dependence of θ[d]	+	–
Hematocrit-dependence of θ[e]	+	±
θ under in vivo conditions recommended by 2017-ERS Standards[a] (PO_2 = 100 mmHg and normal Hb concentration) (mL/min/mmHg/(mL·blood))	0.56 (Guénard et al.[g])	4.5 (Carlsen and Comroe[h])
θ under in vivo conditions recommended by Coffman et al.[f] (PO_2 = 100 mmHg and normal Hb concentration) (mL/min/mmHg/(mL·blood))	1.23 (Reeves and Park[i])	Infinite (Boland et al.[j])

Hb: reduced Hb. [a]2017-ERS Standards [6]. [b]Wise and Houghton [7]. [c]Chakraborty et al. [8] measured for human Hb at a temperature of 20 ~ 25 °C and pH of 7.0. [d]Boland et al. [9]. [e]Borland et al. [10]. [f]Coffman et al. [11]. [g]Guénard et al. [12]. [h]Carlsen and Comroe [13]. [i]Reeves and Park [14]. [j]Borland et al. [15]

gas diffusivity is distinctly varied along diffusion paths from alveolocapillary tissue barrier to erythrocytes. However, it can be assumed that although the absolute diffusivity of a gas differs greatly among alveolocapillary tissue barrier, plasma layer, and interior of the erythrocyte, the relative diffusivity defined by d_X/d_Y (X, Y: two indicator gases) may be approximately the same in any of the diffusion path. This is ascribed to the fact that the d_X/d_Y in a binary diffusion system is approximated by the reciprocal ratio of the square root of molecular weight (MW) of the two gases concerned, which is called the Fick/Graham's law and expressed as:

$$d_X / d_Y \approx (MW_Y)^{1/2} / (MW_X)^{1/2} \tag{16.2}$$

Since the molecular weight of CO and NO is 28 and 30 g/mol, respectively, the d_{NO}/d_{CO} is expected to be 0.97. This value is taken to be the same in any lung tissue forming aqueous-phase diffusion paths, including alveolocapillary tissue barrier, plasma layer, and interior of the erythrocyte (Table 16.1).

The diffusive process of CO and NO in aqueous-phase diffusion paths is prescribed by the Krogh diffusion constant that is defined as ($\alpha \cdot d$). The α for each indicator gas is roughly equal along whole diffusion paths. Furthermore, the relative diffusivity (d_{NO}/d_{CO}) is approximately the same in any diffusion path. Thereby, the relative Krogh diffusion constant of NO against CO, which is defined as $\Upsilon_{NO/CO}$, is expressed by the following equation:

$$\gamma_{NO/CO} = (\alpha \cdot d)_{NO} / (\alpha \cdot d)_{CO} \quad (16.3)$$

$\Upsilon_{NO/CO}$ equals the ratio of membrane diffusing capacity for NO (D_{MNO}) to that for CO (D_{MCO}). Based on the α_{CO} and α_{NO} in water summarized by Wilhelm et al. [23] and the d_{NO}/d_{CO} estimated from the Fick/Graham's law, the 2017-ERS Standards considered that the best value of $\Upsilon_{NO/CO}$ is 1.97 and the D_{MNO}/D_{MCO} is described as:

$$D_{MNO} / D_{MCO} = 1.97 \quad (16.4)$$

However, it should be noted that the Fick/Graham's law is approved to be true only in the gas-phase binary diffusion system but not in the aqueous phase. Wise and Houghton [7] measured d_{CO} and d_{NO} in water and found that they are, respectively, $3.26 \cdot 10^{-5}$ and $4.80 \cdot 10^{-5}$ at 37 °C (Table 16.1), resulting in that d_{NO}/d_{CO} in water is 1.47. The measurement by Wise and Houghton [7] indicates that d_{NO} in water is clearly larger than d_{CO}, which is the opposite of the result derived from the Fick/Graham's law ($d_{NO}/d_{CO} = 0.97$). The peculiar behaviors of NO and CO in water are explained by the fact that although both NO and CO are polar molecules, they carry different electronic dipole moments in aqueous-phase conditions [7, 24]. Assuming that the relationship between d_{NO} and d_{CO} in any diffusion path is the same as that measured in water by Wise and Houghton [7], Kang et al. [24] thought that the most appropriate value of d_{NO}/d_{CO} in aqueous-phase diffusion paths in the lung is 1.47, resulting in that $\Upsilon_{NO/CO}$ in lung tissue is 3.35 (Table 16.1). In the analysis by Kang et al. [24], α_{CO} was taken to be 0.018 and α_{NO} was 0.041. If the $\Upsilon_{NO/CO}$ value predicted by Kang et al. [24] is adopted, the relationship between D_{MNO} and D_{MCO} differs substantially from that described by Eq. (16.4). Supporting the concept of Kang et al. [24], Coffman et al. [11] demonstrated that the exercise-induced increase in D_{MCO} and V_C is better predicted from $\Upsilon_{NO/CO}$ value of 4.4 rather than 1.97. Based on these facts, the author feels that the further in vitro and in vivo validation against the issue of what is the most appropriate value of $\Upsilon_{NO/CO}$ is needed.

The diffusivity of Hb in the erythrocyte should be considered, as well. This is because a slow diffusion of Hb in the erythrocyte induces facilitated diffusion of a gas which is bound to Hb. However, the diffusivity of Hb in the erythrocyte is $4.5 \sim 6.4 \cdot 10^{-8}$ at 37 °C and normal Hb concentration [21], which is only 1.8% of free O_2 in water. Therefore, the facilitated diffusion caused by the ligand-bound Hb in the erythrocyte may be ignored.

2.2 *Chemical Properties of CO and NO Reacting with Hb*

Association rate constant of CO with reduced Hb is 10% of that of O_2, while dissociation rate constant of CO from carboxyhemoglobin (HbCO) is $4 \cdot 10^{-4}$ of O_2 [8], as the result of which the affinity of CO with reduced Hb is 234 times higher than that of O_2 (Table 16.1). Despite this fact, the partial pressure of CO (P_{CO}) in the erythrocyte and thus in the capillary blood is not zero and shows a low but finite value depending on PO_2 surrounding Hb in the erythrocyte [25], which indicates that CO has a nature limited by perfusion, as well. In addition, it should be noted that under physiological conditions (normoxic PO_2 and low P_{CO}), uptake of CO by erythrocytes is predominantly limited by the replacement reaction between CO and HbO_2, which is much slower than the association reaction of CO with reduced Hb [8]. Hence, CO can be regarded as the gas that is limited by the combination of perfusion, aqueous-phase diffusion, and reaction with HbO_2, among them the replacement reaction of CO with HbO_2 being a major limiting step for CO gas transfer. These facts suggest that the estimation of D_{LCO} using Eq. (16.1), which is only validated under conditions with no perfusion limitation, may make an error (see Sect. 16.4.1).

The association rate constant of NO with reduced Hb, which is the oxidative reaction transforming ferrous irons (Fe^{2+}, reduced Hb) to ferric irons (Fe^{3+}, methemoglobin (MetHb)), is of the same order of magnitude as O_2 combining with reduced Hb forming HbO_2, but the dissociation rate constant of NO from Hb is $1.5 \cdot 10^{-6}$ of O_2, resulting in that the equilibrium constant (i.e., the affinity) of NO with Hb is $6.25 \cdot 10^5$ times larger than that of O_2 (Table 16.1) [8]. In in vivo conditions where abundant O_2 is present, NO reacts directly with O_2 in HbO_2 and yields a nitric anion (NO_3^-) and a MetHb (Fig. 16.1). The association and dissociation rate constants of this NO-elicited oxidative reaction with HbO_2 are thought to be almost equivalent to those in the oxidative reaction between NO and reduced Hb. Hence, NO is irreversibly consumed by HbO_2, resulting in that partial pressure of NO (P_{NO}) in the erythrocyte (thus, in the capillary blood) is maintained at zero independent of erythrocyte PO_2 surrounding Hb; namely, there are no free NO molecules to be transported by perfusion in pulmonary capillaries during the single-breath D_{LNO} measurement with a short period of time, which indicates that NO is purely limited by aqueous-phase diffusion but neither by reaction with Hb nor by perfusion and gas transfer of NO from alveolar gas to Hb is correctly described by Eq. (16.1) under a quasi-steady-state condition during a single-breath measurement of D_{LNO}.

3 What Are the Most Appropriate Values of Specific Gas Conductance of CO (θ_{CO}) and of NO (θ_{NO}) in Capillary Blood?

3.1 θ_{CO} in Capillary Blood

The molar ratio of CO against Hb in in vitro rapid-reaction or stopped-flow apparatus is about 1:1 whereas in in vivo single-breath measurements of D_{LCO}, the molar ratio of CO to Hb is about $1{:}1{\cdot}10^4$, implying that CO uptake is largely different between in vitro and in vivo conditions. However, the main limiting process of CO uptake is the replacement reaction of CO with HbO_2 irrespective of the condition in which Hb exists with or without encapsulation by erythrocytes. Thus, the PO_2–$1/\theta_{CO}$ relationships measured with various in vitro methods [3, 26, 27] may work well as those allowing for estimating D_{LCO} in in vivo conditions. Furthermore, the direct in vivo PO_2–$1/\theta_{CO}$ relationship was investigated by Guénard et al. [12]. They searched the best-fit relationship of PO_2–$1/\theta_{CO}$ based on the D_{LCO} values measured at two different PO_2 over a narrow PO_2 range (inspired O_2 concentrations: 15 and 21%). They determined the optimal relationship between capillary PO_2 and $1/\theta_{CO}$ in ten healthy subjects by minimizing the difference in D_{MCO}/V_C. The result obtained by Guénard et al. [12] was quantitatively consistent with most of the relationships of PO_2–$1/\theta_{CO}$ measured in in vitro conditions (Fig. 16.2). Based on these facts, the equation of PO_2–$1/\theta_{CO}$ reported by Guénard et al. [12] was adopted as giving the most appropriate relationship between PO_2 and $1/\theta_{CO}$ in pulmonary capillaries by the 2017-ERS Standards. The author has no objection against this selection and postulates that θ_{CO} value is 0.56 mL/min/mmHg/(mL·blood) at PO_2 of 100 mmHg and normal Hb concentration in blood, which is calculated from the equation proposed by Guénard et al. [12].

3.2 θ_{NO} in Capillary Blood: Finite or Infinite?

Carlsen and Comroe [13] measured the association kinetics of CO and NO bound to free Hb and Hb packed in erythrocytes using a rapid-reaction, constant-flow apparatus, in which the artifact evoked by the stagnant water layer surrounding the erythrocyte is sufficiently removed. They found only a small difference in association kinetics of CO between free Hb and packed Hb, suggesting that CO uptake by erythrocytes is mainly prescribed by the reaction of CO with Hb. On the other hand, they identified a large difference in association kinetics of NO between free Hb (very fast) and packed Hb (very slow), indicating that NO uptake by erythrocytes is limited by the aqueous-phase diffusion in the erythrocyte interior. The θ_{NO} of human biconcave erythrocytes calculated from the Carlson's experiment is 4.5 mL/min/mmHg/(mL·blood). Although several values of θ_{NO} have been reported to date, the

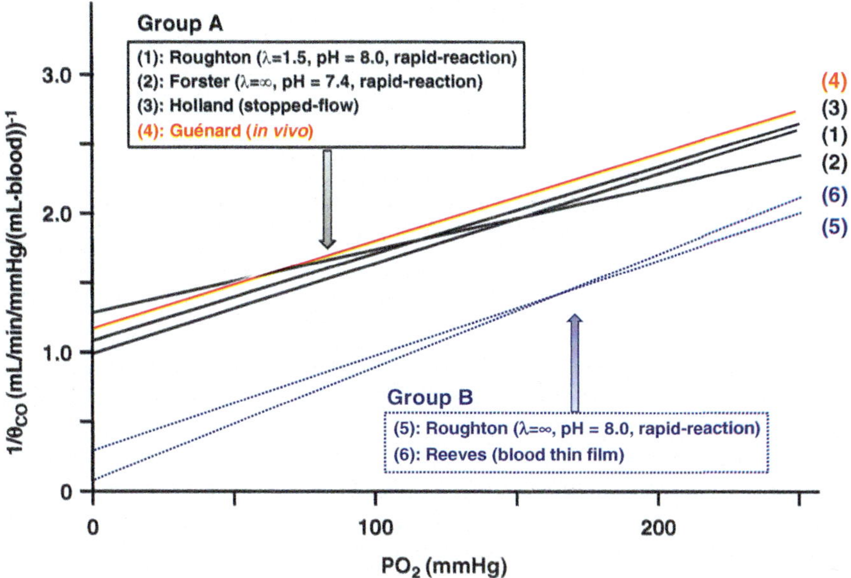

Fig. 16.2 Relationship between PO_2 and $1/\theta_{CO}$. Group A includes relationships demonstrated by (1): Roughton and Forster (1957) [3], (2): Forster (1987) [26], (3): Holland (1969) [27], and (4): Guénard et al. (2016) [12]. Group B includes relationships demonstrated by (5): Roughton and Forster (1957) [3] and (6): Reeves and Parks (1992) [14]. Measurement of Reeves and Parks was made neither with rapid-reaction nor with stopped-flow but with a special procedure using blood thin film with no flow. The findings of Reeves and Park have not been replicated. Lines belonging to each group overlap each other without a large difference. However, lines assigned to group A and those to group B are radically different. Of the lines belonging to group A, the line reported by Guénard et al. [12] was accepted by 2017-ERS Standards as giving the most valid relationship for PO_2-depnedent change in $1/\theta_{CO}$ in in vivo conditions. λ denotes the ratio of CO permeability of erythrocyte membrane to that of erythrocyte interior introduced by Roughton and Foster [3]. However, λ has been certified to be almost infinite, i.e., no gas transfer resistance through erythrocyte membrane

θ_{NO} reported by Carlsen and Comroe [13] is only the estimate directly measured with a reliable method. Thus, the 2017-ERS Standards recommended to use the constant value of 4.5 as the most appropriate value of θ_{NO} in in vivo conditions. Further disavowal against reaction limitation during NO uptake by erythrocytes was ascertained by Borland et al. [9], who identified that D_{LCO} is PO_2- and hematocrit (Ht)-dependent, whereas D_{LNO} is PO_2-independent but Ht-dependent. The PO_2-independent D_{LNO} implies that NO uptake is not limited by chemical reaction between NO and Hb. The Ht-dependent D_{LNO} indicates that NO uptake is impeded with decreasing density of erythrocytes (i.e., anemia), which increases plasma layer thickness. However, due to a larger Krogh diffusion constant of NO than that of CO (Table 16.1), the relative contribution of diffusion limitation elicited by the

Ht-dependent change in plasma layer thickness is anticipated to be small during NO transfer. Indeed, Borland et al. [10] certified that the Hb adjustment for D_{LNO} is unnecessary until Hb concentration reduces to ~4 g/dL. Based on these facts, the constant and finite value of θ_{NO} has been considered to be valid under a wide variety of in vivo conditions.

However, the opposition against the finite value of θ_{NO} has been filed by a couple of investigators [8, 11, 15, 24]. Of them, the theoretical analysis reported by Boland et al. [15] deserves special attention. Carlson and Comroe [13] measured NO uptake through the erythrocyte in a rapid-reaction apparatus under an equimolar condition of NO and Hb, i.e., NO and Hb are 0.15 mmol/L and 0.11 mmol/L, respectively, just after mixing. On the other hand, when measuring the single-breath D_{LNO} in in vivo condition, the subject inhales a gas mixture generally containing 40 ppm of NO for 10 s. Thereby, the NO concentration in the alveolocapillary tissue barrier will be $5.7 \cdot 10^{-5}$ mmol/L just after the beginning of the D_{LNO} measurement. Hb concentration in capillary blood is 20 mmol/L, which indicates that the molar ratio of NO against Hb is $1{:}3.5 \cdot 10^{5}$ in in vivo measurement of D_{LNO} (i.e., the great excess in Hb concentration), which differs largely from that in in vitro condition where the molar ratio of NO against Hb is 1:1. Solving the one-dimensional diffusion–reaction equation for NO uptake by erythrocytes under the steady-state condition, Boland et al. [15] found that NO is absorbed only by the Hb present in the vicinity of erythrocyte surface, i.e., the "surface capture" of NO in in vivo D_{LNO} measurement. On the other hand, in in vitro measurement of NO uptake in a rapid-reaction apparatus, NO should penetrate more deeply into the erythrocyte interior for seeking for Hb that can bind NO, i.e., "deep penetration" of NO in in vitro NO uptake measurement. These considerations suggest that the intracellular diffusion path is short (thus, diffusion resistance is small) when D_{LNO} is measured in in vivo condition, while the intracellular diffusion path is long (thus, diffusion resistance is large) when NO uptake is measured in in vitro condition. Based on these facts, Boland et al. [15] concluded that the aqueous-phase diffusion resistance against NO uptake yielded by the erythrocyte is virtually zero and the θ_{NO} is practically infinite in in vivo condition during single-breath measurement of D_{LNO}. Qualitatively the same conclusion was attained by Kang et al. [24], who analyzed NO uptake by erythrocytes in cooperation with the three-dimensional model of capillaries and erythrocytes under the steady-state condition. Chakraborty et al. [8] expanded the theoretical analysis for NO uptake by discoidal erythrocytes under the condition with no unstirred plasma layer surrounding them and concluded that the in vivo θ_{NO} should be $4.2 \cdot 10^{6}$ mL/min/mmHg/(mL·blood) in most cases of practical interest, the value being virtually infinite. Supporting the aforementioned theoretical results, Coffman et al. [11] demonstrated that while estimating D_{MCO} from D_{LCO} and D_{LNO} measures, the exercise-elicited increase in D_{MCO} is well predicted by infinite θ_{NO} but not by finite θ_{NO} of 4.5. These theoretical and clinical observations suggest that although it is not known at present whether θ_{NO} in in vivo conditions is actually infinite or a large finite value, it is certainly conceivable to exceed at least 4.5. Since the determination of the most appropriate value of θ_{NO} during D_{LNO} measurements is one of the most critical matters while estimating D_{MCO} and V_C from D_{LCO} and D_{LNO} measures, the author sincerely feels that this matter should be vigorously addressed again.

4 Considerations for Various Assumptions Introduced by Roughton and Forster for Describing Diffusing Capacity

The unique assumptions that are needed for establishing the classical Eq. (16.1) proposed by Roughton and Forster [3] are summarized as follows: (1) CO uptake through alveolocapillary tissue barrier, plasma layer, and Hb layer within the erythrocyte is predominantly prescribed by aqueous-phase diffusion and/or reaction with Hb but not by capillary perfusion; (2) no gradient of P_{CO} in the plasma; (3) no parallel heterogeneity in the distribution of D_{LCO} (to be exact, that of D_{LCO} to alveolar volume (V_A) defined as the Krogh factor [28]); and (4) no series (stratified) heterogeneity of CO in the alveolar gas phase. It is necessary to ascertain each of these assumptions one by one.

4.1 Contribution of Perfusion-Limited Gas Transfer to Diffusing Capacity

When an indicator gas has a nature limited by capillary perfusion besides aqueous-phase diffusion and reaction with Hb, the following equation should be used for estimating the total resistance for gas transfer from alveolar gas to Hb inside the erythrocyte (Fig. 16.3):

$$1/D_L = 1/D_M + 1/D'_B \tag{16.5}$$

$$D'_B = D_B + \beta \cdot Q \tag{16.6}$$

where D'_B is the effective blood component of D_L, which is given by the sum of D_B (true blood component of D_L yielded only by erythrocytes) and perfusion (Q) component of D_L, i.e., ($\beta \cdot Q$). ($\beta \cdot Q$) indicates the perfusion-related gas conductance, in which β is the capacitance coefficient of a gas in the capillary blood, corresponding to the effective solubility of a gas [29]. Eq. (16.6) implies that the transfer efficiency of perfusion-limited gas in the capillary is significantly augmented under conditions where effective gas solubility and/or perfusion are increased. As argued in the previous section, NO transfer from alveolar gas to Hb in erythrocytes is predominantly limited by aqueous-phase diffusion imposed by effective alveolocapillary membrane and erythrocyte interior. However, the contribution of reaction with Hb and perfusion is negligible in NO transfer. On the other hand, CO transfer from alveolar

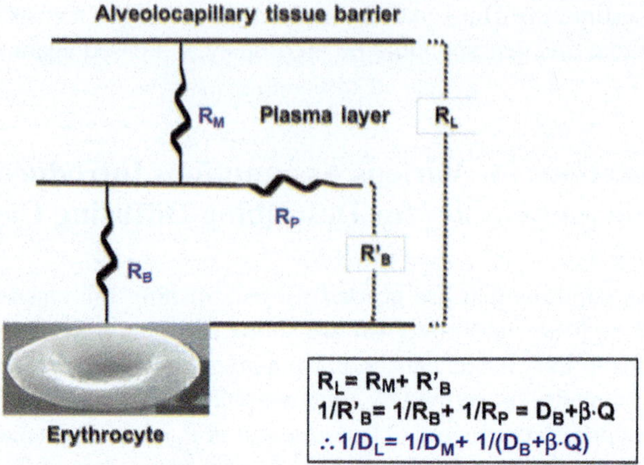

Fig. 16.3 Contribution of perfusion-limited gas transfer under steady-state condition. R_M: gas transfer resistance prescribed by the alveolocapillary tissue barrier and plasma layer, which equals $1/D_M$. R_B: gas transfer resistance prescribed by erythrocytes, which equals $1/D_B$. R_P: gas transfer resistance prescribed by capillary perfusion, which equals $1/(\beta \cdot Q)$, where β is capacitance coefficient of an indicator gas in blood and Q is capillary perfusion. Gas entering into capillary blood is divided into two paths, i.e., that transported by capillary perfusion (limited by R_P) and that captured by erythrocytes (limited by R_B). Thereby, R_B and R_P are taken to be connected in parallel, resulting in that gas transfer resistance, R'_B, prescribed by R_B and R_P is described as $1/R'_B = 1/R_B + 1/R_P$. Replacing $1/R'_B$ by D'_B, $1/R_B$ by D_B, and $1/R_P$ by $(\beta \cdot Q)$, $D'_B = D_B + (\beta \cdot Q)$. Since R_M and R'_B are connected in series, total gas transfer resistance, R_L, is described by $R_M + R'_B$. Hence, $1/D_L = 1/D_M + 1/(D_B + (\beta \cdot Q))$

gas to Hb in erythrocytes is limited by aqueous-phase diffusion through effective alveolocapillary membrane and erythrocyte interior as well as reaction with Hb, in which reaction with Hb plays a major role in limiting CO transfer. However, there is a certain possibility that CO transfer is also influenced by perfusion to some extent. This is inferred from the fact that HbCO concentration in the blood increases by 0.6 ~ 0.7% after a single-breath measurement of D_{LCO} is terminated [30], indicating that the P_{CO} in capillary blood (i.e., free CO molecules) corresponding to the equilibrium P_{CO} of HbCO dissociation curve also increases perceptibly during a single-breath measurement of D_{LCO}. The free CO molecules thus generated are processed by the perfusion, thus increasing the D_B outwardly. To estimate the extent of perfusion limitation in overall CO transfer from alveolar gas to Hb in erythrocytes, the value of $D_{LCO}/(\beta_{CO} \cdot Q)$ was calculated using the diffusion–perfusion equation proposed by Piiper and Scheid [29]. $1/D_{LCO}$ is the total resistance for CO transfer generated during the passage of CO from alveolar gas to Hb in erythrocytes, while $1/(\beta_{CO} \cdot Q)$ is the resistance generated by capillary perfusion (Fig. 16.3). The $(1/(\beta_{CO} \cdot Q))/(1/D_{LCO})$ value, i.e., the $D_{LCO}/(\beta_{CO} \cdot Q)$, is 0.01 ~ 0.02, resulting in that the perfusion resistance is only 1 ~ 2% of the total resistance for CO transfer. Hence, the author concludes that CO transfer is influenced by capillary perfusion but its contribution to overall CO transfer is small, resulting in that CO transfer from alveolar gas to Hb in erythrocytes can be reasonably described by Eq. (16.1). Regarding this issue, the persuasive experiment was performed by Boland et al. [9]. Using a membrane oxygenator circuit in which gas flow, perfusion, and blood volume were changed accordingly, they examined the contribution of perfusion to CO and NO transfer under a condition that mimics the single-breath maneuver. They found that the transfer of CO and NO from the gas phase to Hb in erythrocytes is independent of blood flow over a 25-fold variation from 0.1 to 2.5 L/min, which confirms that the transfer of CO and NO is practically not perfusion-limited in the single-breath maneuver. However, it should be noted that perfusion-related CO transfer plays a significant role when measuring the steady-state D_{LCO}, in which a small amount of CO (0.1% CO in air) is inhaled in minutes [25, 31].

4.2 *Contribution of Incomplete Mixing in Capillary Blood to Diffusing Capacity*

If the mixing in the capillary blood is incomplete, $1/D_L$ may be expressed by the equation as [3]:

$$1/D_L = (1+F)/D_M + 1/D_B \tag{16.7}$$

where F is a constant defining the extent of mixing in the capillary blood. When F = zero, mixing is complete, i.e., there is no partial pressure gradient of an indicator gas in the capillary plasma. However, when F has a finite value, it indicates the presence of incomplete plasma mixing-elicited partial pressure gradient of a gas, which results in the underestimation of D_M by a factor of $1/(1 + F)$. The mixing of capillary plasma is ascribed to convection generated by erythrocyte motions in the capillary. Even today, however, we do not know how complete (or incomplete) such mixing is. Therefore, it is very difficult to estimate the actual effect of incomplete plasma mixing (i.e., the determination of F) on D_{LCO} and D_{LNO} measures.

4.3 *Contribution of Parallel Heterogeneities to Diffusing Capacity*

To estimate the effect of parallel heterogeneities on gas transfer of CO and NO during the single-breath measurement of D_{LCO} and D_{LNO} in a simple way, the lung is assumed to be configured by two acini ($i = 1, 2$) with varied properties of gas transfer, including inspired/expired alveolar tidal volume (V_{AT}), alveolar volume (V_A), and diffusing capacity (D) for CO (D_{CO}) and for NO (D_{NO}). In the current study, the "whole lung" in this simple model is termed the "virtual lung" (Fig. 16.4a). As the acinus is regarded as the functional gas exchange unit, the distributions of V_{AT}, V_A,

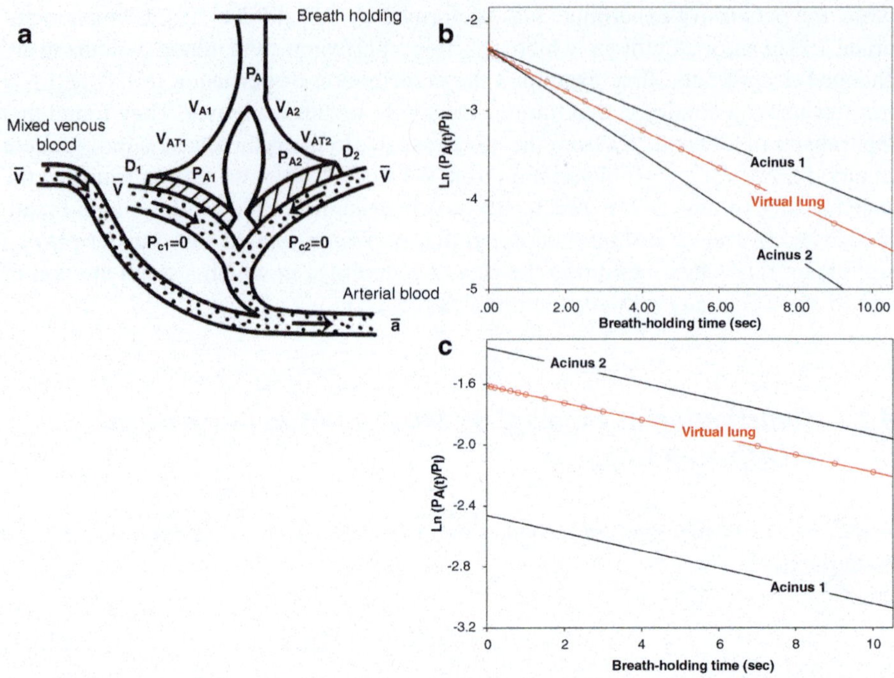

Fig. 16.4 Effects of parallel inhomogeneities on D_{LCO} in a lung model with two distinct acini. (**a**) Acinus is assumed to configure functional gas exchange unit with constant alveolar ventilation, capillary perfusion, and diffusing capacity (i.e., diffusive process through effective alveolocapillary membrane as well as erythrocyte interior and reactive process in erythrocytes). Two acini ($i = 1, 2$) with varied properties for gas transfer are considered in this virtual lung model. Inspired/expired alveolar tidal volume ($(V_{AT})_i$), alveolar volume ($(V_A)_i$), and diffusing capacity ($(D)_i$) for CO are taken to be different between two acini. Inspired CO gas pressure (P_I) is assumed to be identical in two acini. $(P_A)_i$ is gas pressure of CO in each acinus, while P_A is CO gas pressure of mixed expired gas from two acini. Perfusion heterogeneity is not considered and partial pressure of CO in each capillary network ($(P_c)_i$) is taken to be zero. Although results obtained only for CO transfer are depicted here, these are also applied for NO transfer because the qualitative effects of various gas transfer properties are nearly identical between CO and NO. (**b**) CO uptake with time in acini 1, 2, and virtual lung under condition with heterogeneous distribution of $(D/V_A)_i$ but homogeneous distributions of $(V_{AT}/V_A)_i$ and (V_{ATi}/V_{AT}). Conditions of calculation are as follows: $(D/V_A)_1 = 0.24$, $(D/V_A)_2 = 0.4$, $(V_{AT1}/V_A) = (V_{AT2}/V_A) = 0.16$, and $(V_{AT1}/V_{AT}) = (V_{AT2}/V_{AT}) = 0.5$. Although the initial point of logarithmic value of P_{ACO}/P_{ICO} just after the beginning of breath-holding is the same between respective acini and virtual lung, CO uptake rate differs between them. Furthermore, CO uptake process in the virtual lung is represented by a curved line rather than a straight line. Therefore, simulation of this curved line by a straight line underestimates the slope, resulting in that apparent D_L of virtual lung calculated under assumption with no functional heterogeneity of D_L/V_A is underestimated by 6.3% compared to true D_L expressed by $(D_1 + D_2)$. (**c**) CO uptake with time in acini 1 and 2, and virtual lung with heterogenous distributions of both $(V_{AT}/V_A)_i$ and (V_{ATi}/V_{AT}) but homogeneous distribution of $(D/V_A)_i$. Conditions of calculation are: $(D/V_A)_1 = (D/V_A)_2 = 0.08$, $(V_{AT1}/V_A) = 0.08$, $(V_{AT2}/V_A) = 0.24$, $(V_{AT1}/V_{AT}) = 0.25$, and $(V_{AT2}/V_{AT}) = 0.75$. Logarithmic values of P_{ACO}/P_{ICO} in both acini and virtual lung decay linearly with the same slope. Logarithmic values of P_{ACO}/P_{ICO} just after the beginning of breath-holding are different between both acini and virtual lung. Apparent D_L of virtual lung estimated under assumption with no functional heterogeneity is identical to true D_L given by $(D_1 + D_2)$

and D within the acinus are homogeneous, whereas they differ between the acini. Perfusion is homogeneously distributed in the acinus but differently distributed between the acini. Back pressure of CO and NO in any pulmonary microcirculation is supposed to be zero. The tissue components forming D include the surface area and thickness of the alveolocapillary tissue barrier (D_M) and that of capillary blood volume (V_C). As discussed in detail in Sect. 5.3, the relationship between D_{NO} and D_{CO} (D_{NO}/D_{CO}) in the acinus is expressed by a function of anatomically determined D_M/V_C there. Thus, D_{NO} in an intended acinus is described by k·(D_{CO}), where k is a constant, indicating that in a certain acinus, the magnitude of NO-related gas transfer processed by D_{NO} is k-fold that of CO-related gas transfer by D_{CO}. However, there is no qualitative difference between D_{NO}- and D_{CO}-related gas transfer. Thereby, the concentration profile of CO and that of NO in alveolar gas of each acinus during breath-holding are represented correctively in a single equation as:

$$\left[P_{A(t)}/P_I\right]_i = \left(V_{AT}/V_A\right)_i \bullet \text{EXP}\left[-\left(D/V_A\right)_i \bullet \left(P_B - P_{H2O}\right) \bullet t\right] \quad (16.8)$$

in which P_I denotes inspired P_{CO} or P_{NO} at the beginning of breath-holding, while $[P_{A(t)}]_i$ (i = 1, 2) is alveolar P_{CO} or P_{NO} at the time t during breath-holding in respective acini. P_I is assumed to be constant over the lung. P_B and P_{H2O} denote atmospheric and saturated water vapor pressure, respectively. $(V_{AT})_i$ and $(V_A)_i$ are alveolar tidal volume and alveolar gas volume distributed to each acinus. $(D)_i$ is D_{CO} or D_{NO} in an intended acinus. Thereby, the time-dependent decay of P_{ACO} or P_{ANO} in the virtual lung is described by the following equation:

$$\left[P_{A(t)}/P_I\right]_m = \left(V_{AT1}/V_{AT}\right)\cdot\left[P_{A(t)}/P_I\right]_1 + \left(V_{AT2}/V_{AT}\right)\cdot\left[P_{A(t)}/P_I\right]_2 \quad (16.9)$$

$[P_{A(t)}/P_I]_m$ is mean $P_{A(t)}/P_I$ of the virtual lung consisting of the two acini. $V_{AT1} + V_{AT2} = V_{AT}$. $V_{A1} + V_{A2} = V_A$. Eqs. (16.8) and (16.9) suggest the possibility that overall decay of CO or NO with time in the virtual lung is influenced by the three parallel heterogeneities, including the distribution of diffusing capacity to alveolar volume $(D/V_A)_i$, that of alveolar tidal volume to alveolar volume $(V_{AT}/V_A)_i$, and that of alveolar tidal volume in each acinus to total alveolar tidal volume (V_{ATi}/V_{AT}). Of these three parallel heterogeneities, the regional difference in D/V_A is particularly important. The D/V_A is the rate constant of gas uptake, i.e., the Krogh factor [28], in each acinus. The $Ln[P_{A(t)}/P_I]$ against breath-holding time (t) shows the straight line but the different slope in respective acini. Thereby, the $Ln[P_{A(t)}/P_I]_m$ against t of the virtual lung reveals the curved line especially at an initial part of breath-holding due to the different D/V_A in the two acini (Fig. 16.4b). Simulation of this curved line by the straight line as if the lung is homogeneous concerning D/V_A underestimates the apparent D_L of the virtual lung in comparison with the true D_L represented by ($D_1 + D_2$). This indicates that the parallel heterogeneity of D/V_A renders a negative impact on D_{LCO} and D_{LNO}.

The heterogeneous distribution of alveolar tidal volume to alveolar volume $(V_{AT}/V_A)_i$ and that of regional alveolar tidal volume to total tidal volume (V_{ATi}/V_{AT}) are mainly caused by the heterogeneous distribution of convective flow in the lung. Although these convention-associated heterogeneities are substantially reduced by the breath-holding maneuver, part of them are thought to remain. The $(V_{AT}/V_A)_i$ heterogeneity acts as the factor deciding the distribution of inspired gas to each acinus at the beginning of breath-holding with no effect on the gas uptake rate, resulting in that the apparent D_L estimated under assumption with no functional heterogeneity in the virtual lung is not distorted from the true D_L. Although the regional difference in alveolar tidal volume to total tidal volume (V_{ATi}/V_{AT}) fixes the weight of expired gas from each acinus on constituting the mean P_A in alveolar gas sample collected just after the breath-holding is discontinued, this heterogeneity has no effect on the gas uptake rate, as well, thus yielding no difference between the apparent and true D_L. This holds true even if the combined heterogeneities of $(V_{AT}/V_A)_i$ and (V_{ATi}/V_{AT}) are present (Fig. 16.4c). Therefore, the author concludes that only the parallel heterogeneity of $(D/V_A)_i$ acts as the factor modifying the gas uptake rate and underestimates the D_L.

It should be noted that the aforementioned story hypothesizes that the heterogeneity of inspired gas pressure (P_I) is instantly removed just after the beginning of breath-holding. However, this is physiologically not correct and the different P_I in different acini should be taken as the fourth factor eliciting the parallel heterogeneity. The regional heterogeneity of P_I is yielded by the difference in coupling of convection and diffusion in different acini. Supporting this consideration, Magnussen et al. [32] found increasing D_{LCO} (i.e., overestimation of D_{LCO}) with decreasing breath-holding time in patients with bronchial asthma. Their observation cannot be explained from any heterogeneous distribution of $(D/V_A)_i$, $(V_{AT}/V_A)_i$, or (V_{ATi}/V_{AT}) but is interpreted as the evidence suggesting that the heterogeneity of P_I over the lung does not fully fade away when breath-holding time is short.

Of the four parallel heterogeneities argued above, the heterogeneous distribution of alveolar tidal volume to alveolar volume and that of alveolar tidal volume to total alveolar tidal volume exert exactly the same impacts on acinar gas-phase transport of CO and NO. This is because these heterogeneities are imposed by an uneven distribution of convective flow that carries any gas as a lump. Hence, their effects do not differ between CO and NO. The regional difference in P_I is subtle for CO and NO because the difference in the effect of coupling of convection and diffusion in any acinus is small between CO and NO due to a small difference in gas-phase diffusivity of the two gases. The effect of D/V_A heterogeneity on CO and NO gas transfer, which is related to the anatomical alterations of various components forming alveolar septa, can also be conceived to be qualitatively the same between CO and NO. Thereby, the four parallel heterogeneities are expected to exert nearly the same impacts on CO and NO transfer in the lung periphery. These facts send an important message that the effect of all parallel heterogeneities on the pulmonary diffusing capacity may approximately be removed if the D_{LNO}/D_{LCO} is used as the parameter.

4.4 Contribution of Stratified Heterogeneities to Diffusing Capacity

As discussed in detail in Chap. 6 in this book, the gas-phase diffusion-dependent stratification generated during inspiration plays little role in prescribing overall gas mixing within human acinus under the steady-state condition. However, the inspiration-related, diffusion-elicited stratified heterogeneity has the effect that is not ignored under artificial breathing such as breath-holding or rebreathing. In COPD patients, D_{LCO} decreased with shortening of breath-holding time, sometimes reaching a negative value when breath-holding time is very short [33]. This phenomenon is the opposite of the finding investigated by Magnussen et al. [32], who showed increasing D_{LCO} with shorter breath-holding time in patients with bronchial asthma, which was explained by the parallel heterogeneity of inspired gas pressure (P_I) being not eliminated during a short breath-holding time. On the other hand, decreasing D_{LCO} with shortening of breath-holding time in patients with COPD is not explicable from any parallel inhomogeneity. Alternatively, it should be explained from the incomplete gas mixing between inspired gas and resident gas, which is caused by the augmented stratified heterogeneity in patients with COPD having the destruction of acinar airway and airspace. Recently, the important role of stratified heterogeneity on rebreathing D_{LNO} was confirmed by Hsia et al. [34] in an immature canine model with elongated airways. They found that the D_{LNO} measured in the background gas with He and O_2 was higher, while that measured in the background gas with SF_6 and O_2 was lower, than that in the background gas with N_2 and O_2. More recently, Linnarsson et al. [35] examined the effect of atmospheric pressure on single-breath D_{LNO}. They found that the D_{LNO} measured at 4.0 atm was significantly lower than that at 1.0 atm, indicating that diffusive transport of NO in the lung periphery is inversely related to the gas density. Their experimental results revealed that the quantitative effect of gas-phase diffusion-related stratification on the single-breath D_{LNO} is ~5% in normal subjects. Based on these findings, the author concludes that the diffusion-dependent gas-phase stratification exerts a detrimental impact on acinar gas mixing under artificial breathing of rebreathing or breath-holding, though its effect is small.

Although the effect of gas-phase stratified heterogeneity is slightly different between CO and NO because of their gas-phase diffusivity being 1:0.97 (calculated from the Fick/Graham's law that can be applied for analyzing gas behavior in gas-phase binary diffusion system), the author considers that such small variation in gas-phase diffusivity may not induce a significant difference in the effect of stratified heterogeneity on CO and NO transfer. Therefore, when the D_{LNO}/D_{LCO} ratio is introduced, it may almost exclude the effect of series stratified heterogeneity on the pulmonary diffusing capacity.

As an interim summary, the author wants to mention that we should always be aware of the possibility that D_{LCO} and D_{LNO} measured for patients having a variety of lung or cardiac diseases are underestimated by parallel heterogeneity of D/V_A, overestimated by parallel heterogeneity of inspired gas pressure, and underestimated by series stratified heterogeneity. However, the contribution of these functional heterogeneities is thought to be nearly identical to D_{LCO} and D_{LNO} as far as they are measured simultaneously. Therefore, the parameter of D_{LNO}/D_{LCO} may cancel the effect of various functional heterogeneities, indicating that the D_{LNO}/D_{LCO} allows us to straightforwardly identify the anatomical abnormalities of alveolar septa without correcting the functional heterogeneities irrespective of the types of disease.

5 Precautions on D_{LCO} and D_{LNO} Measured in Clinical Practice

5.1 *Effects of Different $\Upsilon_{NO/CO}$ and θ_{NO} on Partitioned D_{MCO} and V_C*

The issue of what is the most appropriate value for the relative Krogh diffusion constant of NO against CO ($\Upsilon_{NO/CO}$) has not been conclusively resolved. Furthermore, the proposition concerning the most appropriate value of θ_{NO}, i.e., finite or infinite, is also uncertain at present. These two values are essential while estimating the membrane components of diffusing capacity (D_{MCO}) and the capillary blood volume (V_C) from D_{LCO} and D_{LNO} measures. Namely, D_{MCO} and V_C estimated from D_{LCO} and D_{LNO} are expected to considerably vary depending on $\Upsilon_{NO/CO}$ and/or θ_{NO}. Therefore, using D_{LCO} and D_{LNO} values measured for a large number of normal subjects [6], the author attempted to estimate the D_{MCO} and V_C based on the different values of $\Upsilon_{NO/CO}$ and θ_{NO} as shown in Table 16.1. The author found that D_{MCO} (and D_{MNO}) decreases, whereas V_C increases, by a factor of two when $\Upsilon_{NO/CO}$ increases from 1.97 to 3.35 (Table 16.2). Qualitatively a similar trend was investigated when θ_{NO} is changed from 4.5 to infinity, i.e., 31 ~ 47% decrease in D_{MCO} while 32 ~ 72% increase in V_C. Furthermore, the relative resistance imposed by the effective alveolocapillary membrane, i.e., $(1/D_{MCO})/(1/D_{LCO}) = D_{LCO}/D_{MCO}$, increases from 22% to 48% when $\Upsilon_{NO/CO}$ increases from 1.97 to 3.35 under constant θ_{NO} at 4.5 (Fig. 16.5). Similarly, D_{LCO}/D_{MCO} increases from 22% to 41% when θ_{NO} increases from 4.5 to infinity at constant $\Upsilon_{NO/CO}$ of 1.97. When $\Upsilon_{NO/CO}$ and θ_{NO} increase simultaneously from 1.97 to 3.35 and from 4.5 to infinity, respectively, D_{LCO}/D_{MCO} increases from 22% to 70%. These results suggest that CO transfer through alveolar septa is predominantly limited by the blood component ($1/D_B$) when the assumption taken by the 2017-ERS Standards is correct (i.e., $\Upsilon_{NO/CO}$: 1.97, θ_{NO}: 4.5). However, when the assumption that $\Upsilon_{NO/CO}$ is 3.35 and θ_{NO} is infinite is correct, the main limiting step for CO transfer is ascribed to the membrane component ($1/D_M$). If the latter is accepted, it indicates that in contrast to the typical story about D_{LCO}, D_{LCO} is insensitive to injury of pulmonary microcirculation but is more sensitive to injury of alveolocapillary tissue

Table 16.2 D_M and V_C estimated from various $\Upsilon_{NO/CO}$ and θ_{NO}

	D_{MCO}	D_{BCO}	D_{MNO}	D_{BNO}	V_C
(1) $\Upsilon_{NO/CO}$: 1.97 θ_{CO}: 0.56 θ_{NO}: 4.5	135	38	266	305	68
(2) $\Upsilon_{NO/CO}$: 1.97 θ_{CO}: 0.56 θ_{NO}: infinite	72	50	142	Infinite	90
(3) $\Upsilon_{NO/CO}$: 3.35 θ_{CO}: 0.56 θ_{NO}: 4.5	61	57	205	460	102
(4) $\Upsilon_{NO/CO}$: 3.35 θ_{CO}: 0.56 θ_{NO}: infinite	42	98	142	Infinite	175

D_{LCO}: 29.6 mL/min/mmHg, D_{LNO}: 142 mL/min/mmHg. These values were adopted from 2017-ERS Standards [6]. θ_{CO}: calculated at constant capillary PO_2 of 100 mmHg using equation proposed by Guénard et al. [12]. D_{MCO} and D_{MNO}: mL/min/mmHg, D_{BCO} and D_{BNO}: mL/min/mmHg, V_C: mL

barrier, resulting in that D_{LCO} has qualitatively the same trait as D_{LNO} regarding the point of what process, i.e., $1/D_M$ or $1/D_B$, is predominantly reflected. These facts certainly indicate that it is premature to accept that D_{MCO}, V_C, and transfer resistance estimated with the method of the 2017-ERS Standards are reasonable and valid. Hence, until the assumption about $\Upsilon_{NO/CO}$ and θ_{NO} is confirmed, we must judge the pathophysiological abnormalities of alveolar septa in patients with various types of lung and cardiac disease simply from the five raw data, including D_{LCO}, D_{LCO}/V_A, D_{LNO}, D_{LNO}/V_A, and D_{LNO}/D_{LCO}.

5.2 *Normal Range and Covariates Prescribing D_{LNO}/D_{LCO}*

Several groups of investigators [36–40] attempted to decide the reference equations for predicting the mean (M), upper limit-of-normal (ULN), and lower limit-of-normal (LLN) of simultaneously measured D_L-related parameters by taking sex, age, and height (representative of lung volume) as the explanatory variables. The author would like to briefly abstract some important messages demonstrated in the literature: (1) age exerts a negative impact on D_{LCO}, D_{LCO}/V_A, D_{LNO}, and D_{LNO}/V_A, and (2) height or lung volume has a positive impact on D_{LCO} and D_{LNO} while it has a negative impact on D_{LCO}/V_A and D_{LNO}/V_A. Zavorsky et al. [41] demonstrated that age-, sex-, and height-corrected D_L-related parameters measured under a resting condition did not differ between morbidly obese and nonobese subjects, indicating that body weight may not act as an anthropometric variable while establishing the

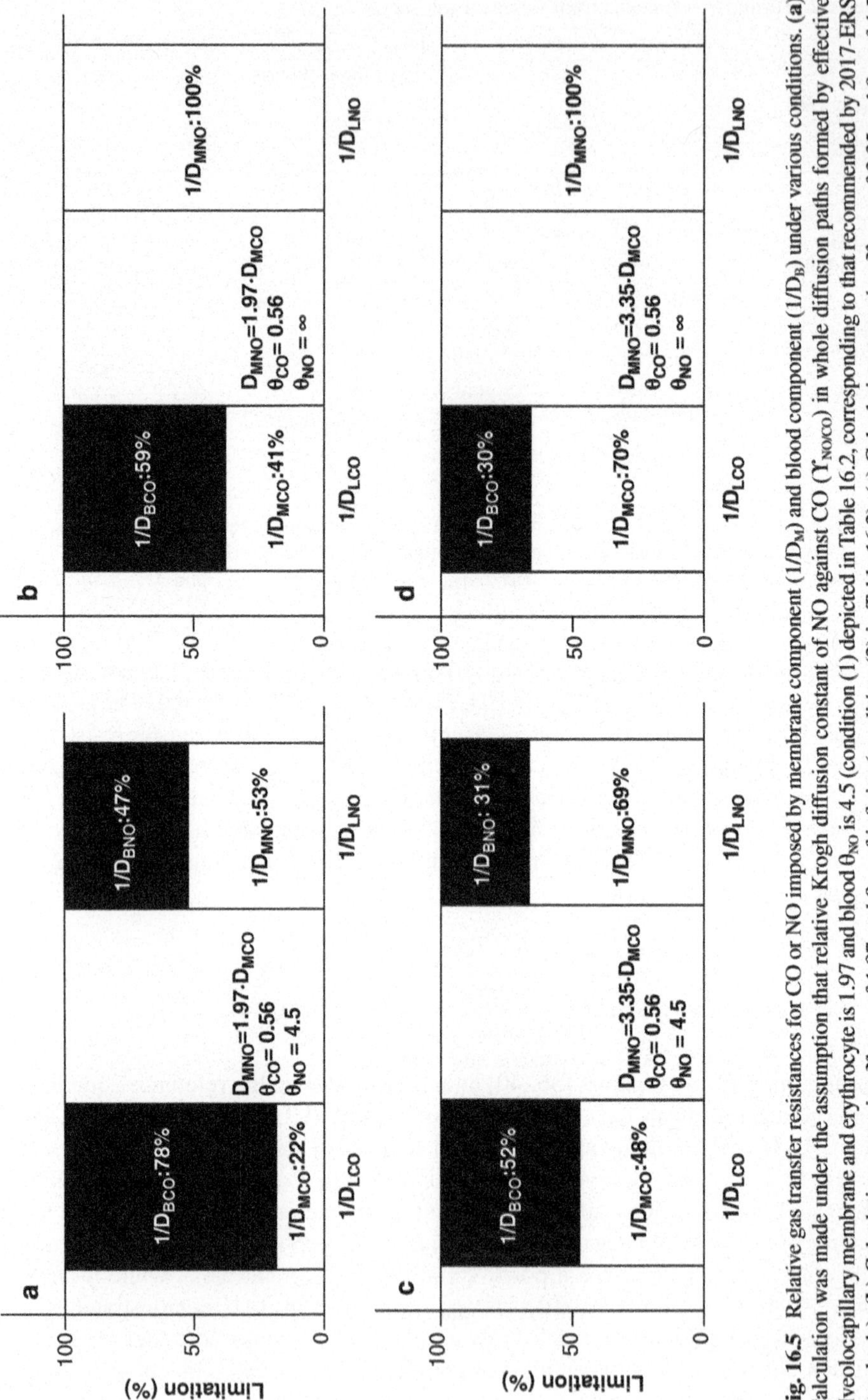

Fig. 16.5 Relative gas transfer resistances for CO or NO imposed by membrane component ($1/D_M$) and blood component ($1/D_B$) under various conditions. **(a)** Calculation was made under the assumption that relative Krogh diffusion constant of NO against CO ($\Upsilon_{NO/CO}$) in whole diffusion paths formed by effective alveolocapillary membrane and erythrocyte is 1.97 and blood θ_{NO} is 4.5 (condition (1) depicted in Table 16.2, corresponding to that recommended by 2017-ERS Standards). **(b)** Calculation assuming $\Upsilon_{NO/CO}$ of 1.97 and θ_{NO} of infinity (condition (2) in Table 16.2). **(c)** Calculation assuming $\Upsilon_{NO/CO}$ of 3.35 and θ_{NO} of 4.5 (condition (3) in Table 16.2). **(d)** Calculation assuming $\Upsilon_{NO/CO}$ of 3.35 and θ_{NO} of infinity (condition (4) in Table 16.2). All calculations were made under the assumption that θ_{CO} is constant at 0.56. Depending on $\Upsilon_{NO/CO}$ and θ_{NO}, relative resistances by $1/D_M$ and $1/D_B$ for CO and NO are considerably varied

reference equations for D_L-related parameters. Analyzing the single-breath D_{LCO} and D_{LNO} values measured for young sub-Saharan Africans (n = 32) and young European Caucasians (n = 32), Simaga et al. [40] demonstrated that D_{LCO} and D_{LNO} are lower in Africans than those in Europeans with no difference in D_{LCO}/V_A and D_{LNO}/V_A between them, which was explained by a lower lung volume with respect to height in Africans. Interestingly, they found higher D_{LNO}/D_{LCO} in Africans, which was explained by a more restricted pulmonary vascular bed in Africans. The findings reported by Simaga et al. [40] indicate that ethnicity should be introduced as one of the explanatory variables, as well.

The reference value of D_{LNO}/D_{LCO} and the effect of covariates modifying it are not consistent in the literature reported so far. Van der Lee et al. [36] demonstrated increasing D_{LNO}/D_{LCO} with advancing age or with increasing lung volume in normal Dutch subjects (n = 124 aged 25 ~ 55 years), whereas Aguilaniu et al. [37] reported that D_{LNO}/D_{LCO} in normal French subjects is independent of any covariate, including sex, age, or height (n = 303 aged 18 ~ 94 years). No correlation of D_{LNO}/D_{LCO} with age was also suggested in North American normal subjects (n = 130 aged 18 ~ 85 years) by Zavorsky et al. [38]. On the other hand, Munkholm et al. [39] found a small but significant negative correlation between D_{LNO}/D_{LCO} and age with a slope of −0.0025/year (p = 0.0003) in normal Caucasians recruited from the Copenhagen General Population Study (n = 282 aged 18 ~ 97 years). In addition, they found a negative correlation between D_{LNO}/D_{LCO} and height with a slope of −0.0028/cm, though this slope was not statistically significant (p = 0.058). The sex-dependent modification of D_{LNO}/D_{LCO} was not identified in their study. As such, it cannot be said that the relationship between D_{LNO}/D_{LCO} and covariates, including sex, age, and height (or lung volume), has been conclusively confirmed. Thereby, the author attempted to analyze the effect of sex, age, or height on D_{LNO}/D_{LCO} using the data reported by the 2017-ERS Standards (n = 490 aged 18 ~ 93 years) (Fig. 16.6). In normal males, the relationship between D_{LNO}/D_{LCO} and age is represented by the curved line but its negative slope is very small at any age, indicating that in a clinical practice, D_{LNO}/D_{LCO} can be regarded as an age-independent parameter in males over a wide range of age. However, in normal females, D_{LNO}/D_{LCO} is considerably lowered in the age over 60 years, indicating that differing from males, age exerts a relatively large negative impact on D_{LNO}/D_{LCO} in elderly females. On the other hand, D_{LNO}/D_{LCO} in both males and females is conspicuously lowered with increasing height, which is also represented by the curved line. The relationship of D_{LNO}/D_{LCO} with lung volume is qualitatively similar to that with height because height is the decisive factor for lung volume. Our analytical results are qualitatively, but not quantitatively, consistent with those reported by Munkholm et al. [39] in respect of a negative correlation between D_{LNO}/D_{LCO} with age or height. However, the reference equation considered in the present study is unsatisfactory because it was contrived by dividing the multiple regression equation of D_{LNO} by that of D_{LCO} but not directly constructed from the raw data of D_{LNO} and D_{LCO} measures. Therefore, the author thinks that it is urgently required to establish the reference equation that decides M, ULN, and LLN of D_{LNO}/D_{LCO} by introducing sex, age, height, and, if possible, ethnicity as covariates.

5.3 Relationship Between D_{LNO}/D_{LCO} and D_{MCO}/V_C

Transforming Eq. (16.1) for CO and that for NO, D_{LNO}/D_{LCO} can be summarized in the following equation:

$$D_{LNO}/D_{LCO} = \gamma_{NO/CO} \cdot \left[1 + (D_{MCO}/V_C)/\theta_{CO}\right] / \left[1 + \gamma_{NO/CO} \cdot (D_{MCO}/V_C)/\theta_{NO}\right] \quad (16.10)$$

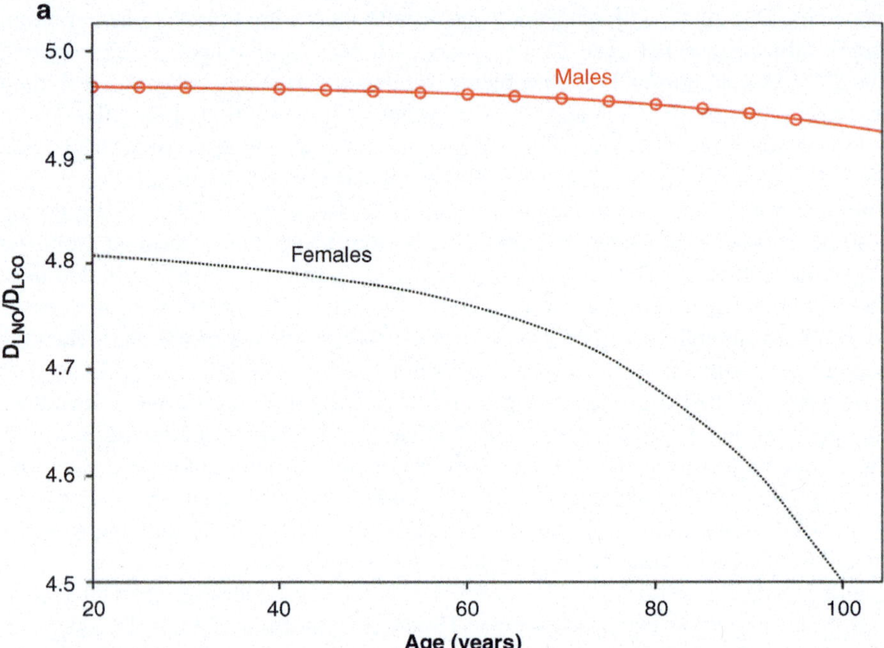

Fig. 16.6 Relationships between D_{LNO}/D_{LCO} and various covariates, including sex, age, and height. D_{LNO}/D_{LCO} values under various conditions were calculated using reference equations of D_{LCO} and D_{LNO} decided for normal males ($n = 248$) and females ($n = 242$), which were summarized in 2017-ERS Standards [6]. (**a**) Age-dependency of D_{LNO}/D_{LCO} in males (red line) and females (black dotted line) in age ranging between 20 and 95 years but at a constant height of 170 cm. Regression analysis revealed that slope with age is −0.0004/year for males over a wide range of ages. Equation covering the whole age range in males was given by $D_{LNO}/D_{LCO} = 4.97–3.94 \cdot 10^{-4} \cdot (\text{age}) + 8.07 \cdot 10^{-6} \cdot (\text{age})^2–8.37 \cdot 10^{-8} \cdot (\text{age})^3$. On the other hand, regression analysis was separated at the age of 60 years in females. Slope with age below 60 years was −0.0013/years, while that over 60 years was −0.007/year. Equation covering the whole age range in females was described as $D_{LNO}/D_{LCO} = 4.88–0.0048 \cdot (\text{age}) + 1.06 \cdot 10^{-4} \cdot (\text{age})^2–9.53 \cdot 10^{-7} \cdot (\text{age})^3$. (**b**) Height-dependency of D_{LNO}/D_{LCO} in males (red line) and females (black dotted line) in height ranging between 130 and 200 cm but at constant age of 50 years. Regression analysis revealed that the average slope with height is −0.021/cm in males and − 0.014/cm in females. Equation covering the whole range of height is given by $D_{LNO}/D_{LCO} = 9.38–0.041 \cdot (\text{height}) + 8.84 \cdot 10^{-5} \cdot (\text{height})^2$ in males and $D_{LNO}/D_{LCO} = 10.45–0.055 \cdot (\text{height}) + 1.25 \cdot 10^{-4} \cdot (\text{height})^2$ in females

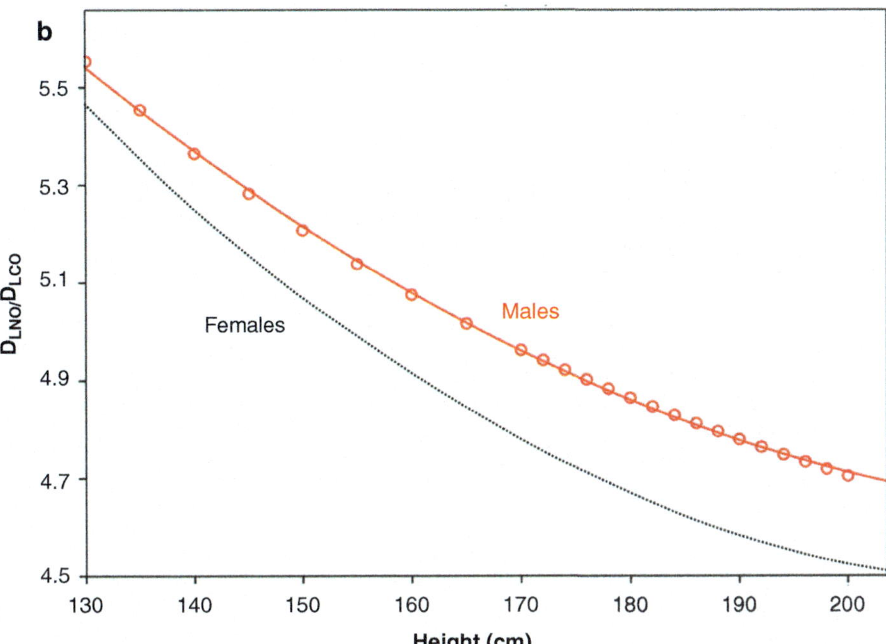

Fig. 16.6 (continued)

Equation (16.10) is the general equation that describes the relationship between D_{LNO}/D_{LCO} and D_{MCO}/V_C under any value of θ_{NO}. Equation (16.10) is not linear but is approximated by a straight line while D_{MCO}/V_C is less than 2.0 (Fig. 16.7), which corresponds to the maximum value of D_{MCO}/V_C calculated from a variety of possible values of $\Upsilon_{NO/CO}$ and θ_{NO} (Table 16.2). The slope of the line, which determines the sensitivity of how preciously the change in D_{MCO}/V_C is reflected by that in D_{LNO}/D_{LCO}, varies depending mainly on θ_{NO} but marginally on $\Upsilon_{NO/CO}$. As depicted in Fig. 16.7a, The extent of change in D_{LNO}/D_{LCO} against that in D_{MCO}/V_C is larger as θ_{NO} is larger. When θ_{NO} is infinite, the relationship between D_{LNO}/D_{LCO} and D_{MCO}/V_C is described by the simple linear equation as:

$$D_{LNO} / D_{LCO} = \gamma_{NO/CO} \bullet \left[1 + \left(D_{MCO} / V_C\right) / \theta_{CO}\right] \tag{16.11}$$

The same equation as Eq. (16.11) was derived by Hughes and van der Lee [42]. Equation (16.10) certifies that D_{LNO}/D_{LCO} is directly correlated with D_{MCO}/V_C. Decrease in D_{MCO}, including heterogeneous distribution of D_{MCO}, lowers D_{MCO}/V_C, thus leading to a decrease in D_{LNO}/D_{LCO}. On the other hand, a decrease in V_C, including heterogeneous distribution of V_C, increases D_{MCO}/V_C, resulting in an increase in D_{LNO}/D_{LCO}. As such, D_{LNO}/D_{LCO} acts as a good, clinical indicator that

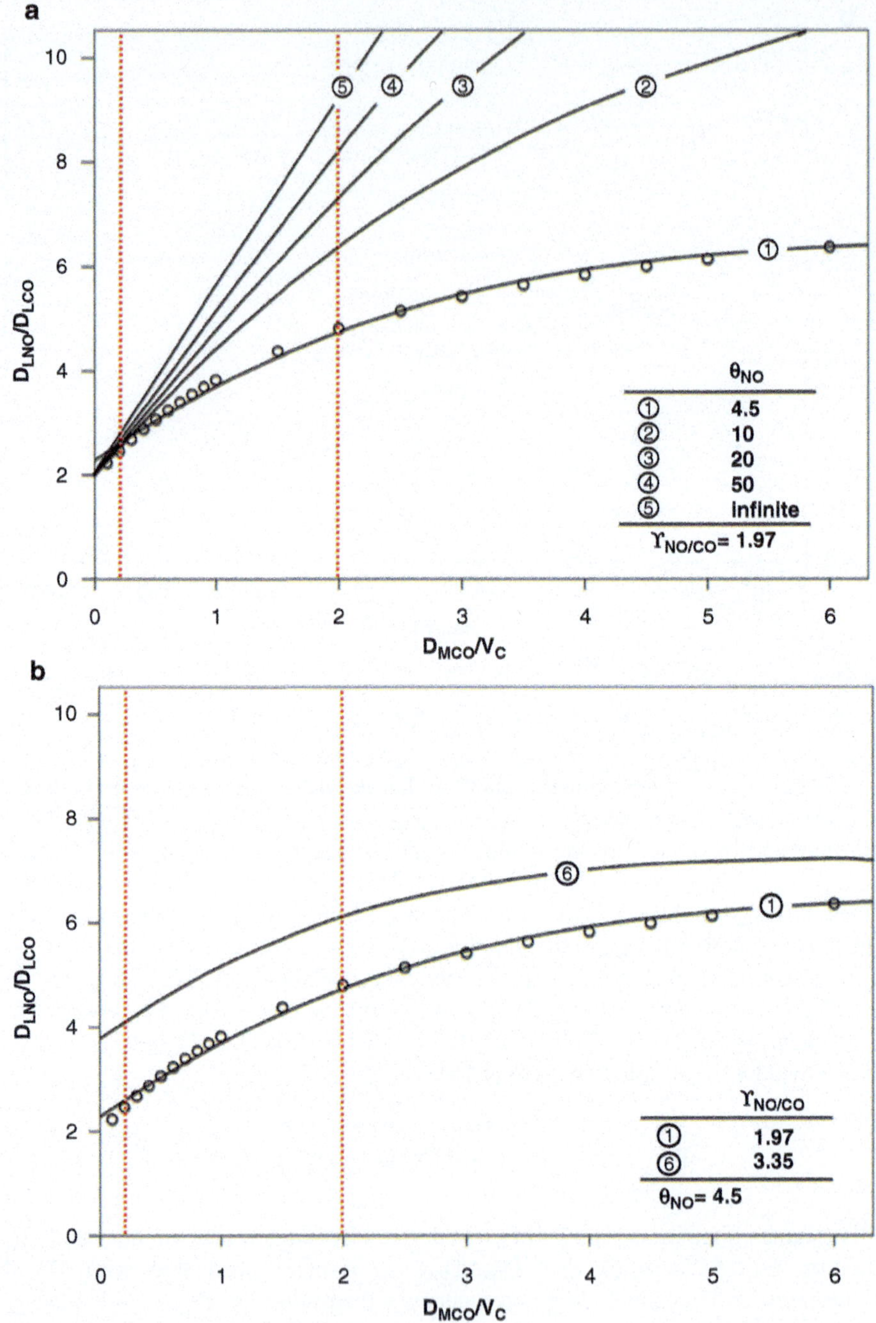

Fig. 16.7 Relationship between D_{MCO}/V_C and D_{LNO}/D_{LCO}. (**a**) Calculations were made using general equation (Eq. 16.10) under conditions in which θ_{NO} is changed from 4.5 to infinity but $\Upsilon_{NO/CO}$ is fixed at 1.97. (**b**) Calculations were made using Eq. 16.10 under conditions with fixed θ_{NO} at 4.5 but $\Upsilon_{NO/CO}$ changing from 1.97 to 3.35. Red dotted lines denote possible ranges of D_{MCO}/V_C in in vivo conditions, i.e., a minimum of 0.2 and a maximum of 2.0 (see Table 16.2)

differentiates lung diseases with decreasing D_{MCO} (thus, decreasing D_{MCO}/V_C and D_{LNO}/D_{LCO}), including interstitial lung diseases and destruction of alveolar architectures (COPD and other cystic diseases) from those with decreasing V_C (thus, increasing D_{MCO}/V_C and D_{LNO}/D_{LCO}), including the destruction of pulmonary microcirculation (idiopathic pulmonary arterial hypertension), microvascular obliteration (e.g., thrombotic microangiopathy (TMA), pulmonary veno-occlusive disease (PVOD), and pulmonary capillary hemangiomatosis (PCH)), and microvascular dilatation (hepatopulmonary syndrome (HPS)). Of note, microvascular angiitis such as ANCA-related microvascular angiitis (e.g., microscopic polyangiitis (MPA), granulomatosis with polyangiitis (GPA), and eosinophilic granulomatosis with polyangiitis (EGPA)) is frequently complicated with alveolar hemorrhage that liberates free Hb into alveolar space, which absorbs CO and NO there, thus increasing D_M (D_{MCO} and D_{MNO}) apparently. This peculiar phenomenon leads to a synergetic action of microvascular injury-associated decrease in V_C and hemorrhage-elicited increase in D_{MCO}, which augments D_{MCO}/V_C and D_{LNO}/D_{LCO} largely. Although in our previous paper [43], the author stated that D_{LNO}/D_{LCO} may be difficult to be applied in a clinical practice due to the uncertainty regarding its physiological implication. Now, the author has changed the mind and agrees to the opinion that D_{LNO}/D_{LCO} plays an important role in predicting the anatomical relationship between D_{MCO} and V_C.

6 Learning from D_{LCO} and D_{LNO} Measures in Various Types of Lung and Cardiac Disease

As extensively argued above, the validity of D_{MCO} (D_{MNO}) or V_C partitioned with the method of the 2017-ERS Standards is not decisive at present. Therefore, the author would like to use five raw measures, including D_{LCO}, D_{LCO}/V_A, D_{LNO}, D_{LNO}/V_A, and D_{LNO}/D_{LCO}, but not D_{MCO} (D_{MNO}) and V_C, for interpreting the pathophysiological abnormities in patients with various types of disease. However, when the change in D_{MCO} (D_{MNO}) or V_C was estimated at various conditions based on the simultaneous measurement of D_{LCO} and D_{LNO}, it was used as the meaningful, pathophysiological parameter. This is because the change in D_{MCO} (D_{MNO}) or V_C may qualitatively (but not quantitatively) reflect the pathophysiological truth independent of the assumption concerning $\Upsilon_{NO/CO}$ and θ_{NO}. Otherwise specified, D_L-related metrics featured in this section were simultaneously measured by means of the single-breath maneuver.

6.1 Smokers and Chronic Obstructive Lung Disease (COPD)

In heavy smokers ($n = 236$) and subjects with COPD ($n = 36$), van der Lee et al. [44] demonstrated that although D_{LCO}, D_{LCO}/V_A, D_{LNO}, and D_{LNO}/V_A decreased in these subjects, D_{LNO}/V_A had the highest sensitivity for detecting CT-based emphysema. D_{LCO}/V_A and D_{LNO}/V_A were better than D_{LCO} and D_{LNO} for detecting emphysema.

Furthermore, the sensitivity of D_{LCO}/V_A and D_{LNO}/V_A for detecting emphysema was higher than the spirometric parameter of FEV_1/FVC. In this study, D_{MCO} and V_C were not estimated.

Moinard and Guénard [45] considered that the specific feature of early COPD is the reduction in V_C, which reflects the patchy destruction of acinar microcirculation, while that of advanced COPD is the simultaneous reduction in D_{MCO} and V_C, which reflects the diffuse destruction of acinar architectures, including acinar airways, alveolocapillary tissue barrier, and acinar microcirculation. Although their consideration is potentially interesting from a clinical point of view, we should wait to accept their consideration until the reliable method for determining D_{MCO} and V_C is established.

6.2 Cystic Fibrosis (CF)

Measuring various D_L-associated parameters in patients with CF ($n = 21$), Dressel et al. [46] found that D_{LCO}, D_{LCO}/V_A, D_{LNO}, and D_{LNO}/V_A decreased significantly in these patients. In addition, they identified that the severity of CF judged from CT scores, including bronchiectasis, mucus plugging, peribronchial thickening, and parenchymal diseases, was more sensitively detected by D_{LNO} than D_{LCO}. Furthermore, D_{LCO}/V_A was not correlated with the CT scores of CF. In this study, D_{MCO} and V_C were estimated under the assumption that $\Upsilon_{NO/CO}$ and θ_{NO} equal 1.97 and infinity, respectively. The V_C value thus obtained seemed to be not different from the expected normal V_C. However, the judgment on the validity of D_{MCO} and V_C reported in this paper is not possible at present.

6.3 Pulmonary Sarcoidosis (SAR)

Phansalkar et al. [47] estimated various D_L-associated parameters in patients with pulmonary SAR ($n = 25$), in which D_{MCO} and V_C were calculated under the assumption that $\Upsilon_{NO/CO}$ is 2.42 and θ_{NO} is infinite. They identified that although the D_{MCO} and V_C obtained for patients with SAR without overt fibrosis under the rebreathing maneuver were much lower than those measured for normal controls, the reduction in V_C in these patients was significantly larger than that in D_{MCO}. Their findings were interpreted as evidence for the microvascular damage playing a primary role in the pathology of SAR. Phansalkar et al. presumed that in SAR with little fibrosis, patchy alveolar deposition of noncaseating granulomas selectively obliterates pulmonary capillaries, which is accompanied by variable inflammation that may damage alveolocapillary tissue barrier. Unfortunately, however, their findings have not been ascertained yet from a morphological point of view. Furthermore, the assumption of $\Upsilon_{NO/CO}$ and θ_{NO} employed in their study has not been certified.

6.4 Usual Interstitial Pneumonia-Idiopathic Pulmonary Fibrosis (UIP-IPF) and Nonspecific Interstitial Pneumonia (NSIP)

Barisione et al. [48] investigated various D_L-related parameters in patients with UIP-IPF ($n = 30$) and those with NSIP ($n = 30$). D_{MCO} and V_C values were calculated under the presupposition that $\Upsilon_{NO/CO}$ or θ_{NO} is 1.97 or 4.5, respectively. Additionally, they calculated D_{MCO} and V_C under the presupposition of θ_{NO} being infinite but $\Upsilon_{NO/CO}$ being fixed at 1.97. The extent of fibrotic change and that of alveolar interstitial inflammation (GGA: ground-glass attenuation) were quantified in terms of CT images taken at six axial levels between the aortic arch and 2 cm below the dome of right hemidiaphragm. They found that the relative volume of fibrosis was larger in UIP-IPF, while the relative volume of GGA was larger in NSIP. The augmentation of CT-decided fibrotic change was more closely correlated with the decrement in D_{LNO} than that in D_{LCO}, indicating that the D_{LNO} measurement provides a more sensitive information on the fibrotic change than the D_{LCO} in either UIP-IPF or NSIP. On the other hand, no correlation between the relative volume of GGA and D_{LCO} as well as D_{LNO} was found, indicating that GGA does no act as an insult for lowering D_{LCO} and D_{LNO}. Although the absolute values of D_{MCO} and V_C were substantially different between those obtained from the assumption of θ_{NO} at 4.5 and those from the assumption of θ_{NO} at infinity, both values showed qualitatively the same tendencies, i.e., the D_{MCO} at θ_{NO} of 4.5 and that at θ_{NO} of infinity were decreased as the extent of fibrosis was augmented. However, this was not the case for V_C, i.e., neither the V_C at θ_{NO} of 4.5 nor that at θ_{NO} of infinity was correlated with the extent of fibrosis in patients with UIP-IPF or NSIP (strictly, a weak correlation in UIP-IPF but no correlation in NSIP), indicating that the fibrotic change predominantly injures alveolo-capillary tissue barrier represented by decreasing D_{MCO}, which may be disproportionated with the injury of pulmonary microcirculation represented by decreasing V_C. D_{LNO}/D_{LCO} values in patients with UIP-IPF or NSIP were significantly decreased in comparison with those in control subjects, indicating that decrease in D_{MCO}, thus, decrease in D_{MCO}/V_C, is reflected by a decrease in D_{LNO}/D_{LCO}. This indicates that D_{LNO}/D_{LCO} acts as a good, clinical predictor for D_{MCO}/V_C in patients with fibrosis.

6.5 Systemic Sclerosis (SSc) with and Without Pulmonary Arterial Hypertension (PAH)

Applying the method proposed by the 2017-ERS Standards, a couple of investigators [49, 50] investigated the behaviors of D_{LCO} and D_{LNO} in patients with SSc with and without PAH. Degano et al. [49] divided 572 patients with SSc into four groups, including group A with neither interstitial lung disease (ILD) nor PAH ($n = 313$), group B with ILD but no PAH ($n = 201$), group C with both ILD and PAH ($n = 35$), and group D with PAH but no ILD ($n = 23$). They identified the following findings. (1) D_{LCO} and D_{LNO} values in patients with PAH (groups C, D) were much lower than those with ILD alone (group B). However, there was no difference in D_{LCO} and D_{LNO} between patients with both ILD and PAH (group C) and those with PAH alone (group D), suggesting that decreased D_{LCO} and D_{LNO} in SSc patients are attributed primarily to PAH. Hence, Degano et al. considered that D_{LCO} and D_{LNO} had an identical diagnostic value for detecting PAH. (2) D_{LCO} and D_{LNO} in patients with ILD alone (group B) were appreciably lower than those in patients with neither ILD nor PAH (group A), indicating that D_{LCO} and D_{LNO} are also influenced by ILD to approximately the same degree, though the degree of influence by ILD is smaller than that by PAH. (3) Although D_{LNO}/D_{LCO} showed an increasing trend in patients with PAH (groups C, D) in comparison with those with ILD alone (group B), there was no statistical difference between them. However, it should be noted that the patients assigned to group C, who had both ILD and PAH, were included in the analysis of D_{LNO}/D_{LCO}. In the patients belonging to group C, it is expected that the coexistence of ILD that decreases D_{LNO}/D_{LCO} due to the reduction in D_{MCO}/V_C conceals the PAH-associated increase in D_{MCO}/V_C that increases D_{LNO}/D_{LCO}. Thereby, the author considers that the findings observed by Degano et al. do not deny the clinical importance of D_{LNO}/D_{LCO} for detecting the microvascular injury causing PAH. (4) D_{MCO} and V_C were partitioned from D_{LCO} and D_{LNO} according to the method recommended by the 2017-ERS Standards. Degano et al. demonstrated that D_{MCO}, V_C, and D_{MCO}/V_C had much lower diagnostic value for PAH detection than D_{LCO} and D_{LNO} measures, indicating no merit of the partition into D_{MCO} and V_C for identification of PAH. (5) Multivariate analysis revealed that D_{LCO} (but not D_{LNO}), FVC, age at onset, and PAH were significantly associated with survival. Among them, D_{LCO} had the highest hazard ratio. Through such meticulous analysis, Degano et al. concluded that compared with D_{LCO} alone, simultaneous measurements of D_{LCO} and D_{LNO} do not boost the capability to detect PAH in SSc patients. Furthermore, Degano et al. issued a message that the story one likes to present, i.e., D_{LNO} is sensitive to the injury of alveolocapillary tissue barrier, while D_{LCO} is specifically sensitive to that of pulmonary microvasculature, may not stand true. However, it should be noted that it is difficult to judge the validity of D_{MCO} and V_C estimated in the study of Degano et al. because of the uncertainty about $\Upsilon_{NO/CO}$ and θ_{NO} that were assumed to be 1.97 and 4.5, respectively.

Barisione et al. [50] performed a study similar to that by Degano et al. [49]. Barisione et al. enrolled 96 patients with SSc with and without lung restriction and found that somewhat differing from the observation by Degano et al., D_{LNO} has a higher sensitivity to detecting the CT-confirmed ILD than D_{LCO}. However, in line with the observation by Degano et al., the D_{MCO} and V_C partitioning by the method of the 2017-ERS Standards does not provide the clinically useful information on whether the different behavior of D_{LCO} and D_{LNO} is due primally to ILD or microvascular injury. Barisione et al. considered that decreased D_{LCO} in the absence of lung restriction does not allow to suspect PAH without ILD, which is consistent with the conclusion of Degano et al. [49]. The author considers that the clinical uselessness of D_{MCO} and V_C shown by Degno et al. and Barisione et al. does not imply that these two parameters have no pathophysiological significance at all, rather it indicates that the validity of $\Upsilon_{NO/CO}$ and θ_{NO} adopted by the 2017-ERS Standards should be evaluated again.

6.6 *Diffuse Parenchymal Lung Disease (DPLD) and Pulmonary Arterial Hypertension (PAH)*

In patients with DPLD ($n = 41$), including sarcoidosis, NSIP, IPF, and so forth, van der Lee et al. [51] demonstrated that D_{LCO}, D_{LCO}/V_A, D_{LNO}, and D_{LNO}/V_A were significantly reduced in comparison with normal controls. Furthermore, they exhibited that in patients with PAH ($n = 26$) caused by primary pulmonary hypertension (PPH), chronic thromboembolic pulmonary hypertension (CTEPH), and SSc-elicited pulmonary hypertension, D_{LCO}, D_{LCO}/V_A, D_{LNO}, and D_{LNO}/V_A were decreased to the same extent as observed for patients with DPLD. However, they identified that D_{LNO}/D_{LCO} in patients with PAH were significantly higher than those in patients with DPLD and healthy normal subjects. Their findings suggest that although the single value of either D_{LCO} or D_{LNO} cannot differentiate between DPLD and PAH, the ratio, i.e., the D_{LNO}/D_{LCO}, has the ability to separate PAH from DPLD.

6.7 *Lung Diseases After Allogenic Stem Cell Transplantation (Allo-SCT)*

Bourgeois et al. [52] searched the pretransplant parameters that can predict the pulmonary complications and outcomes after allo-SCT ($n = 103$). Using the pretransplant D_{LCO}, D_{LNO}, and D_{LNO}/D_{LCO} as the explanatory variables, they analyzed the association between these D_L-related parameters and the 2-year cumulative incidence of severe pulmonary complication (SPC) defined as any pulmonary complication responsible for hospitalization, ARDS, and pulmonary complication-related mortality. In univariate analysis, there was a significant association between lower value of pretransplant $\%D_{LNO}$ and higher risk of SPC and ARDS. Although D_{LNO}/D_{LCO} did not show an association with SPC and ARDS, it was significantly correlated with pulmonary complication-related mortality. Furthermore, the lymphoid types of disease acted as the decisive factor for SPC, ARDS, pulmonary complication-related mortality, and non-relapse mortality. Based on these findings, Bourgeois et al. concluded that in addition to the types of disease, D_{LNO} and D_{LNO}/D_{LCO} are more useful to predict pulmonary complications than D_{LCO} after allo-SCT.

6.8 *Bronchiolitis Obliterans Syndrome (BOS) After Lung Transplantation*

Winkler et al. [53] cross-sectionally analyzed the association between various pulmonary function tests (spirometry, lung volumes, airway resistance, and simultaneous measurement of D_{LCO} and D_{LNO}) and incidence as well as the severity of BOS after either single or bilateral lung transplantation ($n = 61$, single: 19, bilateral: 42). The underlying lung diseases required for lung transplantation were IPF ($n = 22$), COPD ($n = 16$), CF ($n = 11$), SAR ($n = 3$), and others ($n = 9$). D_{MCO} and V_C were partitioned according to the method recommended by the 2017-ERS Standards. The discrimination study performed based on all pulmonary function parameters revealed that $\%D_{LNO}$ remained as only a significant variable separating patients with early BOS from those with no BOS. Other parameters, including spirometric parameters, lung volumes, airway resistance, $\%D_{LCO}$, $\%D_{MCO}$, and $\%V_C$, did not separate early BOS from no BOS. As for the prediction of advanced BOS, FEV_1/FVC, and $\%FEV_1$ among spirometric parameters, all lung volumes except for %TLC, airway resistance, and D_{LNO}/D_{LCO} among D_L-related parameters revealed the significant power for recognizing the difference between advanced BOS and early or no BOS.

6.9 *Atrial Septal Defect (ASD) Before and After ASD Closure*

Nassif et al. [54] measured the D_L-related parameters before and 6 months after the closure of atrial septal defect (ASD, $n = 29$), which improves the left-to-right shunting blood flow through the interatrial septum. In this study, D_{MCO} and V_C were estimated in compliance with the 2017-ERS Standards. Since the changes in D_{MCO} and V_C before and after the closure of ASD were investigated, the author assumed that they would qualitatively reflect the truth for closure-elicited changes in alveolocapillary tissue barrier and pulmonary microcirculation. After ASD closure, D_{LCO}, D_{LCO}/V_A, and D_{LNO}/V_A decreased significantly, though the decrease in D_{LNO} was marginal. Although decreasing change in V_C was significant, no change in D_{MCO} was identified. D_{LNO}/D_{LCO} increased after closure and its increase was inversely correlated with change in V_C. These findings led Nassif et al. to conclude that the ASD closure-elicited rectification of pulmonary hemodynamics, i.e., the decrease in V_C in concomitance with a decrease in pulmonary blood flow, is the primary factor for explaining their findings. Their observation demonstrated two important issues: (1) reduction in pulmonary blood flow does not change the alveolocapillary surface area available for gas transfer of CO and NO (i.e., no change in D_{MCO} before and after ASD closure), and (2) D_{LNO}/D_{LCO} may act as an important indicator for predicting the relationship between D_M and V_C (i.e., D_M/V_C).

7 Conclusion Remarks

The measurement of D_{LNO} is more convenient and simpler than that of D_{LCO} because D_{LNO} is minimally affected by PO_2 and Hb concentration in the pulmonary microcirculation. D_{LCO} and D_{LNO} share 80 ~ 90% of the pathophysiological information in a certain lung disease, indicating that the sensitivity of these two parameters to diagnosing a lung disease is almost equal, though it is shown to be a little higher in D_{LNO}. The most appealing issue of adding the D_{LNO} measurement is that the single simultaneous measurement of D_{LCO} and D_{LNO} allows for estimating D_{MCO} and V_C, while two measurements of D_{LCO} at different inspired PO_2 are needed when the classic Roughton–Forster's D_{LCO} procedure is applied. Unfortunately, however, there are a couple of serious problems that should be certainly addressed before trusting the significance of D_{MCO} and V_C partitioned from D_{LCO} and D_{LNO}. The relative Krogh diffusion constant of NO against CO in various diffusion paths from alveolar gas to Hb in the erythrocyte, which is denoted by $\Upsilon_{NO/CO}$ (D_{MNO}/D_{MCO}), is one of the key problems. In addition, the specific gas conductance of NO in the blood (θ_{NO}) is another key problem. Although the 2017-ERS Standards recommended to use $\Upsilon_{NO/CO}$ of 1.97 and θ_{NO} of 4.5, there are many oppositions against these recommended values. Hence, the author's personal opinion at present is that one should wait for partitioning D_{LCO} and D_{LNO} into D_{MCO} and V_C until reliable values of $\Upsilon_{NO/CO}$ and θ_{NO} are finalized. Instead of D_{MCO} and V_C, the author encourages one to use D_{LNO}/D_{LCO} ratio, which is certified to act as the representative of D_{MCO}/V_C under any pathophysiological conditions irrespective of θ_{NO}, i.e., finite or infinite. Of note, the effects of parallel and series heterogeneities on D_{LNO}/D_{LCO} are expected to be marginal. Supporting the important role of D_{LNO}/D_{LCO} derived from theoretical analyses, patients with IPF, NSIP, various lung diseases after allogeneic stem cell transplantation, and bronchiolitis obliterans syndrome after lung transplantation showed decreasing D_{LNO}/D_{LCO}, reflecting the enhanced injury of alveolocapillary tissue barrier (i.e., decreasing D_{MCO} and/or increasing heterogeneity of D_{MCO} distribution) in these diseases. On the other hand, patients with PAH showed increasing D_{LNO}/D_{LCO}, which reflects the augmented injury of pulmonary microcirculation (i.e., decreasing V_C or increasing heterogeneity of V_C distribution). Similarly, patients with atrial septal defect (ASD) exhibited increasing D_{LNO}/D_{LCO} after ASD closure. This indicates that the rectification of pulmonary hemodynamics after ASD closure significantly decreases V_C and/or its heterogeneous distribution. As such, D_{LNO}/D_{LCO} well explains the pathophysiological abnormalities in alveolocapillary tissue barrier and pulmonary microcirculation without knowing actual estimates of D_{MCO} and V_C and without correcting the effects of parallel and series heterogeneities. These facts certainly indicate that it is of clinical importance to establish the reliable reference equation of D_{LNO}/D_{LCO} as early as possible for expanding the application of simultaneous measurements of D_{LCO} and D_{LNO} to the differential diagnosis of the patients with diverse injuries of alveolar septa.

References

1. Krogh M. The diffusion of gases through the lungs of man. J Physiol. 1915;49:271–300. https://doi.org/10.1113/jphysiol.1915.sp001710.
2. Ogilvie C, Forster R, Blakemore W, Morton JW. A standardized breath-holding technique for the clinical measurement of the diffusing capacity of the lung for carbon monoxide. J Clin Invest. 1957;36:1–17. https://doi.org/10.1172/JCI103402.
3. Roughton F, Forster R. Relative importance of diffusion and chemical reaction rates in determining rate of exchange of gases in the human lung, with special reference to true diffusing capacity of pulmonary membrane and issue of blood in the lung capillaries. J Appl Physiol. 1957;11:290–302. https://doi.org/10.1002/ardp.19572901307.
4. Guénard H, Varene N, Vaida P. Determination of lung capillary blood volume and membrane diffusing capacity in man by the measurements of NO and CO transfer. Respir Physiol. 1987;70:113–20. https://doi.org/10.1016/s0034-5687(87)80036-1.
5. Borland CD, Higenbottam TW. A simultaneous single breath measurement of pulmonary diffusing capacity with nitric oxide and carbon monoxide. Eur Respir J. 1989;2:56–63.
6. Zavorsky GS, Hsia CC, Hughes JM, Borland CD, Guénard H, van der Lee I, et al. Standardisation and application of the single-breath determination of nitric oxide uptake in the lung. Eur Respir J. 2017;49:1600962. https://doi.org/10.1183/13993003.00962-2016.
7. Wise DL, Houghton G. 1968. Diffusion coefficient of neon, krypton, xenon, carbon monoxide and nitric oxide in water at 10–60 °C. Chem Eng Sci. 1968;23:1211–6. https://doi.org/10.1016/0009-2509(68)89029-3.
8. Chakraborty S, Balakotaiah V, Bidani A. Diffusing capacity reexamined: relative roles of diffusion and chemical reaction in red cell uptake of O_2, CO, CO_2, and NO. J Appl Physiol. 2004;97:2284–302. https://doi.org/10.1152/japplphysiol.00469.2004.
9. Borland CDR, Dunningham H, Bottrill F, Vuylsteke A. Can a membrane oxygenator be a model for lung NO and CO transfer? J Appl Physiol. 2006;100:1527–38. https://doi.org/10.1152/japplphysiol.00949.2005.
10. Borland CDR, Dunningham H, Bottrill F, Vuylsteke A, Yilmaz C, Merrill DD, et al. Significant blood resistance to nitric oxide transfer in the lung. J Appl Physiol. 2010;108:1052–60. https://doi.org/10.1152/japplphysiol.00904.2009.
11. Coffman KE, Chase SC, Taylor BJ, Johson BD. The blood transfer conductance for nitric oxide: infinite vs. finite θ_{NO}. Respir Physiol Neurobiol. 2017;241:45–52. https://doi.org/10.1016/j.resp.2016.12.007.
12. Guénard HJ, Martinot JB, Martin S, Maury B, Lalande S, Kays C. In vivo estimates of NO and CO conductance for haemoglobin and for lung transfer in humans. Respir Physiol Neurobiol. 2016;228:1–8. https://doi.org/10.1016/j.resp.2016.03.003.
13. Carlsen E, Comroe JH Jr. The rate of uptake of carbon monoxide and nitric oxide by normal human erythrocytes and experimentally produced spherocytes. J Gen Physiol. 1958;42:83–107. https://doi.org/10.1039/an9588300042.
14. Reeves RB, Park HK. CO uptake kinetics of red cells and CO diffusing capacity. Respir Physiol. 1992;88:1–21. https://doi.org/10.2307/2074765.
15. Borland C, Patel S, Zhu Q, Vuylsteke A. Hypothesis: Why θ_{NO} could be finite *in vitro* but infinite *in vivo*. Respir Physiol Neurobiol. 2017;241:58–61. https://doi.org/10.1016/jrespir.2017.02.013.
16. Power GG. Solubility of O_2 and CO in blood and pulmonary and placental tissue. J Appl Physiol. 1968;24:468–74. https://doi.org/10.1152/jappl.1968.24.4.468.
17. Sendroy J, Robert T, Dillon RT, Van Slyke DD. Studies of gas and electrolyte equilibria in blood XIX. The solubility and physical state of uncombined oxygen in blood. J Biol Chem. 1934;105:597–632.
18. Christoforides C, Hedley-Whyte J. Effect of temperature and hemoglobin concentration on solubility of O_2 in blood. J Appl Physiol. 1969;27:592–6. https://doi.org/10.1152/jappl.1969.27.5.592.

19. Goldstick TK, Ciuryla VT, Zuckerman L. Diffusion of oxygen in plasma and blood. Adv Exp Med Biol. 1976;75:183–90. https://doi.org/10.1007/978-1-4684-3273-2_23.
20. Chick H, Lubrzynska E. The viscosity of some protein solutions. Biochem J. 1914;8:59–69. https://doi.org/10.1042/bj0080059.
21. Kreuzer F, Hoofd LJ. Facilitated diffusion of oxygen in the presence of hemoglobin. Respir Physiol. 1970;8:280–302. https://doi.org/10.1016/0034-5687(70)90037-X.
22. Stein TR, Martin JC, Keller KH. Steady-state oxygen transport through red blood cell suspensions. J Appl Physiol. 1971;31:397–402. https://doi.org/10.1152/jappl.1971.31.3.397.
23. Wilhelm E, Battino R, Wilcock RJ. Low-pressure solubility of gases in liquid water. Chem Rev. 1977;77:219–62. https://doi.org/10.1021/cr60306a003.
24. Kang M-Y, Grebenkov D, Guénard H, Katz I, Sapoval B. The Roughton-Forster equation for D_{LCO} and D_{LNO} re-examined. Respir Physiol Neurobiol. 2017;241:62–71. https://doi.org/10.1016/j.resp.2016.12.014.
25. Yamaguchi K, Kawai A, Mori M, Asano A, Takasugi T, Umeda A, et al. Distribution of ventilation and of diffusing capacity to perfusion in the lung. Respir Physiol. 1991;86:171–87. https://doi.org/10.1016/0034-5687(91)90079-X.
26. Forster RE. Diffusion of gases across the alveolar membrane. In: Fishman AP, Farhi LE, Tenney SM, Geiger SR, editors. Handbook of physiology, the respiratory system, gas exchange. Bethesda MD: Am Physiol Soc; 1987, sect. 3, vol. IV, chap. 5. p. 71–88.
27. Holland RA. Rate at which CO replaces O_2 from O_2Hb in red cells of different species. Respir Physiol. 1969;7:43–63. https://doi.org/10.3817/0369003043.
28. Hughes JMB, Pride NB. Examination of the carbon monoxide diffusing capacity (D_{LCO}) in relation to its K_{CO} and V_A components. Am J Respir Crit Care Med. 2012;186:132–9. https://doi.org/10.1164/rccm.201112-2160CI.
29. Piiper J, Scheid P. Blood-gas equilibration in lungs. In: West JB, editor. Pulmonary gas exchange. Vol. I. Ventilation, blood flow, and diffusion. New York: Academic Press; 1980. p. 131–71.
30. Zavorsky GS. The rise in carboxyhemoglobin from repeated pulmonary diffusing capacity tests. Respir Physiol Neurobiol. 2013;186:103–8. https://doi.org/10.1016/j.resp.2013.01.001.
31. Yamaguchi K, Mori M, Kawai A, Takasugi T, Oyamada Y, Koda E. Inhomogeneities of ventilation and the diffusing capacity to perfusion in various chronic lung diseases. Am J Respir Crit Care Med. 1997;156:86–93. https://doi.org/10.1164/ajrccm.156.1.9607090.
32. Magnussen H, Holle JP, Hartmann V, Schoenen JD. D_{CO} at various breath-holding times: comparison in patients with chronic bronchial asthma emphysema. Respiration. 1979;37:177–84. https://doi.org/10.1159/000194024.
33. Piiper J, Scheid P. Diffusion and convection in intrapulmonary gas mixing. In: Fishman AP, Farhi LE, Tenney SM, Geiger SR, editors. Handbook of physiology, section 3: the respiratory system, Vol. IV: gas exchange. Bethesda, MD: American Physiological Society; 1987. p. 51–69.
34. Hsia CCW, Yan X, Dane DM, Johnson RL Jr. Density-dependent reduction of nitric oxide diffusing capacity after pneumonectomy. J Appl Physiol. 2003;94:1926–32. https://doi.org/10.1152/japplphysiol.00525.2002.
35. Linnarsson D, Hemmingsson TE, Frostell C, Van Muylem A, Kerckx Y, Gustafsson LE. Lung diffusing capacity for nitric oxide at lowered and raised ambient pressures. Respir Physiol Neurobiol. 2013;189:552–7. https://doi.org/10.1016/j.resp.2013.08.008.
36. van der Lee I, Zanen P, Stigter N, van den Bosch JM, Lammers JWJ. Diffusing capacity for nitric oxide: reference values and dependence on alveolar volume. Respir Med. 2007;101:1579–84. https://doi.org/10.1016/j.rmed.2006.12.001.
37. Aguilaniu B, Maitre J, Glénet S, Gegout-Petit A, Guénard H. European reference equations for CO and NO lung transfer. Eur Respir J. 2008;31:1091–7. https://doi.org/10.1183/09031936.00063207.
38. Zavorsky GS, Cao J, Murias JM. Reference values of pulmonary diffusing capacity for nitric oxide in an adult population. Nitric Oxide. 2008;18:70–9. https://doi.org/10.1016/j.niox.2007.10.002.

39. Munkholm M, Marott JL, Bjerre-Kristensen L, Madsen F, Pedersen OF, Lange P, et al. Reference equations for pulmonary diffusing capacity of carbon monoxide and nitric oxide in adult Caucasians. Eur Respir J. 2018;52:1500677. https://doi.org/10.1183/13993003.00677-2015.
40. Simaga B, Forton K, Motoji Y, Naeije R, Faoro V. Lung diffusing capacity in sub-Saharan Africans versus European Caucasians. Respir Physiol Neurobiol. 2017;241:23–7. https://doi.org/10.1016/j.resp.2017.01.003.
41. Zavorsky GS, Kim DJ, McGregor ER, Starling JM, Gavard JA. Pulmonary diffusing capacity for nitric oxide during exercise in morbid obesity. Obesity (Silver Spring). 2008;16:2431–8. https://doi.org/10.1038/oby.2008.402.
42. Hughes LMB, van der Lee I. The $T_{L,NO}/T_{L,CO}$ ratio in pulmonary function test interpretation. Eur Respir J. 2013;41:453–61. https://doi.org/10.1183/09031936.00082112.
43. Yamaguchi K, Tsuji T, Aoshiba K, Nakamura H. Simultaneous measurement of pulmonary diffusing capacity for carbon monoxide and nitric oxide. Respir Investig. 2018;56:100–10. https://doi.org/10.1016/j.resinv.2017.12.006.
44. van der Lee I, Gietema HA, Zanen P, van Klaveren RJ, Prokop M, Lammers JW, et al. Nitric oxide diffusing capacity versus spirometry in the early diagnosis of emphysema in smokers. Respir Med. 2009;103:1892–7. https://doi.org/10.1016/j.rmed.2009.06.005.
45. Moinard J, Guénard H. Determination of lung capillary blood volume and membrane diffusing capacity in patients with COLD using the NO-CO method. Eur Respir J. 1990;3:318–22.
46. Dressel H, Filser L, Fischer R, Marten K, Müller-Lisse U, de la Motte D, et al. Lung diffusing capacity for nitric oxide and carbon monoxide in relation to morphological changes as assessed by tomography in patients with cystic fibrosis. BMC Pulm Med. 2009;9:30. https://doi.org/10.1186/1471-2466-9.
47. Phansalkar AR, Hanson CM, Shakir AR, Johnson RL, Hsia CC. Nitric oxide diffusing capacity and alveolar microvascular recruitment in sarcoidosis. Am J Respir Crit Care Med. 2004;169:1034–40. https://doi.org/10.1164/rccm.200309-1287OC.
48. Barisione G, Brusasco C, Garlaschi A, Baroffio M, Brusasco V. Lung diffusing capacity for nitric oxide as a marker of fibrotic changes in idiopathic interstitial pneumonias. J Appl Physiol. 2016;120:1029–38. https://doi.org/10.1152/japplphysiol.00964.2015.
49. Degano B, Soumagne T, Delaye T, Berger P, Perez T, Guillien A, et al. Combined measurement of carbon monoxide and nitric oxide lung transfer does not improve the identification of pulmonary hypertension in systemic sclerosis. Eur Respir J. 2017;50:1701008. https://doi.org/10.1183/13993003.01008-2017.
50. Barisione G, Garlaschi A, Occhipinti M, Baroffio M, Pistolesi M, Brusasco V. Value of lung diffusing capacity for nitric oxide in systemic sclerosis. Physiol Rep. 2019;7:e14149. https://doi.org/10.14814/phy2.14149.
51. van der Lee I, Zanen P, Grutters JC, Snijder RJ, van den Bosch JMM. Diffusing capacity for nitric oxide and carbon mon- oxide in patients with diffuse parenchymal lung disease and pulmonary arterial hypertension. Chest. 2006;129:378–83. https://doi.org/10.1378/chest.129.2.378.
52. Bourgeois AL, Malard F, Chevallier P, Urbistandoy G, Guillaume T, Delaunay J, et al. Impact of pre-transplant diffusion lung capacity for nitric oxide (D_{LNO}) and of D_{LNO}/pre-transplant diffusion lung capacity for carbon monoxide (D_{LNO}/D_{LCO}) ratio on pulmonary outcomes in adults receiving allogeneic stem cell transplantation for hematological diseases. Bone Marrow Transplant. 2016;51:589–92. https://doi.org/10.1038/bmt.2015.284.
53. Winkler A, Kahnert K, Behr J, Neurohr C, Kneidinger N, Hatz R, et al. Combined diffusing capacity for nitric oxide and carbon monoxide as predictor of bronchiolitis obliterans syndrome following lung transplantation. Respir Res. 2018;19:171. https://doi.org/10.1186/s12931-018-0881-1.
54. Nassif M, van Steenwijk RP, van der Lee I, Sterk PJ, de Jongh FHC, Hogenhout JM, et al. Impact of atrial septal defect closure on diffusing capacity for nitric oxide and carbon monoxide. ERJ Open Res. 2019;5:00260–2018. https://doi.org/10.1183/23120541.00260-2018.

The manufacturer's authorised representative in the EU is Springer Nature Customer Service Centre GmbH, Europaplatz 3, 69115 Heidelberg, Germany. If you have any concerns regarding our products, please contact ProductSafety@springernature.com

Printed and bound by CPI Group (UK) Ltd, Croydon, CR0 4YY
15/07/2026
02167634-0004